AF327827

NEW CONCEPTS IN BLOOD FORMATION AND CELL GENERATION
in Malignant and Benign Tissues

Volume IV

Blood and Blood Vessels Developing Locally from Every Tissue

Part One: Benign Tissues
Part Two: Malignant Tissues

HEMPROVA GHOSH McDONALD, M.D., F.C.A.P.

Director, Diagnostic and Cell Research Institute
Waco, Texas

The author gratefully acknowledges the assistance of
Mr. Bernard Louis Machovsky, Waco, Texas,
in the preparation of this manuscript.

DIAGNOSTIC AND CELL RESEARCH INSTITUTE

WACO, TEXAS

2004

Published by :
Diagnostic and Cell Research Institute
2713 N. 43rd Street
Waco, Texas 76710

Library of Congress Control Number: 89090849

ISBN: 0-9627824-3-2

Copyright (C) 2004 by Hemprova Ghosh McDonald. All Right Reserved.
This book is protected by copyright. No part of it may be reproduced in any form
or by any means without written permission of the author.

PRINTED IN THE UNITED STATES OF AMERICA
BY DAVIS BROTHERS PUBLISHING COMPANY, LTD., WACO, TEXAS

DEDICATION

This volume and those published in the past, and to be published in the future are dedicated to the Almighty God, who has given power to my eyes to see, the ability of my mind to analyze, and has provided in me a great desire and perseverance to continue these basic cellular studies year after year.

PREFACE

I have realized that during the investigative studies of tissues and cells under the microscope, we not only use our eyes, but more essentially our minds and power of discrimination, which may tell us from where and how these tissues or cells make their appearances. I feel one should also critically examine with inquisitive minds the surrounding cellular and acellular structures. Such critical examinations may require hours of time, spread over a long period, not only observing but also mentally formulating the possible origin and developmental processes of the tissues or cells in question. Often, we suddenly see something for the first time, even though it has been present all the time in front of our eyes, but we are ignorant of its existence and importance. For example, the blue fibers or blackish reaction of red cells, both of which are highly important for generation of red and white blood cells,[1] presented here in figures 7.9-7.14.

In addition, we usually have many preconceived ideas about the development of various acellular and cellular structures (such as red cells, varieties of white blood cells, platelets, reticulocytes, megakaryocytes, and other cells) which we have learned from various textbooks and from teachers. Because of these preconceived ideas, we often fail to see the actual cellular structures or changes that may be very obvious. In this book, I present various morphological and developmental processes which can now be easily found after I have presented them through microscopic examinations of blood, bone marrow, and other tissues. One's mind should remain open to receive these phenomena, which may be contradictory to what one is accustomed to learning. Such open-mindedness helps us to understand at least some of the various ways of nature's performance.

When I examine tissue sections I see the cells and tissues at certain morphological stages of their lives. I try to visualize where they came from, what cellular changes they are undergoing, and what their possible direction would be in the future. We should look at tissue sections not as static substances, but as dynamic living substances which have been stopped at a certain point in their life processes. When examining the photomicrographs in this book, I hope that the readers will view them as constantly changing tissues and cells suspended at one point in time. I find it amazing to see how cells differentiate, dedifferentiate and redifferentiate into the same cell type, or other cell types.[2]

The study of tissue sections and smears has been my passion, especially since it keeps me challenged to find out what individual tissues or cells are doing. I hope that this book and my earlier publications will be helpful to readers in solving some morphological problems which were previously unaddressed or ignored.

---

1. Described in detail in Volume III (McDonald, 2001), Chapters 7 and 8.
2. As is demonstrated in tissue sections in this Volume and Volumes I, II and III (McDonald, 1989, 1995 and 2001).

ACKNOWLEDGMENTS
AND A SHORT LIFE HISTORY
OF THE AUTHOR

I am grateful to Dr. J. C. Ray, MD (Berlin), the late Director and Founder of the Indian Institute for Biochemistry and Medical Research, Calcutta, India. He had created in me a research spirit by his own example which has stayed with me throughout my life. Together with my late first husband Dr. Nirendra Nath Ghosh, who was also my classmate in Medical College, Calcutta, we had a chance to work with Dr. Ray in immunological research in kala-azar and amebiasis.*

After my first husband's death in early 1950 from chronic myelogenous leukemia (which he valiantly fought for seven years) I came to the United States in 1951 with a Danforth Foundation fellowship kindly offered by the late Dr. E. V. Cowdry, Director of the Cancer Research Division of Washington University in St. Louis, Missouri. I am thankful to Dr. Cowdry and also to the late Dr. Lauren V. Ackerman, Head of the Department of Surgical Pathology at the same institution, who instructed me in surgical pathology.

After I had finished my general pathology residency from St. Louis City Hospital, Dr. Ackerman gave me the opportunity to carry on, independently, a small scale experimental cancer chemotherapy program which I set up using a few chemicals (some of which had an effect on plant tumors) on breast carcinomas in C3H mice. It was during this time that I first noticed blood formation in nonmedullary tissues. Since then I have continued my cellular and tissue research with much of the work being performed on human surgical and autopsy slides in my own laboratory, Diagnostic and Cell Research Institute, Waco, Texas, which now after my retirement from clinical practice has been shifted to my home address.

Throughout my cellular research career of five decades, many persons have assisted me immensely in different ways. These persons have assisted me with experiments, kept records, prepared slides, made valuable suggestions, and given me technical and secretarial help. I am grateful to the following persons for their assistance in the publication of this and earlier volumes: Dixie McGregor and Lois Grawe, who helped me in my early work on C3H mice in St. Louis; Michael Soto, Susan Vartdol, Kelly Moulds, Jo Ann Lee, Sean DeLue, Jeff Deloach and Melissa Chan have been very helpful during their premedical college years at Baylor University, Waco, Texas.

I gratefully acknowledge the assistance of Mr. Bernard Louis Machovsky of Waco, Texas, a laboratory technologist and my primary assistant and co-worker for the last thirty years. Mr. Machovsky has an extensive knowledge of my research materials,

---

*1. A preliminary note on the complement-fixation reaction in Kala-azar with a specific antigen as an aid to diagnosis. Hemprova Ghosh, N.N. Ghosh, and J.C. Ray. *Ann Biochem & Exp Med* 5 : 153-158, 1945.

*2. Complement-fixation reaction in sera of rabbits actively immunized with living culture of Leshmania donovania. Hemprova Ghosh and N. N. Ghosh. *Ann Biochem & Exp Med* 7 : 1-2, 1947.

*3. Agglutination reaction in sera of rabbits immunized with different strains of Leishmania donovani. Hemprova Ghosh and N. N. Ghosh. *Ann Biochem & Exp Med* 7 : 3-6, 1947.

*4. A preliminary note on complement-fixation reaction in amoebiasis. N. N. Ghosh and Hemprova Ghosh. *Ann Biochem & Exp Med* 8 : 3-10, 1948.

*5. Further studies on complement-fixation reaction in Kala-azar with a specific antigen as an aid to diagnosis. Hemprova Ghosh, N. N. Ghosh, and J. C. Ray. *Ann Biochem & Exp Med* 9 : 173-178, 1949.

research activities, and findings. His assistance in the preparation of this and earlier volumes is greatly appreciated.

I would also like to thank Dr. Rajat Murkerjee (who visited me twice) and Dr. Sujal Chandra, both pathologists from Calcutta, India, who spent some time with me to get first-hand experience by studying the actual slides and photographs used in my research. Both of them are now carrying on research projects related to the new way of studying cells and tissues at the Cancer Center Welfare Home and Research Institute, Calcutta, India.

Finally, I wish to express my heartfelt thanks and appreciation to my beloved husband of over forty years, Hendley A. McDonald, MD, an internist, who has been a source of encouragement all these years and who has borne most of the expenses of my publications.

I greatly appreciate the permission for reproduction given by the following editors and publishers of some of my previously published articles: *Texas Medicine,* formerly *Texas State Journal of Medicine* (figs. 4.5 and 4.8); *Experimental Pathology* (figs. 4.6-4.9); *Journal of Clinical Pathology* (fig. 4.9); *British Journal of Cancer* (figs. 4.17-4.21), *Journal of the American Medical Women's Association* (fig. 12.10); and *The Woman Physician* (figs. 24.11 and 24.12).

Hemprova G. McDonald, M.D.
Waco, Texas
2004

TABLE OF CONTENTS

PREFACE 4

ACKNOWLEDGMENTS AND A SHORT LIFE HISTORY OF THE AUTHOR 5

INTRODUCTION 17

PRECONSIDERATIONS ON THE APPROACH OF STUDY 23

MATERIALS AND METHOD OF STUDY 25

    1. Materials Used in the Study of Blood Formation and Cell Generation 25

    2. Method of Study 25

Part One. Benign Tissues

(Chapters 1 - 21)

Chapter 1. EMBRYOLOGY [Including Chick and Human Embryos, and Chick Embryos in Tissue Culture] (figs. 1.1-1.9) 29

    1. Embryological Studies of Blood Formation in Chicks and Humans 29

    2. Red Cell Formation in Chick Embryos, including Yolk Sac Membrane and Liver (figs. 1.1-1.4) 30

       In the yolk sac membrane, red cells developing through cells somewhat similar to commonly known erythropoietic series cells; Red cells developing directly as hemoglobin globules within formative blood capillaries developing from liver parenchyma

    3. Blood Formation in Human Embryonic Tissues (figs. 1.5-1.7) 31

       Red cells and blood capillaries developing from mesenchymal tissue and cardiac muscle; Red cells and erythrogenic nucleated cells developing directly from lysing liver cells

    4. Red Cell Development *In Vitro* from Chick Embryo Nonmedullary Tissues (figs. 1.8-1.9) 33

       Red cells developing from liver cells and cardiac muscle cells *in vitro*

Chapter 2. SKIN [Including Epidermis, Dermis, Papillary Dermis and Benign Skin Lesions] (figs.2.1-2.25) 34

    1. Origin of Blood, Blood Vessels and Inflammatory Cells from Epidermal, Dermal and Papillary Dermal Tissue; and the Transformation of Lower Epidermis into Papillary Dermis (figs. 2.1-2.15) 34

    2. Capillary Hemangioma (figs. 2.16-2.17) 39

       Blood and blood vessel formation from dermal collagen

    3. Hemorrhagic Dermatofibroma (figs. 2.18-2.20 39

       Development of red blood cells and hemosiderin pigment from fibrocytes

    4. Nevus (figs. 2.21-2.23) 41

       Transformation of nevus cells into red blood cells, blood capillaries and lymphocytes

    5. Senile Ecchymosis (figs. 2.24-2.25) 42

       Development of red blood cells from collagen bundles of upper dermis

Chapter 3. ADIPOSE TISSUE [Including Adipose Tissue from Pericolon, Lipoma, and Throat] (figs. 3.1-3.16) 43

Adipose tissue forming blood, blood capillaries, erythrogenic SN cells, RVR cells, red hemoglobin-like substance, and plasma cells

Chapter 4. MUSCLES [Including Cardiac, Skeletal, and Smooth Muscles] (figs. 4.1-4.40) 48

    1. Cardiac Muscle [Heart] (figs. 4.1-4.21) 48

Development of red cells and blood capillaries from cardiac muscle fibers; Development of red cells and intracellular fat from chick embryonic cardiac muscle *in vitro*; Degenerated muscle element giving rise to fibrinoid material and various types of Aschoff cells, Types A, B and C; origin and multiplication of Anitschkow myocytes from cardiac muscle fibers, cytogenesis of Types A and B Aschoff cells; Acute myocardial infarction causing coagulation necrosis and the subsequent developments; Prior to nuclear regeneration, frequently seen changes in color and consistency of bright-red infarcted myofibers into bluish-gray, ground glass-like substance; and subsequent development of lymphocyte-like cells, developing blood capillaries, endothelial cells, and spindle-shaped nuclei; Massive red cell formation directly from fresh infarcted muscle fibers; Development of acute inflammatory cells and blood capillaries from disappearing myofibers; Possible transformation of ischemic cardiac muscle into blood and a large blood vessel; Independently growing mammary carcinoma emboli within the cardiac cavities and coronary vessels, but not connected with the heart or vessel walls, questioning the theory that stromal tissue originates only from host tissue

    2. Skeletal Muscle (figs. 4.22-4.33) 56

Direct origin of red cells as hemoglobin globules and blood capillaries from skeletal muscle; Red blood cells and capillaries developing from embryonic skeletal muscle; Origin of reactive/inflammatory cells (erythrogenic SN cells, lymphocytes, multinucleated giant cells, eosinophils, erythrogenic plasma cells' and histiocytes) from skeletal muscle; Without evidence of mitosis, production of various inflammatory cells (including lymphocytes, segmented nuclear cells, fibrocytes and unclassified cells) as a reaction of skeletal muscle fibers against advancing cancer; Dissolution of individual muscle fibers with retention of sarcolemma giving the false appearance of adipose tissue

    3. Smooth Muscle [Including Smooth Muscle from Prostate, Stomach, Colon and Urinary Bladder] (figs. 4.34-4.40) 61

Blood, blood capillary, and reactive plasma cell development directly from smooth muscle; Development of erythrogenic SN cells, endothelium, and erythrogenic eosinophils from smooth muscle

Chapter 5. ENDOCRINE GLANDS [Including Thyroid, Parathyroid, Adrenal, Pituitary and Pancreas] (figs. 5.1-5.50) 64

    1. Thyroid (figs. 5.1-5.24) 64

Red cell development directly as hemoglobin globules from the colloid of thyroid follicles; Blood vessel formation by blood-forming follicles with endothelium replacing the disappearing epithelium; Multiplying glandular epithelial cells dissolving and forming colloid and red cells, and development of blood capillaries from the interfollicular space; Origin of red blood cells and erythrogenic nucleated cells from colloid of follicles; Development of hemosiderin pigmented cells and blood capillaries from colloid of thyroid follicles; Transformation of colloid and glandular epithelium into collagen and development of blood and blood vessels from this collagen; Origin of lymphocytes and lymphatic tissue from thyroid tissue in Hashimoto's disease

2. Parathyroid (figs. 5.25-5.29) 72

Development of red blood cells and blood vessels from glandular tissue; Development of blood and blood capillaries from hyalinized stromal tissue developed from glandular epithelium

3. Adrenal Gland (figs. 5.30-5.37) 73

Development of blood, blood vessels, lymphocytes, erythrogenic SN cells and plasma cells from adrenal tissue; Sickle cell development from adrenal cells

4. Pituitary [Hypophysis Cerebri] (figs. 5.38-5.43) 76

Three zones of the pituitary; Oxyphil cells developing red blood cells and blood capillaries and chromophobe cells developing red blood cells

5. Pancreas (figs. 5.44-5.50) 79

Transformation of acinar structures into Islet of Langerhans; Formation of red blood cells and blood capillaries from Islets of Langerhans and from acinar glandular structures

Chapter 6. CARTILAGE AND BONE [Blood and Bone Marrow Development from Cartilage and Bone] (figs. 6.1-6.5) 81

Original bone marrow cells developing from lysing cartilage cells producing sinuses filled with original marrow cells and marrow fluid; Red cells developing directly as hemoglobin globules, and tiny undifferentiated nucleated cells arising from dissolved cartilage cells in marrow cavities; Dissolution of bone in the development of blood and vascular channels

Chapter 7. BLOOD AND BONE MARROW (figs. 7.1-7.20) 84

Red cells developing directly as hemoglobin globules from erythrogenic mesh, and crystallization of erythrogenic plasma gel; Red cells and nucleated cells developing from clotted blood plasma; Smeared fibrinoid bodies developing from coagulated marrow plasma, and transforming into naked nuclear bodies, and then developing into so-called hematopoietic nucleated marrow cells; Large fibrinoid particles forming red cells and segmented nuclear cells; Participation of blue fibers in the development of red cells and white blood cells, mainly segmented nuclear cells; Combination of blue fibers and blackish reaction of red cells in development of white blood cells; Blackish reaction of red cells in bone marrow giving rise to commonly known erythropoietic, granulopoietic series cells including segmented nuclear cells, eosinophils, lymphocytes and megakaryocytes; Reticulocytes developing from minute clumps of reticulum of finely-beaded fibrils (SGF, the name given by Griffith, et al., 1970) of unknown origin and surrounding blood plasma; Colorless hemolysis as a means of disappearance of red cells; Platelets developing from red blood cells and plasma in peripheral blood and from red blood cells and marrow plasma in bone marrow; Megakaryocytes showing no tendency towards platelet development

Chapter 8. LYMPH NODES AND LYMPHATIC CHANNELS (figs. 8.1-8.9) 91

1. Development and Transformation of Lymphocytes (figs. 8.1-8.6) 91

Red cell development from erythrogenic mononuclear cells of lymph node; Red blood cell and blood capillary development from lymphoid tissue; Grape cells (Russell body cells) developing red cells in lymph node

2. Development of Lymphatic Vessels (figs. 8.7-8.9) 94

Development of lymphocytes and lymphatic channels from papillary dermal tissue

Chapter 9. THYMUS (figs. 9.1-9.4) 96
Red cell and blood capillary development from lymphocytes of thymus

Chapter 10. SPLEEN (figs. 10.1-10.9) 98
Origin of red cells from reticular cells of the spleen, eosinophils, by coalescence of eosinophilic granules, lymphocytes, segmented nuclear cells, and plasma cells; Origin of reactive cells from fibrous connective tissue of spleen; Direct development of sickle cells from splenic tissue

Chapter 11. LUNG (figs. 11.1-11.8) 101
Emphysematous lung; Origin of red cells and erythrogenic SN cells within alveolar lumens from edema fluid; Red cell development from alveolar septal cells; Development of hemosiderin pigment as an abortive attempt at red cell development by so-called histiocytes/phagocytes; Origin of red cells from histiocytes/phagocytes

Chapter 12. LIVER (figs. 12.1-12.28) 104
Liver cell nuclei and cytoplasm producing red cells; Red blood cell and blood vessel development from liver parenchyma; Development of red cells and a central vein from dissolving liver cells; Red cell development from chick embryonic liver and from chick embryonic liver *in-vitro*; Red cells and erythrogenic nucleated cells directly developing from lysing liver cells of human embryonic tissue; Red cell development from liver sinusoids through fine erythrogenic mesh; Liver cell cords developing red cells; Development of irregularly shaped red cells from liver tissue; Large central veins with developing red cells arising from lysing liver parenchyma; Development of sickle cells from liver tissue; Blood and blood vessel formation from lysing liver parenchyma; Rapid and extensive mobilization of liver tissue in the formation of a large blood vessel packed with developing red cells

Chapter 13. GALL BLADDER (figs. 13.1-13.9) 116
Reactive/inflammatory cells arising from gall bladder wall and forming red blood cells and blood capillaries; Red cell development from locally derived eosinophils, SN cells, and lymphocytes; Red blood cells and blood capillaries developing from dissolving smooth muscle, connective tissue and desquamating epithelial cells

Chapter 14. STOMACH [Inflammatory Lesions of the Gastric Wall Including Mucosa, Submucosa and Smooth Muscle Coat] (figs. 14.1-14.14) 119
The development of red blood cells, blood vessels and chronic erythrogenic inflammatory cells in the gastric mucosa; and red cell development from erythrogenic segmented nuclear cells; Origin of blood and blood vessels in gastric submucosa; Red blood cell development from epithelial cells of gastric gland and directly from the gastric lining epithelium; Red cell development through Russell body cells (grape cells) in the gastric mucosa; Red blood cell, endothelium and eosinophil development directly from smooth muscle coat of gastric wall; Development of erythrogenic SN cells and endothelium from smooth muscle coat of gastric wall (see figure 4.38 in Chapter 4, subsection 3, Smooth Muscle); Presence of RVR cells in edematous subserosa of stomach (see figures 21.18-21.22 in Chapter 21, subsection 3, RVR Cells)

Chapter 15. RECTAL TISSUE IN INTERNAL HEMORRHOIDS (figs. 15.1-15.4) 126
Blood and blood vessel formation from rectal mucosa and underlying tissue.

Chapter 16. KIDNEY (figs. 16.1-16.13) 128

Transformation of renal tubules into blood and blood vessels and origin of lymphocytes from renal tubular cells; The development of red cells through the stages of vacuolar red cell (ghost red cell) formation; Plasma cell development from renal tubular epithelium; Transformation of glomeruli into blood and blood vessels; Direct development of sickle cells from renal tissue

Chapter 17. PROSTATE (figs. 17.1-17.3) 133

Red blood cell and blood capillary formation from partly hyalinized stromal tissue of prostate and from glandular epithelial cells; Development of red cells from smooth muscle of the prostate (see figure 4.34 in Chapter 4, Smooth Muscle)

Chapter 18. BREAST (figs. 18.1-18.9) 135

1. Mammary Gland (figs. 18.1-18.3) 135

Red cells arising from glandular epithelium and from erythrogenic gel derived from glandular epithelium

2. Intraductal Papilloma (figs. 18.4-18.9) 136

Typical intraductal papilloma of the breast; Red cell and blood vessel development from dissolving tumor tissue

Chapter 19. BRAIN [Cerebrum] (figs. 19.1-19.8) 139

Development of red cells directly from the cerebrum, as well as through erythrogenic SN cells arising from cerebral tissue; Eccentric enlargement of a vascular lumen by erythrogenic SN cells; Direct development of sickle cells from brain cells

Chapter 20. EYE [Retina and Choroid] (figs. 20.1-20.4) 142

Development of blood and blood capillaries from choroid and plexiform layers of the retina

Chapter 21. INFLAMMATORY OR REACTIVE CELLS OF
NONMEDULLARY ORIGIN (figs. 21.1-21.34) 144

1. Inflammatory Exudates [Red Cell and Inflammatory Cell Development from Desquamated Epithelium of Vagina, Nasal Lining and Gastric Mucosa] (figs. 21.1-21.9) ... 144

In desquamated vaginal epithelium red cell formation through vacuolated mesh; Granular changes towards development of red cells from vaginal epithelium and intercellular substance; Stages of red cell development from linearly locally formed SN cells in vaginal exudate; Development of SN cells from vaginal epithelium; Origin of eosinophils from nasal epithelium and from gastric lining mucosa; Red cell development from desquamating epithelial cells of the gall bladder wall (see fig. 13.9 in Chapter 13, Gall Bladder)

2. Granulation Tissue [Blood Formation in Granulation Tissue Consisting of Inflammatory Cells of Local Origin] (figs. 21.10-21.17) 148

Development of red cells and blood capillaries from inflammatory (reactive) cells of local origin; Red cell development from SN cells and grape cells

3. Redefined Vasoformative Cells of Ranvier (RVR cells), or *Fish-Head* Cells, as an Alternate Source for Development of Red Cells, White Blood Cells, and Blood Capillaries in Nonmedullary Tissues (figs. 21.18-21.32) 150

Presence of RVR cells in edematous subserosa of stomach; Capillary formation and red cell development from RVR cells; Presence of RVR cells in edematous labium majora granulation tissue; Blood, blood vessel, hemoglobin globule, SN cell and eosinophil formation from RVR cells; Development of RVR cells from adipose tissue, and reactive cells and blood capillary development from RVR cells (see figures 3.10 and 3.11 in Chapter 3, Adipose Tissue)

4.   Example of Vascularization of a Pericolic Abcess (figs. 21.33-21.34) 155

Red blood cell and blood vessel development from the break down of poorly differentiated plasma cells of an inflammatory lesion

Part Two. Malignant Tumors

(Chapters 22 - 36)

Chapter 22. CANCERS OF THE SKIN (figs. 22.1-22.19) 158

   1. Basal Cell Carcinoma (figs. 22.1-22.6) 158

Deep and widespread ulcerative basal cell carcinoma of the skin; Transformation of tumor cells into plasmacytoid cells; Red cell development directly from basal cell carcinoma; Transformation of plasmacytoid cells into erythrogenic SN cells; Direct and rapid transformation of spindle-shaped cancer cells into red cells

   2. Squamous Cell Carcinoma of the Skin and Esophagus (figs. 22.7-22.12) 160

Squamous cell carcinoma of the skin transforming into reactive/inflammatory cells; Development of red cells from keratinizing squamous cell carcinoma; Blood and blood vessel development from squamous cell carcinoma of the esophagus; Plasma cell development as a reaction to nearby cancer

   3. Malignant Melanoma (figs. 22.13-22.19) 162

     A. Melanoma of skin (figs. 22.13-22.16) 162

Blood and nucleated cell development from the keratin layer of melanoma of the skin; Malignant changes in squamous epithelium with the development of melanoma cells producing red cells; Blood and blood vessel formation directly from dissolving melanoma cells

     B. Melanoma metastatic to lymph node (fig. 22.17) 164

Red cell development from metastatic melanoma cells

     C. Melanoma metastatic to lung (figs. 22.18-22.19) 164

Red cells developing from metastatic melanoma cells; Plasma cells developing from metastatic melanoma cells

Chapter 23. LIPOSARCOMA (Malignant Tumor of Adipose Tissue)
   (figs. 23.1-23.3) 166

Blood and blood vessel development from liposarcoma cells

Chapter 24. BREAST CANCER (figs. 24.1-24.17) 168

   1. Mammary Carcinoma in Mice (figs. 24.1-24.12) 168

Mammary carcinoma with prominent, extensive, and rapid blood formation associated with tissue destruction and necrosis; Development of blood and blood cysts (so-called hemorrhage and hemorrhagic cysts common in C3H mice) from necrotic tumor tissue;

The simultaneous origin of vascular content, red cells and plasma, and the vessel wall from local cancer tissue; The possible transformation of cancer cells into benign cells and the development of collagenous connective tissue stroma

2. Mammary Carcinoma in Humans (figs. 24.13-24.17) 173

The origin of red cells and blood capillaries directly from cancer cells undergoing stromal changes with collagen formation; The development of red cells and blood capillaries from locally developed plasma cells of cancer cell origin; The mechanism of the local spread of mammary carcinoma into adipose tissue demonstrated by the progressive malignant changes in the latter induced in some way by the nearby cancer

Chapter 25. THYROID CANCERS (figs. 25.1-25.4) 175

Development of hemoglobinized colloid and red cells from dissolving cancer tissue; Normal thyroid tissue developing hemoglobinized colloid as a reaction to nearby adenocarcinoma; Blood and blood vessel development from follicular carcinoma of the thyroid

Chapter 26. ADENOCARCINOMA OF SALIVARY GLAND AND REACTIONS IN NEARBY NORMAL GLANDULAR TISSUE (figs. 26.1-26.6) 177

Poorly differentiated papillary adenocarcinoma of the salivary gland; Blood and blood vessel development from glandular epithelium as a reaction to nearby poorly differentiated papillary adenocarcinoma; Transformation of glandular tissue into plasma cells as a reaction to nearby cancer

Chapter 27. LUNG CANCERS (figs. 27.1-27.7) 181

1. Bronchogenic Squamous Cell Carcinoma of the Lung (figs. 27.1-27.2) 181

Development of inflammatory cells, blood and blood vessels from tumor cells

2. Adenocarcinoma (figs. 27.3-27.4) 182

Plasma and red cell development from liquefying hemorrhagic necrosed tumor tissue

3. Mesothelioma (figs. 27.5-27.7) 183

Fibrinoid changes in mesothelioma with development of red cells and vessel formation

Chapter 28. PANCREATIC CANCERS (figs. 28.1-28.8) 185

1. Adenocarcinoma of Pancreas (figs. 28.1-28.4) 185

Adenocarcinoma of the pancreas showing hyalinization with marked erasement of pancreatic tissue structures; Formation of inflammatory cells and spindle cells and red cells from pancreatic tissue

2. Islet Cell Tumor of Pancreas (figs. 28.5-28.6) 186

Blood development from necrosed tumor tissue

3. Anaplastic Carcinoma Head of Pancreas Metastatic to Liver (figs. 28.7-28.8) 187

Development of blood from necrosed liver tissue as a reaction to nearby carcinoma

Chapter 29. ADENOCARCINOMA OF LIVER (figs. 29.1-29.5) 188

Red cell and blood vessel development from malignant epithelium; Stroma development from cancer cells; Invasion of normal liver tissue by cancer cells; Normal liver tissue near the approaching cancer developing blood and a blood vessel

Chapter 30. TUMORS OF THE GASTROINTESTINAL TRACT (figs. 30.1-30.21) 190

   1. Adenocarcinoma of Stomach (figs. 30.1-30.4) 190

   Blood and blood vessel development through locally developed erythrogenic inflammatory cells

   2. Reticulum Cell Sarcoma of Stomach (figs. 30.5-30.9) 191

   Reticulum cell sarcoma with blood and blood vessel formation; Reticulum cell sarcoma of stomach with amebiasis

   A. Reticulum cell sarcoma of stomach metastatic to perigastric lymph nodes (figs. 30.10-30.11) 193

   Red cell development from metastatic reticulum cell sarcoma

   B. Perigastric adipose tissue near reticulum cell sarcoma of stomach (fig. 30.12) 193

   Blood and blood vessel development from perigastric adipose tissue as a reaction to approaching cancer

   3. Colon Carcinoma (figs. 30.13-30.21) 194

   Blood and prominent blood vessel development from highly necrotic tumor tissue; Red cell development from SN cells of necrotic tumor tissue origin; Development of mucin from this tumor; Connective tissue stroma development from epithelial tumor tissue; Red cell and blood vessel development directly from cancer tissue

Chapter 31. CARCINOID TUMORS (figs. 31.1-31.42) 198

   1. Primary Carcinoid of the Ileum (figs. 31.1-31.16) 198

   Typical carcinoid tumor; Hemorrhagic necrosis of tumor tissue in production of red hemoglobin-like substance; Hyalinization and liquefaction of tumor tissue in development of red hemoglobinized substance, red cells and blood vessels; Transformation of carcinoid cells into fibrous stroma; Development of fibrous connective tissue from hyalinized carcinoid tissue; Red cell development directly from carcinoid tissue; Extension of carcinoid into smooth muscle; Development of calcifications by carcinoid tissue

   2. Primary Carcinoid of the Colon (figs. 31.17-31.20) 203

   Lacy pattern of carcinoid of colon; Development of red cells, plasmacytoid cells, and stromal tissue from primary carcinoid of the colon

   3. Carcinoid Metastatic to Lymph node (figs. 31.21-31.23) 204

   Carcinoid of the ileum metastatic to mesenteric lymph node; Red cell development from metastatic carcinoid; Stroma development from metastatic carcinoid

   4. Carcinoid Metastatic to Liver (figs. 31.24-31.33) 205

   Blood and blood vessel development from metastatic carcinoid in the liver; Stroma development from metastatic carcinoid in the liver; Extension of metastatic carcinoid into normal liver tissue; Development of linear calcifications by carcinoid metastatic to liver

   5. Reactions in Intestinal Wall of Ileum to Nearby Carcinoid (figs. 31.34-31.42) 209

   Fibrinoid degeneration of muscle coat and submucosal tissue with development of blood and blood vessels; Development of calcifications in uninvolved intestinal wall near carcinoid

Chapter 32.   ADRENAL CORTICAL TUMOR   (figs. 32.1-32.2) 212
Red cell development directly from cancer cells

Chapter 33.   RENAL CANCERS   (figs. 33.1-33.12) 213
1.   Adenocarcinoma (Hypernephroma) (figs. 33.1-33.3) 213
Development of red cells from liquefied tumor cell product
2.   Papillary Adenocarcinoma (figs. 33.4-33.6) 214
Striking example of profuse red blood cell formation by papillary adenocarcinoma of the
kidney
3.   Wilms' Tumor (figs. 33.7-33.12) 216
Erythrogenic changes in normal kidney tissue as a reaction to nearby Wilms' tumor;
Blood and blood vessel development directly from tumor tissue

Chapter 34.   ADENOCARCINOMA OF THE PROSTATE   (figs. 34.1-34.8) 219
1.   Prostatic Carcinoma (figs. 34.1-34.5) 219
Typical adenocarcinoma of prostate; Perineural invasion by adenocarcinoma of prostate;
Origin of vacuolar red cells (ghost red cells) and blood vessel formation from malignant
epithelium
2.   Prostatic Carcinoma Metastatic to Bone Marrow (figs. 34.6-34.8) 221
Development of red cells from adenocarcinoma of prostate metastatic to bone marrow

Chapter 35.   BONE CANCERS   (figs. 35.1-35.6) 222
1.   Giant Cell Tumor (figs. 35.1-35.3) 222
Blood and blood vessel formation from giant cell tumor of the bone
2.   Bone with Metastatic Bronchogenic Carcinoma (figs. 35.4-35.6) 223
Red cells developing directly from metastatic cancer cells

Chapter 36.   BRAIN CANCERS   (figs. 36.1-36.6) 225
1.   Astrocytoma (figs. 36.1-36.2) 225
Blood and blood vessel formation from astrocytoma; Red cell development from tumor
tissue
2.   Glioblastoma Multiforme (figs. 36.3-36.6) 226
Blood and blood vessel development from glioblastoma multiforme

Chapter 37.   ACTIVE CELLULAR LYSIS, A PHENOMENON OF
GROWTH PROCESSES 228

APPENDIX

MY OBSERVATIONS AND COMMENTS ON VOLUME I 231

DIAGRAM I. [Red Cell Formation Directly as Hemoglobin Globules from Malignant and Benign Tissues, Including Reactive (Inflammatory) Cells] (Reproduced from Volume I) 236

MY OBSERVATIONS AND COMMENTS ON VOLUME II 237

MY OBSERVATIONS AND COMMENTS ON VOLUME III 239

GLOSSARY 244

REFERENCES 249

LIST OF ORIGINAL SCIENTIFIC PAPERS CONTRIBUTED BY THE AUTHOR 254

INDEX 257

INTRODUCTION

My interest in cellular study started in the mid fifties when I was still in surgical pathology training under Dr. Lauren V. Ackerman at Barnes Hospital, Washington University, Saint Louis, Missouri. Along with surgical pathology training, I was also involved in a small scale experimental cancer chemotherapy program, using breast carcinomas in C3H mice. One particular strain of spontaneous breast carcinoma in C3H mice is very *hemorrhagic*. While examining autopsy specimens of this tumor, I felt that the red blood cells were appearing through crystallization of the red *hemorrhagic* fluid. After studying this idea for a while, I showed the slides to Dr. Ackerman and explained my theory. He studied the slides for a few minutes and then shook his head in amazement. From then on I have critically studied all tissue slides with great interest for the local development of cells and tissues.

In the mid 1950s, while I was an instructor in surgical pathology at Barnes Hospital, the department routinely received as surgical specimens small pieces of atrial appendages surgically removed during mitral commissurotomy procedures performed to dilate stenosed mitral valves. [This deformity frequently occurs as a sequel of untreated or poorly treated rheumatic fever associated with inflammation of cardiac valves.] While carefully examining these specimens I came to the eye-opening conclusion that rheumatic fever is primarily a cardiac muscle disease and not one of the collagen diseases (as was widely accepted at the time). I noticed in these specimens an intimate relationship between Aschoff bodies or Aschoff nodules (the histological diagnostic feature of rheumatic fever) and cardiac muscle, rather than with collagen as was generally accepted. This finding supports the idea that Aschoff bodies arise from degenerated cardiac muscle. After

more additional studies of rheumatic fever, the preliminary report of my findings was first presented at the Annual Meeting of the American Association of Pathologists and Bacteriologists in 1957 (McDonald, see Ghosh 1957), and subsequent work on this subject was published in 1963, 1975, 1978 and 1995.

Since then, nearly five decades of continuous microscopic cellular study have unveiled many new important findings in benign and malignant human and animal tissues (mostly breast carcinomas in C3H mice). Subsequent zealous studies have provided me with some understanding about the basic continuous processes involved in cellular life. These basic processes include cell generation, differentiation, growth, transformation (transdifferentiation), dissolution, dedifferentiation and redifferentiation.

Many of these findings have been published over the last fifteen years in the first three volumes of my series *New Concepts in Blood Formation and Cell Generation, in Malignant and Benign Tissues* (McDonald 1989, 1995 and 2001). These findings are demonstrated in Volume I in mammary carcinomas in mice and humans, basal and squamous cell carcinomas, malignant melanoma, carcinomas of colon, kidney, and prostate; with brief examples in benign gastric wall, spleen, adrenal cortex, renal medulla, liver, epidermis, dermis, adipose tissue, bone marrow, and in cardiac, skeletal and smooth muscles. In Volume II, *Cardiac Muscle*, my findings are demonstrated in great detail in the myocardium of adult and embryonic tissues of humans and animals, and also in humans in chronic ischemic conditions, acute rheumatic fever, and coronary occlusion with myocardial infarction. In Volume III, *New Discoveries in Hematology*, findings are presented in embryonic tissues in chicks and humans, chick embryonic tissues *in vitro*, and also in cartilage and bone of rabbits and humans. However, the majority of the findings in Volume III are presented in human bone marrow and peripheral blood, with a few brief examples in liver, thymus, spleen, brain, cardiac muscle, lung, lymph nodes, submucosal tissue, adrenal gland, and kidney in cases of sickle cell anemia. (Brief summaries of these volumes are provided in My Observations and Comments on Volumes I, II and III at the end of this book.)

During the last four decades I have not accepted any grants from the government or private sources, thereby, retaining my freedom to study and pursue the research topics of my interest. These research topics include the study of various ways of medullary and nonmedullary[1] production of *blood* cells and their inclusion in developing vascular channels (angiogenesis) formed from local tissues (both benign and malignant tissues in humans and animals). These *blood* cells include not only red blood cells but also various white blood cells (inflammatory or reactive cells).

I have studied the erythrogenic capacity of these inflammatory or reactive cells, and the development of endothelium (angiogenesis) from the surrounding local tissues in the formation of blood and lymphatic channels. Some of my time has also been spent on the study of cell origin and transformation in the development of connective tissue stroma from benign and malignant tissues, and the capacity of stromal tissue to form blood and blood vessels. I have also studied the transformation of one type of tissue into another (i.e., benign to malignant, and malignant to benign).

My research also included the study of cellular lysis as an important phenomenon of growth processes, as it provides vital nutrients, spaces for ever-multiplying tumor cells, and possibly self-reproducing substances (as suggested by my demonstration of new growths out of necrotic and liquefied cell products). These findings were originally published in the *British Journal of Cancer* (McDonald, see Ghosh 1959b). Some of my published works on breast carcinomas in mice are summarized in a concise form in the *Indian Journal of Medical Research* (McDonald, 1983).[2]

One of my initial focal points of interest was the following: How are the so-called *hemorrhagic* cysts (which are common in breast carcinomas in C3H mice and which are lined by intact malignant epithelium) formed and filled with blood? The origin of the so-called *hemorrhagic* cysts and the origin of red blood cells inside the formative vascular channels, as well as outside the blood vessel, were studied under light microscope using sections of various tissues stained mainly with routine histochemical stains. The findings, when systematically analyzed, lead to the conclusion that red blood corpuscles develop directly as hemoglobin globules, like secretory granules. These red cells may arise from any tissue or cell (intact, broken down, necrosed, or even liquefied) without passing through the nucleated phases of erythropoiesis (as is generally believed to be the only way of red cell formation). These findings are demonstrated in Volumes I, II and III, and also in this volume.

The above findings on red cell formation are in contradiction to the general belief which was originated and, with support from a few authors, established by the forceful nature of Neumann (who wrote many papers suggesting nucleated erythropoiesis as the only way of red cell formation beginning in 1868, as mentioned by Michels,

---

1.   Nonmedullary - The term preferred by the present author, instead of *extramedullary*, for all organs and tissues other than bone marrow, because these tissues are equally important in the direct local development of red and white blood cells without passing through various nucleated stages as seen in bone marrow.
2.   These topics and additional findings were later demonstrated in great detail in Volume I (McDonald, 1989).

1931a). Neumann propounded the theory that throughout mammalian life, red blood cells arise only through one distinctive series of nucleated phases which occur exclusively in the bone marrow in adult life. Numerous authors have disagreed with Neumann's theory. Michels stated that, owing partly to the tenacity with which Neumann held to his views and to the strength with which he opposed all others, his theory had great impact. Michels also stated that one investigator, Robin (1874), claimed that Neumann's theory was actually "encumbering science."

Mentioned below are the names of some of the authors whose findings disagree with Neumann's conclusion that red blood cells are developed only by passing through a series of nucleated phases and only in bone marrow in adult life. These authors, in agreement with my findings (McDonald, 1989, 1995, 2001, and this volume), observed the possibility of red blood cells arising as hemoglobin globules, like inclusion bodies, without passing through nucleated phases of erythropoiesis. Weber and Kolliker (1845) followed the origin of red blood corpuscles in the cytoplasm of liver cells. Jones (1846) and Drummond (1854) both asserted that nucleated colorless cells (white blood cells) can produce red blood corpuscles without passing through a series of transitional stages. Wedl (1853) observed non-nucleated red blood cell formation in human cancers and normal tissues. Rollet (1862) and Rindfleisch (1863) maintained that red blood corpuscles arise as non-nucleated offspring of other cells, as secretory globules. Heitzmann (1872) believed that protoplasm of hematoblastic substance can break into fragments which are themselves formed into red blood corpuscles. Ranvier (1874) and Schaeffer (1874) observed the origin of red blood corpuscles through endogenous intraprotoplasmic differentiation of vasoformative cells. Latta (1921)

noted red blood corpuscle formation in the connective tissue of the intestine. Jordan (1926) described transformation of lymphocytes directly into erythrocytes. By applying various diagnostic tools, Keasby (1923) concluded that hemoglobin is present in the globules of the globular leukocytes [same as Russell body cells or grape cells (McDonald, 1989)] derived from plasma cells. Duran-Jorda described secretion-like origin of red blood cells (1947), and origin of red blood cells from eosinophils (1948). Recently, Niimi, et al, demonstrated the appearance of erythrocyte-like hemoglobin globules in the visceral yolk sac endodermal cells of mice (Niimi, et al, 2001, 2002b, 2003) and hamsters (Niimi, et al, 2001, 2002a).

In addition, my observations also agree with the findings of Schwann (1847) and Wedl (1853) regarding developmental processes of stellate cells (large cells with cytoplasmic projections in different directions) forming a network of capillaries and producing red cells and plasma in the lumen.[3]

My observations on red cell formation are in agreement with the findings of each of the above-mentioned investigators.[4] Without passing through the nucleated phases of erythropoiesis, the direct origin of red blood corpuscles has been observed in tissues, benign or malignant, epithelial or non-epithelial, including bone marrow. Demonstrated in this and earlier volumes are various cellular and tissue changes in the development of red blood cells and ways of formation of vascular channels enclosing the developing or already developed red cells from various benign and malignant tissues. Based on my observations, I consider red cells formed according to Neumann's theory of erythropoiesis to contribute only to a very small fraction of the total red cell mass produced in the body. Also a vast amount of red cells are formed directly as hemoglobin globules from the acellular liquid plasma of

---

3.   See RVR cells Volume III (figs. 14.11-14.24) and Chapter 21 of this volume (figs. 21.18-21.32).
4.   I have also found additional ways of red cell formation, demonstrated in Volume III (McDonald, 2001).

the bone marrow as demonstrated in Volume I (figs. 130-133), Volume III (figs. 3.1-3.20 and 10.9-10.24), and in this volume (figs. 7.1 and 7.2). Red cell formation from peripheral blood plasma clot is demonstrated in Volume III (figs. 3.21 and 3.23) and in this volume (fig. 7.3).

Not only red blood cells, but also inflammatory or reactive cells such as plasma cells, lymphocytes, segmented nuclear or SN cells,[5] eosinophils, monocytes, histiocytes, etc., derive locally from benign or neoplastic tissues. The generation of these nucleated cells and red cells proceeds locally and directly, without passing through the series of nucleated phases derived through mitosis, as is generally believed to occur in bone marrow. Each kind of these inflammatory or reactive cells shows the capacity to produce single or multiple red cells directly as hemoglobin globules. These cells usually arise locally in great numbers in inflammatory conditions and also in reaction to some exciting factors.

These inflammatory cells may also arise for the purpose of red cell formation, usually in the pathway of formative vascular channels. In eosinophils, red cells are shown to be formed by coalescence of orange-red granules. Both lymphocytes and plasma cells may form red cells in single or multiple number as hemoglobin globules. In addition, plasma cells may produce red cells in multiple numbers directly as hemoglobin globules through the distinctive form of grape cells (commonly known as Russell body cells). Plasma cells may also produce a large number of segmented nuclear cells, and to a lesser extent eosinophils, and these offspring cells in turn may produce red cells directly as hemoglobin globules, usually within formative vascular channels.

Wherever red cell formation occurs in

tissues, usually signs of or attempts at vessel formation may be seen enclosing the newly developed or developing red cells. The participation of plasma cells in the development of endothelium surrounding the newly formed red cells is frequently seen in areas of chronic inflammation with a heavy proliferation of plasma cells. Some of the inflammatory cells may gain their intra-vascular position either by their development within the lumen area of the formative vascular channels or by the annexation of the area containing such cells within the vascular lumen (like the widening of a river) by dissolution of the peripheral tissue and realignment of the new endothelium.

Marked proliferation of reactive cells may occur in adjacent benign tissues in some way induced by the nearby cancer. This inflammatory or reactive cell production within or outside cancer tissue is a protective mechanism against the tumor growth with possible regression of the tumor. On the other hand, the aggressiveness of the cancer may be shown by its influence in causing the adjacent benign tissues to become malignant.[6]

The common belief is that cell multiplication takes place through mitosis, whether it is normal or abnormal. Through my observations, I came to the understanding that mitosis, whether normal or abnormal, is the *unusual* way of cell multiplication. If this common belief were true, we would see very frequent mitosis throughout surgical specimens, which are excised and soon placed in formalin solution for fixation, but that is not the case.

My findings are also in sharp discordance with the popular theories on the development of red and white blood cells, platelets, reticulocytes, megakaryocytes, as well as many other cellular and acellular structures which are not mentioned in literature.

---

5. I prefer this name rather than neutrophils, since SN cells do not always have neutrophilic granules and do not have cytoplasm of their own at origin
6. The possibility of benign tissue becoming cancerous, as a way of cancer spread, is shown in Volume I by the demonstration of malignant changes in adipose tissue adjoining breast carcinoma and partially reproduced in Part Two of this volume (figs. 24.16 and 24.17).

According to Harris and Kellermeyer (1970), authors of *The Red Cell*, "the present working assumption is that primitive, undifferentiated cells (stem cells) comprise a compartment that is capable of maintaining and perpetuating its population by self-renewal as well as maintaining the population of erythroid cells by differentiation." According to Platt (1969, Plate 1), who illustrates (mainly with hand drawn pictures) the commonly believed theory of hematopoiesis, erythroid cells pass through four or five mitotic divisions; beginning with so-called stem cells and following a progression from hemocytoblasts, to pronormoblast, to basophilic normoblasts, to polychromatophilic normoblasts, to normoblasts, to reticulocytes, and then finally producing mature erythrocytes. Yet, mitosis is *rare* in normal bone marrow.[7] Also, these stem cells should be present in significant numbers in the adult bone marrow, however, stem cells are still not widely acknowledged as having been "acceptably identified by either morphologic or biochemical criteria" [Bennett and Cudkowciz (1968), Loutit (1968), Moffatt et al. (1967), Murphey et al. (1969), Yoffey et al. (1967), all cited by Harris and Kellermeyer (1970)].

I also do not believe the popularly accepted theory that after the closed vascular system has developed and the circulation begun in the embryo, new blood vessels always arise by "budding" from pre-existing blood vessels (Maximow and Bloom, 1957). I agree with Clark and Clark (1932), who believe that under normal conditions new capillaries are continually being formed while others may be retracted or absorbed. In addition, the capillary pattern in any part of the body is plastic and capable of alteration in response to changes in the immediate environment.[8] Recently, Lopez, et al, (2003) demonstrated that the mouse embryo makes red blood cell progenitors in every tissue simultaneously with blood vessel morphogenesis. In tumor tissue, Maniotis, et al, (1999) demonstrated vascular channel formation by human melanoma cells *in vivo* and *in vitro* and independent of endothelial cells.

Many recent scientific articles have acknowledged the important role of transdifferentiation[9] in normal growth and regeneration in a large variety of tissues. As Springer, Brazelton, and Blau (2001) state in their commentary:

> It used to be simple. Regenerating liver was made only by liver cells, skin begat skin, endothelium begat endothelium, skeletal muscle came from skeletal muscle precursors, and brain and heart simply did not regenerate. Moreover, new cells made in regenerating tissues like endothelium and smooth muscle were derived from reserve cells in the vicinity of the damage within a given tissue.
>
> In the last several years, virtually all of these long-held assumptions have come into question. There has been a plethora of reports that bone marrow-derived cells can differentiate *in vivo* into cells with properties characteristic of muscle, endothelium, liver, heart, and neuronal cells of the brain. Moreover, even cells derived from some non-bone marrow tissues appear to change their identities and give rise to cells of other tissues. These transformations from one cell type into another suggest a novel level of complexity...

Throughout this volume, and my earlier volumes, I demonstrate many examples of transformation of one distinct cell type into another.

Also, during the last few years the scientific literature has been bombarded with descriptions and examples of extramedullary hematopoiesis[10] in practically every

---

7. As is demonstrated in Volumes I and III, and in this volume in Chapter 7, Blood and Bone Marrow.
8. See Volume I, figs. 7, 17, 61, 89, 90, 92, 106, 107, 110 and 120-1.
9. Transdifferentiation - The direct conversion of one differentiated cell type into another distinct cell type.
10. Extramedullary hematopoiesis - The formation of blood or blood cells outside the bone marrow.

tissue and organ. Many of these authors believe that somehow bone marrow progenitor cells have made their way to these extramedullary sites and begun to multiply, or that reserve primitive pluripotential embryonic progenitor cells in these tissue sites have somehow become activated. Even today the great capacity for local adult cells and tissues to dedifferentiate and redifferentiate[11] or transdifferentiate is not greatly understood or appreciated.

Over the decades the terminology used to describe cells and tissues has varied greatly, and has frequently changed. In this book, in describing the various cellular and acellular structures, I have used the accepted cellular terminology of erythropoiesis and granulopoiesis (in most cases) to make it easier for the reader to comprehend the essence of the discussion without getting entangled in a web of revised terms and nomenclature. Such usage in no way diminishes my firm conviction regarding the way cellular and acellular structures behave.

In Volume I (McDonald, 1989) of this series *New Concepts in Blood Formation and Cell Generation, in Malignant and Benign Tissues*, I describe blood formation and cell generation with blood vessel development mainly in malignant tissues, and only briefly in a few benign tissues. In Volume II, *Cardiac Muscle* (McDonald, 1995), I describe these processes in great detail but exclusively in benign cardiac muscle. Volume III, *New Discoveries in Hematology* (McDonald, 2001), of my series deals mostly with blood formation in bone marrow and, to a small extent, peripheral blood. Brief summaries of these volumes are provided in My Observations and Comments on Volumes I, II and III, included at the end of this book so that one may be familiarized with some of the findings on histocytogenesis which I have already published.

In this volume, I have described blood and blood vessel formation from a wide variety of benign and malignant (non-medullary) tissues and organs, covering practically every tissue and organ. I have also demonstrated the local development of erythrogenic inflammatory cells and their contribution to blood and blood vessel development. Scattered throughout this book are examples of cellular lysis which plays an important role in blood formation and cell generation. This topic is discussed in detail in the last chapter of this book.

I have intended this volume to be a broad and complete overview of my findings in both benign and malignant tissues, and I have tried to include examples of blood and blood vessel formation and cell generation in as many different benign tissues and organs, and in as many tumors, as possible. In compiling this volume, I have made good use of the material presented in my first three volumes. Chapter 1, Embryology (figs. 1.1-1.9), and Chapter 6, Cartilage and Bone (figs. 6.1-6.5) are reproduced from Volume III. Chapter 7, Blood and Bone Marrow (figs. 7.1-7.20) consists of selected photomicrographs from Volume III, covering some of my important new findings presented in that volume. Chapter 4, subsection 1, Cardiac Muscle (figs. 4.1-4.21) consists of photomicrographs taken from Volume II and one of my earlier articles published in *The British Journal of Cancer* (Ghosh, see McDonald, 1959a). Chapter 24, Breast Cancer consists mostly (figs. 24.4-24.17) of photomicrographs taken from Volume I. Also, scattered throughout this volume are other selected photomicrographs from Volume I (figs. 2.1, 2.6, 4.28, 4.29, 4.31, 4.32, 4.38, 5.32, 5.33, 10.1, 12.1, 12.28, 14.1, 16.4, 22.6, 22.15, 22.16, 34.4, 34.7, and 34.8) and Volume III (figs. 3.11, 5.37, 8.2, 12.11-12.13, 12.24, 16.12, 16.13, 19.7, 19.8, 21.18, and 21.22). These photomicrographs are reproduced here, in a single volume, for the convenience of the reader. The three hundred ninety-two new photomicrographs presented here make up the bulk of this volume.

---

11. Regeneration of myocardial fibers through tiny dedifferentiated 'seed state' nuclei arising from lysing cardiac muscle fibers was described by the author in an original paper published in the *Journal of Clinical Pathology* (McDonald, 1975).

PRECONSIDERATIONS ON THE APPROACH OF STUDY

As mentioned in my earlier volumes, cellular study using tissue sections is found to be an excellent way to follow the natural biological processes in the development of various histological structures. Le Gross-Clark (1958) remarked that in tissue sections with simple stain, histological processes can be studied when the cellular structures are not dissociated. In addition to the microscopic examination of tissue sections, critical examinations of many photomicrographs have proven to be very helpful. With an analytical mind, careful and repeated examinations of photomicrographs, hours of 'groping' usually result in the discovery of missing links, however incomplete they may be, in the cellular and tissue growth processes.

I would like to emphasize the fact that our present day technology, remarkable as it is, often helps us overlook the obvious. We should not let ourselves over-rely on technology and curtail our power of observation and analysis using our greatest tool: *the human mind.*

Preconceived ideas and beliefs on various histological processes apparently limit one's accurate view of the phenomenon at hand and may lead to interpretations of the developmental processes which may not be completely accurate. There is much to be learned about the capacity of the living substance for cell generation, differentiation and transformation. Based on circumstantial evidence William E. Horner, Professor of Anatomy at the University of Pennsylvania, pointed out in his book *Special Anatomy and Histology* (1851), the possibility of the development of new elements, in the laboratory of animal systems inside the body, by the all-controlling influence of living principle. One must experience an aura of wonder when he or she tries to imagine the complexity, intricacy and diversity of the enor-

mous biochemical reactions that must be taking place in order to transform the simple looking white and yellow of a fertilized egg into a fully developed chick embryo with feathers, a brain, and everything else, while still within the shell. It would not be out of place to suspect that there may be many new elements, as well as additions to the existing known elements, created by the all-controlling influence to build this fully developed embryo.

It should be remembered that cell patterns seen at one point only identify the characteristics for that time, yet there are many more cells present which do not conform to our classification and challenge our preconceived notions on cell growth patterns, as there are continuous processes of cell differentiation often in different directions and at different speeds. The nucleus, cytoplasm, cell membrane and intercellular substances are all forming, transforming or dissolving in a continuous protoplasmic mass from which similar or completely different kinds of cells and tissue structures may make their appearance, as suggested in this book and in my previous publications.

I am aware of the fact that there must be many gaps I have not yet considered in describing various cellular processes. I am hopeful that with these preliminary findings many inquisitive minds will take up the challenge to unveil some of the additional pathways of cellular and tissue growth, the full extent of which is beyond human capability.

The preconceived idea that every cell has to come from similar cells of embryological lineage and mainly through mitosis has prevented the understanding of cellular and tissue growth patterns to a great extent. Schwann (1847), who made a classical study on cells over a century ago, pointed out the difficulties in formulating any scheme for the classification of cells of different histological structures, by writing:

> Nature is very unwilling to accommodate herself to our schemes. The object of her aim is quite opposed to that of our intellect. She accords and accommodates all contrarieties by gentle transition: the intellect disjoins and seeks everywhere for strongly marked contrast.

In this book and in my earlier publications, I have tried to show my microscopic findings of "gentle transition" in the development of the various cellular and tissue structures constructed with nature's guidance, which is beyond our comprehension.

MATERIALS AND METHOD OF STUDY

1. Materials Used in the Study of Blood Formation and Cell Generation

Numerous photomicrographs were prepared from thousands of slides of autopsy and surgical specimens, and smears and clot sections of blood and bone marrow specimens. These slides were critically examined and analyzed in order to learn the relationships among various cellular and tissue structures concerning dedifferentiation, differentiation, transformation and regeneration. The blood and bone marrow slides were studied in detail to formulate the possible patterns of development of red cells, nucleated cells and various blood and bone marrow cellular and acellular structures. The specimens studied include:

1. Human venous blood, infrequent finger prick blood, and human bone marrow derived from sternal aspirations, as well as from iliac crest bone marrow aspirations and needle biopsies. A majority of these specimens were taken from patients seen by me on hematological consultations. A few were from other sources, including routine blood smears.

2. Human embryonic tissues (received from miscarriage specimens) and chick embryonic tissues studied from early incubation until hatching (used as part of my study of developmental anatomy).

3. Short-term tissue culture products of chick embryo tissues, mainly heart and liver.

4. Adult human tissues obtained from autopsy and surgical specimens.

5. Tissues from various animals such as mice, rats, rabbits and chicks used in both short and long-term experiments.

6. Fresh specimens of cartilage, bone and other tissues from a rabbit sacrificed by inhalation of ether, used to study the developmental processes of bone marrow, blood and other tissues.

2. Method of Study

Many basic cytogenic and hematopoietic processes were studied and analyzed in their natural setting by critically studying thousands of tissue sections, bone marrow aspiration clot and biopsy sections, as well as blood and bone marrow smears under a light microscope. Occasionally phase contrast microscopy was used. Numerous photomicrographs were taken, studied and used in the formulation of various cellular and tissue growth processes. Critical analysis of these photomicrographs again and again with a highly inquisitive mind led to many startling new findings, which were restudied and verified by taking even more photomicrographs. Some discoveries have taken many years of study and examination. It should be remembered that histological sections represent the cross section of various dynamic growth processes in their natural setting. In my method of study I agree with Weiss (1950), who stated:

> We must form the habit...of viewing the shape of a given cell, tissue, or organ not as a static feature, but as a phase, transitory or terminal, in a continuous chain of transformation: as a cross section through a stream of processes proceeding in time. Every morphological criterion is but an expression of antecedent processes that have contributed to its formation.

The observations presented by photomicrographs were based on the study of formalin-fixed tissue sections of various organs, formalin-fixed bone marrow aspiration clot and biopsy sections, as well as peripheral blood and bone marrow aspiration smears. Routinely, Wright stain was applied to cover glass preparations of aspirated bone marrow smears and circulating venous blood smears. The blood and bone marrow smears for the detection of reticulocytes were stained using the New Methylene Blue method. The tissue sections of various organs and bone marrow clots were made at four to six microns thickness, and were routinely stained with hematoxylin and eosin (H&E) and frequently with Giemsa stain. For detection of iron pigment, the stain for Prussian blue reaction was used for bone marrow smears and clot sections.

Infrequently applied was a battery of special histochemical stains such as Mallory's phosphotungstic acid hematoxylin, Mallory's aniline blue-acid fuchsin, Orange G, Verhoeff van Gieson stain, Chromotrope Aniline Blue stain, Masson's trichrome stain, Foot's modification of Bielschowsky's method for reticulum stain, periodic acid Schiff method of McManus, Oil O Red and Benhold's Congo red stain. For confirmation of shapes of certain morphological structures in their entirety, in three dimensions, rarely serial sections were examined.

Since the present day theory of erythropoiesis found in all textbooks originated from the cellular study of the yolk sac membrane of early chick embryos, I studied and saw for myself blood and blood vessel formation from very early embryonic stages until hatching. In addition, simple tissue culture experiments were set up using chick embryo tissues as implants, grown *in vitro* on a cover glass. The medium used was composed of Eagles solution containing Hanks balanced salt solution, vitamins, amino acids, glutamine and horse serum. Both penicillin and streptomycin were added to a certain concentration. The final pH was kept between 7.2 and 7.4 by addition of sodium bicarbonate. This procedure was adapted following a method used at that time at Southwestern Medical Center in Dallas, Texas (Reynolds and Montgomery, 1967). Cover glasses holding the culture products were stained with H&E, Giemsa, Oil O Red (fat stain) and a few others. A short study was also made of red cell and blood vessel development in human embryos from different organs and the placenta (from miscarriage specimens).

PART ONE

BENIGN TISSUES

Chapters 1-21

Chapter 1

EMBRYOLOGY (figs. 1.1-1.9)
(Including Chick and Human Embryos, and Chick Embryos in Tissue Culture)

1. Embryological Studies of Blood Formation in Chicks and Humans

Bone marrow is commonly believed to be the only site of blood formation in adults, even though many authors (Denys, 1886; Drummond, 1854; and Duran-Jorda, 1948) including myself (McDonald, 1989, 1995, 2001 and this volume) have demonstrated that both red and white blood cells arise directly from benign and malignant tissues, as well as from bone marrow and blood plasma. The popular theory of blood formation in embryos is that original blood-forming cells (which are known as stem cells) arise in the yolk sac membrane (Sabin, 1920). These stem cells form various cellular elements and travel to the embryonic liver and then to the bone marrow where hematopoietic series cells finally conclude in the formation of red cells, white blood cells and platelets.

The popular theory of hematopoiesis in adults, with red cells developing only from stem cells through various nucleated cell stages, was derived from microscopic observations of blood formation from the yolk sac membrane of chick embryos. This theory traces red cell formation starting from hypothetical stem cells in bone marrow to hemocytoblast, to erythroblast, to proerythroblast, to basophilic normoblast, to polychromatophilic normoblast, to orthrochromic normoblast, to reticulocyte and finally to erythrocyte.

This theory of hematopoiesis is commonly described in textbooks of hematology, mainly with hand drawn illustrations. The youngest cells, myeloblasts and erythroblasts, have large nuclei with delicate chromatin, prominent nucleoli and a rim of dark blue ribosomal cytoplasm. As the cells mature, they develop their distinctive nuclear and cytoplasmic characteristics. In the red cell series, the round erythroblast nucleus becomes smaller and darker and progressively more pyknotic and the cytoplasm becomes

pinker, reflecting increasing hemoglobin content, until the normoblast stage is recognizable. Finally, the nucleus disappears, producing anuclear reticulocytes and mature red cells.

In the myeloid line, the myeloblast enlarges and gains nonspecific azurophil granules and becomes a promyelocyte. Specific cytoplasmic granules in the myelocytes distinguish them as neutrophilic, eosinophilic and basophilic precursors. Each then undergoes progressive changes in nuclear shape through the metamyelocyte and band form, until the mature neutrophils, eosinophils, or basophils are formed.

During megakaryocyte maturation, cells undergo cycles of DNA synthesis without cell division producing large polypoid mononuclear cells. Thus, megakaryocytes are very large compared to other mature cells (because of their large multi-lobed nuclei and abundant eosinophilic cytoplasm). Production of monocytes, lymphocytes and mast cells also occurs in the marrow, as well as, in nonmedullary tissues.

To satisfy my curiosity, I studied chick embryos, particularly red cell and blood vessel development from the yolk sac membrane and other tissues, including bone marrow. I found that my observations agree only to a small extent with the popular theory of blood development mentioned above. I noticed that in the yolk sac membrane, as well as in the bone marrow, there are cells somewhat comparable to commonly believed erythropoietic series cells. However, at the same time, there are other tissues and organs in chickens, both adult and embryo, from which a vast amount of red cells develop directly without passing through the various nucleated cell stages of erythropoiesis. (It must be noted that chick red cells, in both adult and embryo, are spindle-shaped and contain a small round hyperchromatic nucleus.)

Similar to chick embryos, I found that in humans (both adults and embryos) the majority of red cells develop directly from local tissues, both benign and malignant.[1] Because of the preconceived idea that red cells are formed only by passing through various nucleated stages via mitosis, direct formation of red cells as hemoglobin globules from various tissues escaped popular notice.

The common theory of red cell development states that there are four or five separate mitosis from the hemocytoblast to the red cell stage (as shown by Platt, 1969, Plate 1). On the contrary, there is hardly any mitosis seen in the development of red cells in normal bone marrow. In addition, a profuse number of red cells develop directly from marrow plasma and from other organs, particularly the heart, where a large amount of red cells develop from liquefying cardiac muscle in the formation of heart cavities in the embryonic stage. In the chorionic villi and decidual tissues of human embryos, I found the direct local development of red cells and blood capillaries. Rarely, erythrogenic SN cells (segmented nuclear cells) develop directly from the chorionic villi of the human placenta and transform into red cells during the development of blood capillaries. Finally, a vast amount of blood formation from decidual tissue is noted in the placenta. These findings are shown in detail in chick embryos, human embryos, and chick embryo tissue culture products, and are published in Volume III (McDonald, 2001). To save space and expense, only a few of the embryology photomicrographs from Volume III are included in this chapter.[2]

2. Red Cell Formation in Chick Embryos, including Yolk Sac Membrane and Liver (figs. 1.1-1.4)

These photomicrographs show blood formation from the yolk sac membrane and liver of a chick embryo. Figures 1.1-1.4 are taken from a six-day-old chick embryo, including yolk sac membrane (figs. 1.1-1.3) and liver (fig. 1.4). Figures 1.1, 1.3 and 1.4 below are reprinted from Volume III (McDonald, 2001).[3]

In the yolk sac membrane, red cells developing through cells somewhat similar to commonly known erythropoietic series cells (figs. 1.1-1.3)

Figures 1.1-1.3. These figures are from the yolk sac membrane of a six-day-old chick embryo. Figure 1.1 shows part of a yolk sac membrane with developing blood and vascular cavities in low magnification. Area (A) is seen in higher magnification in figures 1.2 and 1.3. In figure 1.3, normal spindle-shaped red cells are forming just beneath the dissolving vacuolated lining through stages that can be compared to erythropoiesis, with the more mature spindle-shaped red cells seen in the lower right corner. Nucleated erythropoiesis is somewhat demonstrated by these cells. This is how the theory that red cells only develop through a nucleated series originated. However, no mitosis is seen in this figure, contrary to the commonly believed theory. Figure 1.1 H&E x 80; figure 1.2 H&E x 260; and figure 1.3 H&E x 520

Red cells developing directly as hemoglobin globules within formative blood capillaries developing from liver parenchyma (fig. 1.4)

Figure 1.4. Blood sinusoids are originating from liver parenchyma. Liver cells are being transformed into spindle-shaped red cells with tiny round nuclei (A). The area at (A) will be connected with already more-developed blood sinusoids on the right, left and below, all developing from liver parenchyma. The lining endothelium has already developed in some areas (arrows). (See also figure 12.11 in Chapter 12, Liver.) H&E x 520

---

1. As demonstrated by photomicrographs of cardiac muscle, liver, mesenchymal tissue, etc., (McDonald, 1989, 1995, 2001, and this volume).
2. For a detailed discussion of my findings in embryonic tissues see Chapter 1, Volume III (figs. 1.1-1.45).
3. For a detailed discussion of my findings in chick embryonic tissue see Chapter 1, Volume III (figs. 1.1-1.12).

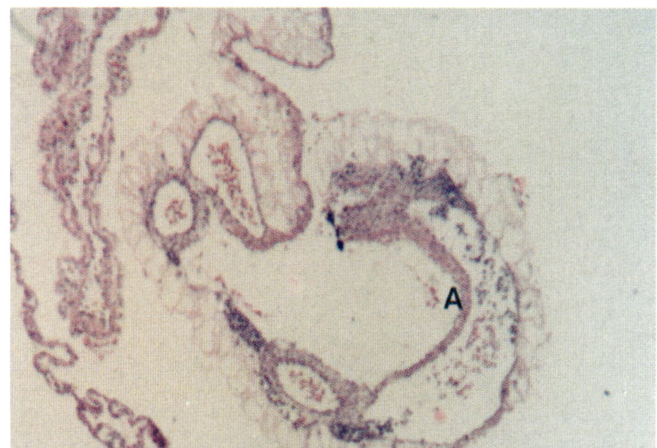

Figure 1.1, H&E x 80

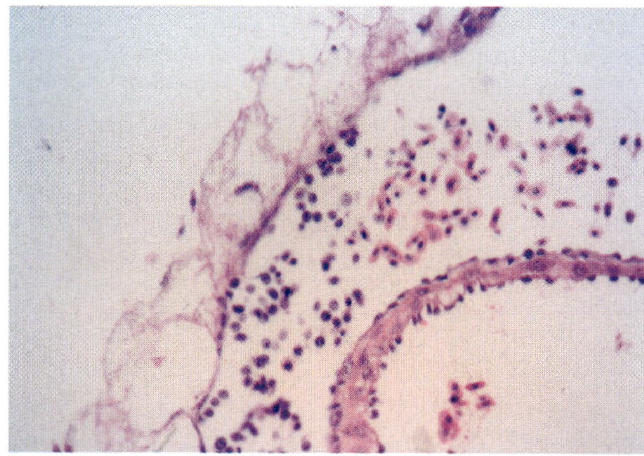

Figure 1.2, H&E x 260

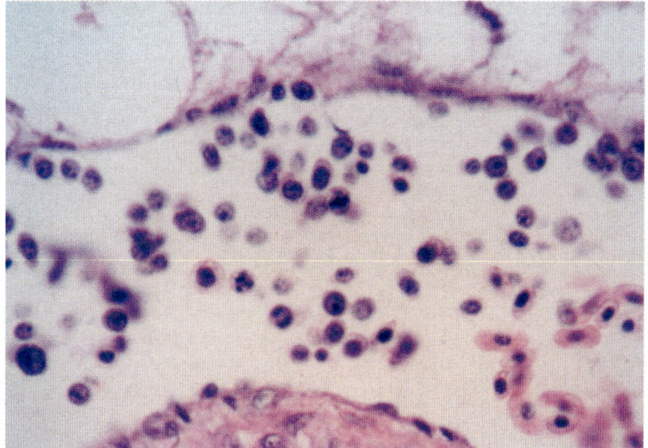

Figure 1.3, H&E x 520

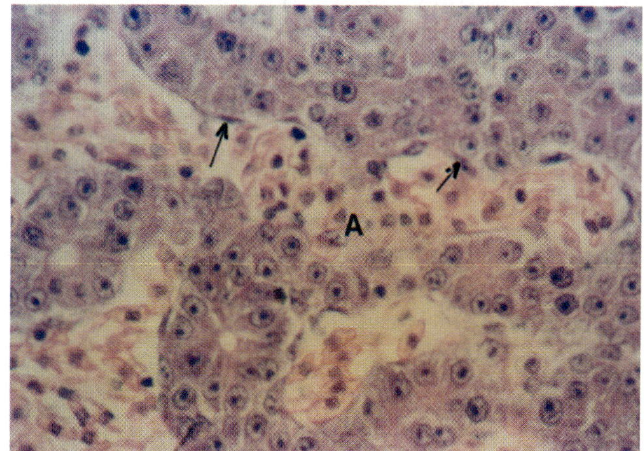

Figure 1.4, H&E x 520

3. Blood Formation in Human Embryonic Tissues (figs. 1.5-1.7)

In these photomicrographs, red cells develop from local liquefying products or intact local tissues. The liquefying process may not always be very prominent.[4] The mechanism of the appearance of tiny hyperchromatic blue-black round bodies (which are present in some red cells) may be similar to Howell-Jolly body formation in adult red cells. Howell-Jolly bodies are commonly found in red cells following acute spleenic infarction (Larrimer, et. al., 1975). In addition, I have shown that red cells can directly develop from cardiac muscle, and many of these red cells may contain Howell-Jolly bodies (fig. 1.6), the significance of which is unknown.

Whenever we see red cells arising through nucleated stages we automatically, but incorrectly, assume that they are following Neumann's (1868) theory of erythropoiesis. Neumann's erythropoietic series cells, which are somewhat similar to the cellular series shown in figure 1.7 in the liver, have little erythrogenic property compared to the direct development of red cells as hemoglobin globules, as shown in figure 1.6 in an area of cardiac muscle lysis. Similar observations have been made by Badarau (1969).

Figures 1.5-1.7 are taken from a formalin-fixed human embryo weighing 7.8 grams that was a product of spontaneous abortion (miscarriage). The tissues were sectioned at four or five microns thickness and stained with hematoxylin and eosin. The following

---

4.   Prominent liquefaction of embryonic cerebellum is demonstrated in Volume III (figs. 1.24 and 1.25).

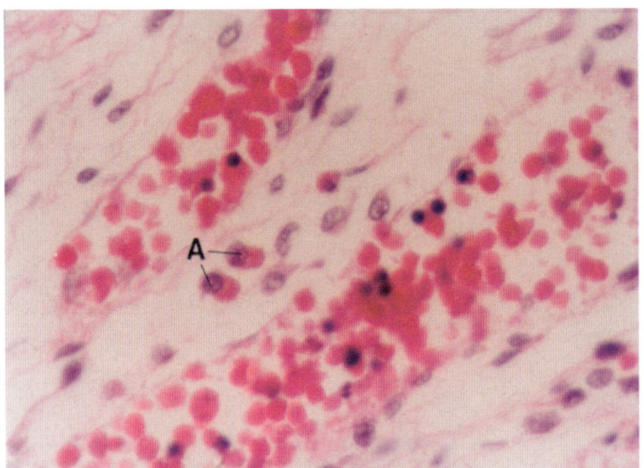

Figure 1.5, H&E x 520

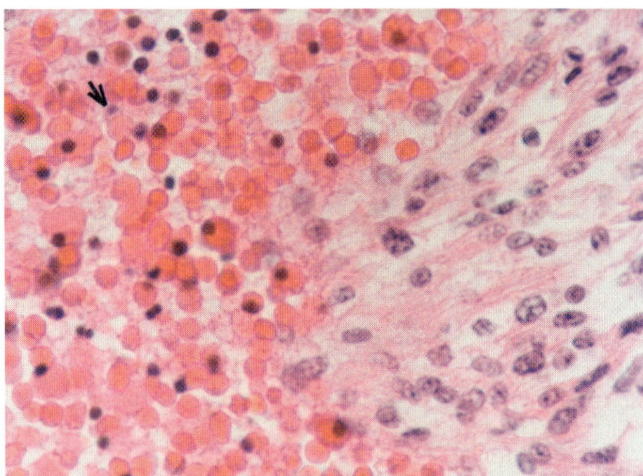

Figure 1.6, H&E x 520

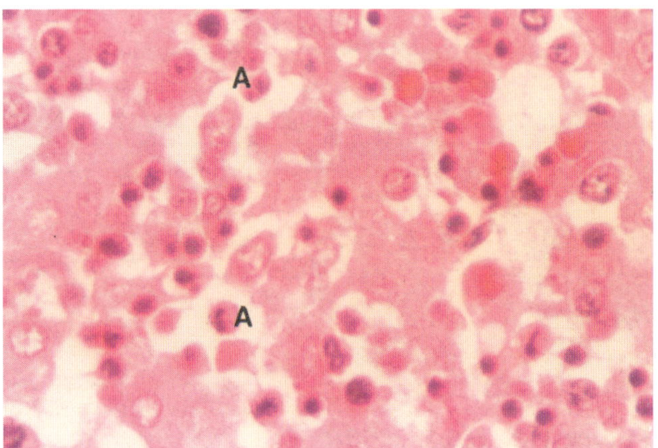

Figure 1.7, H&E x 800

photomicrographs are taken from undifferentiated mesenchymal tissue (fig. 1.5), cardiac muscle (fig. 1.6), and liver (fig. 1.7). Figures 1.5-1.7 below are reprinted from Volume III (McDonald, 2001).[5]

Red cells and blood capillaries developing from mesenchymal tissue (fig. 1.5); and cardiac muscle (fig. 1.6)

Figure 1.5. Blood and blood vessels are developing from the liquefied product of undifferentiated mesenchymal tissue. This figure shows red hemoglobin globules of various sizes (without nuclei) in two columns. A few of the red globules contain tiny hyperchromatic dots known as Howell-Jolly bodies. Note that there is a suggestion of endothelial membrane formation at the margins of these columns towards the development of large blood capillaries. There are two erythrogenic nucleated cells (A) of mesenchymal origin between the two columns of red cells. H&E x 520

Figure 1.6. Massive development of red cells as hemoglobin globules from dissolving cardiac muscle is shown in this figure. The muscle tissue has undergone direct red cell formation (this is commonly mistaken as a hemorrhage) and the muscle fiber is being replaced by developing red cells. Several red cells contain tiny hyperchromatic bluish-black round dots, which are known as Howell-Jolly bodies (arrow). H&E x 520

Red cells and erythrogenic nucleated cells developing directly from lysing liver cells (fig. 1.7)

Figure 1.7. Scattered throughout this figure are many round hyperchromatic nucleated cells which mimic so-called Neumann's erythropoietic series cells. However, red cells are actually forming directly as hemoglobin globules from liquefying liver cells. In the upper right of this figure, a group of red cells is originating from dissolving liver tissue. In addition, there are hyperchromatic small nuclei with very little hemoglobin developing from them. These cells are traditionally compared to Neumann's erythropoietic series cells. (A) points to cells considered erythrogenic SN cells (segmented nuclear cells). H&E x 800

---

5.   For a detailed discussion of my findings in human embryonic tissue see Chapter 1, Volume III (figs. 1.13-1.33).

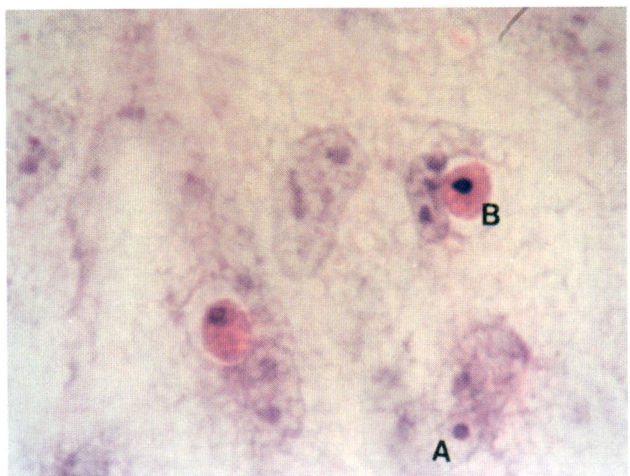

Figure 1.8, H&E x 1300

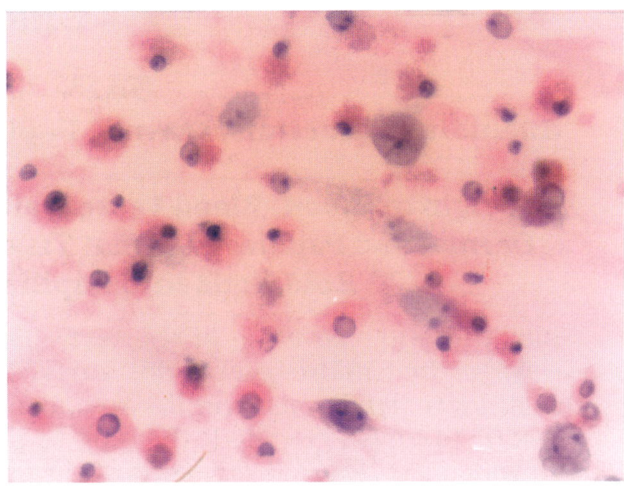

Figure 1.9, H&E x 520

4. Red Cell Development *In Vitro* from Chick Embryo Nonmedullary Tissues
(figs. 1.8-1.9)

Red cell development *in vitro* from both liver and kidney tissues has been previously observed in tissue cultures of tadpoles of the bullfrog *Rana catesbeiana* (Broyles and Deutsch, 1975). In order to study the possibility of red cell development from different tissues *in vitro*, I took advantage of a few slides which are still in good condition from my simple tissue culture experiments done in 1964. These photomicrographs are taken from the available slides of simple tissue culture experiments set up using chick embryo tissues as implants, grown *in vitro* on a cover glass. For more information on the procedure, see Materials and Method of Study.

Cellular proliferation occurs from inoculated tissue fragments grown between two cover glasses and bathed in tissue culture medium. After different periods of incubation, the cover glasses with growth are usually stained with H&E. In a few instances, other staining procedures are employed such as Giemsa and fat stain (Oil O Red).

These photomicrographs are from the following chick embryo tissue cultures: fig-ure 1.8 - a 3 day liver tissue culture of a 6-day-old chick embryo; and figure 1.9 - a 6 day cardiac muscle tissue culture of a 7-day-old embryo. Figures 1.8 and 1.9 below are reprinted from Volume III (McDonald, 2001).[6]

Red cells developing from liver cells (fig.1.8)

Figure 1.8. Within liver cell nuclei, there arise large hemoglobin globules containing single chromatin dots. Initially, red cells start as chromatin dots surrounded by clear space (A), which later becomes hemoglobinized and turns into well-developed red cells (B). (See also figure 12.12 in Chapter 12, Liver.) H&E x 1300

*Red cells developing from cardiac muscle
(fig. 1.9)*

Figure 1.9. Red cells of various sizes with tiny hyperchromatic dots (known as Howell-Jolly bodies) are developing from cardiac muscle cells *in vitro*. (See also figure 4.3 in Chapter 4, subsection 1, Cardiac Muscle.) H&E x 520

---

6. For a detailed discussion of my findings in embryonic tissue *in vitro* see Chapter 1, Volume III (figs. 1.34-1.45).

Chapter 2

SKIN (figs. 2.1-2.25)
(Including Epidermis, Dermis, Papillary Dermis and Benign Skin Lesions)

1. Origin of Blood, Blood Vessels and Inflammatory Cells from Epidermal, Dermal and Papillary Dermal Tissue; and the Transformation of Lower Epidermis into Papillary Dermis (figs. 2.1-2.15)

Skin, the outer covering of the body, is made up of two main layers: the surface epithelial layer (*epidermis*) and the so-called "true skin" (*dermis*). Beneath the skin is a loose *connective tissue* layer, consisting of oblique elastic fibers and containing lymphatics, blood vessels, nerves, and generally fat. In a histological section of normal skin the border between the epidermis and the dermis is irregular because the cone-shaped dermal papillae reach upward and indent the inner surface of the lower epidermis. Because the ridges of the epidermis separating the papillae appear in histological sections as pegs, they are often referred to as rete pegs or rete ridges (Lever, 1961).

In textbooks, the epidermis is usually divided into four layers. Starting with the deepest layer adjoining the dermis and going outwards, these layers are: 1. the basal layer; 2. the stratum malpighii (or *stratified squamous cell layer* which contains the intercellular bridges); 3. the granular layer (which usually contains keratohyaline granules); and 4. the horny *keratin* layer (which usually contains keratin fibrils). When remnants of pyknotic nuclear material are present in the horny layer, it is called the parakeratotic layer [*parakeratosis*, see (B), figure 2.1]. The dermis is made up of the outer *papillary* layer containing the nerve endings, capillaries and glandular structures, and a deeper *reticular* layer, largely made up of fibrous bundles.

Upon critical examination of various cellular structures, the transformations of one type of cell to another (including blood and blood vessels) are found to have originated from different cells of the epidermis and dermis. Also shown here are the various stages of transformation (transdifferentiation) of epidermal cells in the development of dermis from the lower part of the epidermal tissue, thus shifting the level of dermal-epidermal junction outward.

Presented below are the general changes observed in the development of blood and/or blood vessels in epidermal tissue. Also shown is the transformation and proceeding outward of the lower layer of epidermal cells with origin of the papillary dermis with blood and blood vessels. From squamous cells, the development of reactive cells such as lymphocytes, segmented nuclear cells and plasma cells is seen. A lymphatic channel, with lymphocytes within its lumen, is observed originating from newly developing papillary dermis. The development of papillary dermis from lower epidermal cells, and blood and blood vessels originating from the newly developed papillary dermis can also be observed. For cellular changes in cancers of the skin see Chapter 22.

Figures 2.1-2.15 are taken from seven skin specimens surgically excised for benign lesions from seven different adult patients. These figures were selected to show the prominence of different cell layers of the epidermis and dermis, and also in order to demonstrate various cellular and tissue changes in the development of blood, blood capillaries, lymphatic channels, and reactive or inflammatory cells (and their erythrogenic properties).

Figure 2.1 is from a skin lesion of the left great toe showing marked hyperkeratosis, parakeratosis, acanthosis and some ulceration. Figures 2.2, 2.3, 2.5, 2.7, 2.8, 2.10 and 2.13-2.15 are from a skin lesion of the back with seborrheic keratosis, hyperkeratosis, acanthosis, papillomatosis and inflammatory reaction in the dermis. Fig-

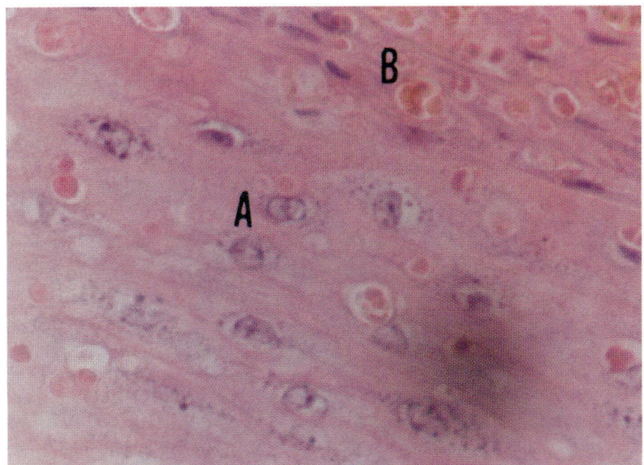

Figure 2.1, H&E x 520

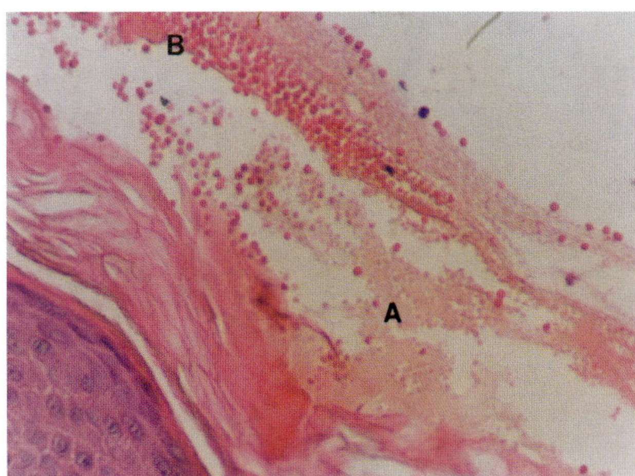

Figure 2.2, H&E x 260

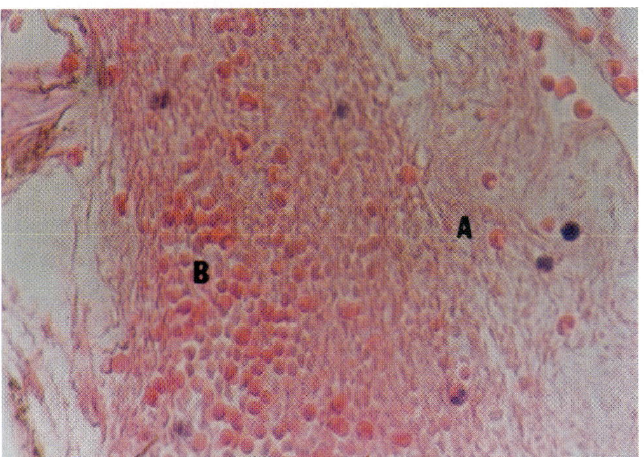

Figure 2.3, H&E x 520

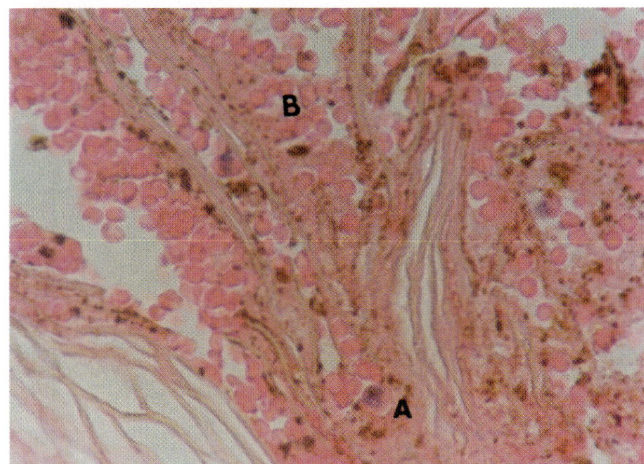

Figure 2.4, H&E x 520

ure 2.4 is from a skin lesion of the right cheek with seborrheic keratosis, marked keratinization, and superficial hemorrhage in the keratin layer. Figure 2.6 is from a skin lesion of the scalp with hyperkeratosis and prominent acanthosis. Figure 2.9 is from a skin lesion of the right ear with seborrheic keratosis in polypoid formation. Figure 2.11 is from a skin lesion from the base of the penis with seborrheic keratosis. Figure 2.12 is from a skin lesion of the right ear with a fibroepithelial polyp and microabscesses. Figures 2.1 and 2.6 are reprinted from Volume I (McDonald, 1989).

Development of blood and blood vessels from epidermal tissue (figs. 2.1-2.10)

Figure 2.1. Red cells appear directly as hemoglobin globules within the squamous cells of the stratum granulosum (A), and the parakeratotic cell layer (B). H&E x 520

Figure 2.2. Partial liquefaction of the thick keratin of the outer epidermis and development of red cells from this liquefying keratin, and its liquefied product, are shown here. The red cells start as minute and faintly stained granules (A), and then become almost full-size and fully hemoglobinized red cells (B). H&E x 260

Figure 2.3. In a higher magnification, these red cells are developing in a mosaic pattern from the keratin fibers, beginning as barely stained tiny particles (A) and maturing to regular-sized red cells (B). H&E x 520

Figure 2.4. Red cells (A) are in the early stages of development. Red cells (B) are further along in their development in size

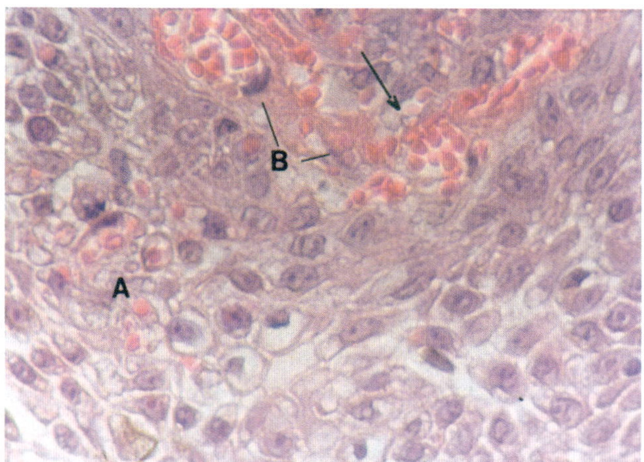

Figure 2.5, H&E x 520

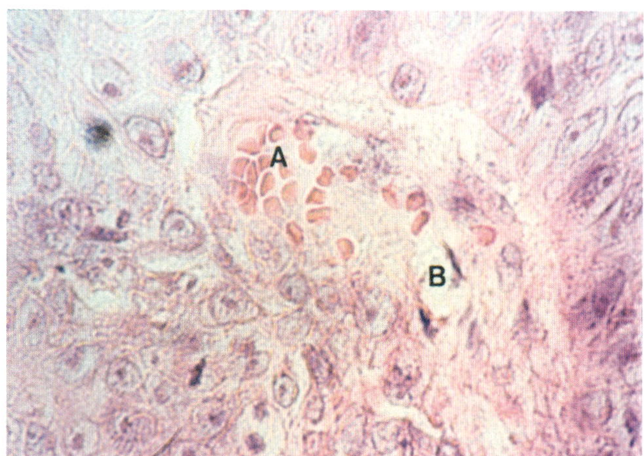

Figure 2.6, H&E x 520

and shape, but not in hemoglobinization. Note that some red cells still show remnants of melanin (brown) pigment. H&E x 520

Figure 2.5. Isolated red cells are arising from the squamous epithelial cells in area (A). In the upper section, red cells are seen developing in clusters (or columns) toward the formation of blood vessels. Endothelial nuclei (B) bordering the clusters of red cell columns are arising from vanishing squamous cells. The arrow points to early, barely hemoglobinized globules. H&E x 520

Figure 2.6. Red cells (A), mostly polyhedral in shape and in a mosaic pattern, are appearing within squamous cells with prominent intercellular bridges. Note that the intercellular bridges pass through squamous cells (including the nuclei), intercellular spaces, and even through red cells in the early developmental stages (can be seen with a magnifying glass). (B) points to liquefied squamous cell product, clear plasma, partially encircled by developing endothelium that arise from peripheral cell substance. H&E x 520

Transformation of epidermis into papillary dermis (figs. 2.7-2.10)

Figure 2.7. The papillary dermis is being formed by the transformation (transdifferentiation) of deeper layers of epidermal cells into loose, fibrillary connective tissue. The arrow points to the remains of one epidermal cell connected to streams of loose wavy fibrils of papillary dermis.

There is a large developing blood capillary probably forming from the epidermal cells, similar to figure 2.6. H&E x 520

Figure 2.8. Surrounded by and developing from squamous epithelium is newly formed papillary dermal tissue, which contains blood and a vascular channel. Note the different stages of blood vessel formation (A and B). At (A), the endothelial cells enclose the lumen partly occupied by erythrogenic gel. At (B), a few well-developed red cells are in the capillary lumen lined by developing endothelial cells. (C) points to an undissolved epithelial cell fragment in the lumen. The arrow points to the remains of dissolving epidermal nuclear structures. H&E x 400

Figure 2.9. A further transformation of lower level epidermal cells (upper marginal area) into papillary dermis containing multiple blood vessels is shown here. Note the similarity among cells situated in the basal layer of the epidermis (A), in the developing papillary dermal area (B), and in the not yet dissolved vascular lumen (C). The origin of vascular content from the local liquefied tissue is evidenced by the presence of protoplasmic cellular remnants, and the varying sizes and various degrees of hemoglobinization of the red blood cells. H&E x 520

Figure 2.10. The commonly known "dermal papillae" (with blood vessels in cross-section) are surrounded by stratified

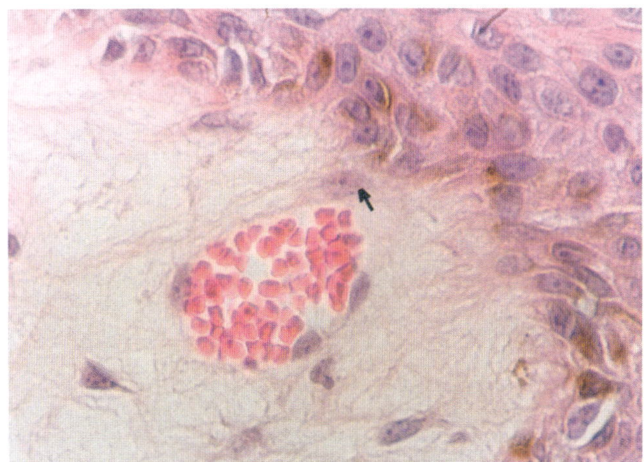

Figure 2.7, H&E x 520

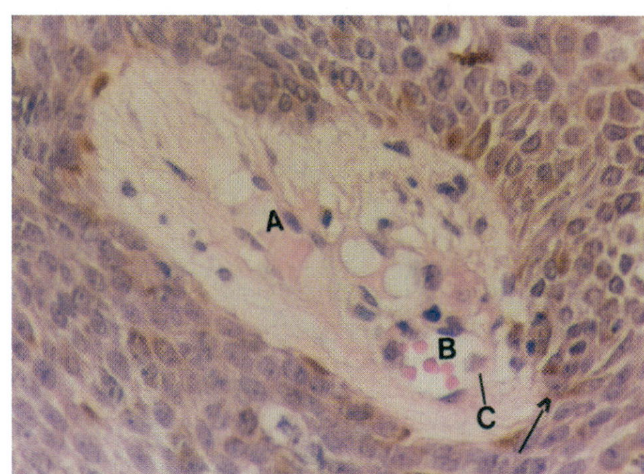

Figure 2.8, H&E x 400

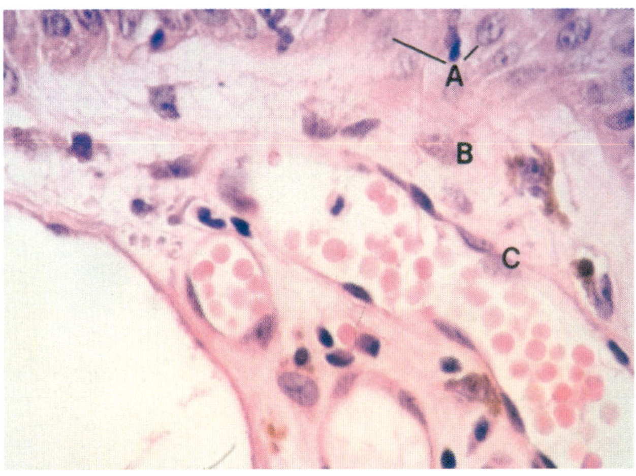

Figure 2.9, H&E x 520

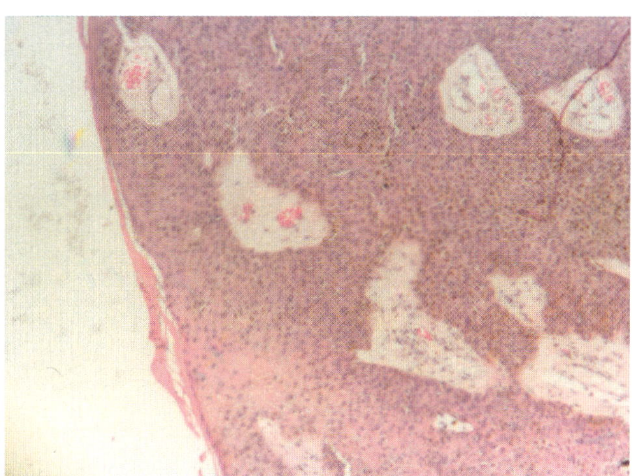

Figure 2.10, H&E x 80

squamous epithelium. The origin of such structures from local squamous epithelium is explained by figures 2.7-2.9. H&E x 80

From squamous epithelial cells the development of reactive/inflammatory cells; lymphocytes (figs. 2.11-2.15); SN cells (figs. 2.13-2.15); and plasma cells (fig. 2.15)

Figure 2.11. The left side of this figure demonstrates transitional stages of lymphocyte development from squamous cell nuclei, with clearing of the surrounding cytoplasm. One such lymphocyte (A) is surrounded by squamous epithelium. Note one red blood cell originating in hyalinized epidermal cells (arrow). H&E x 800

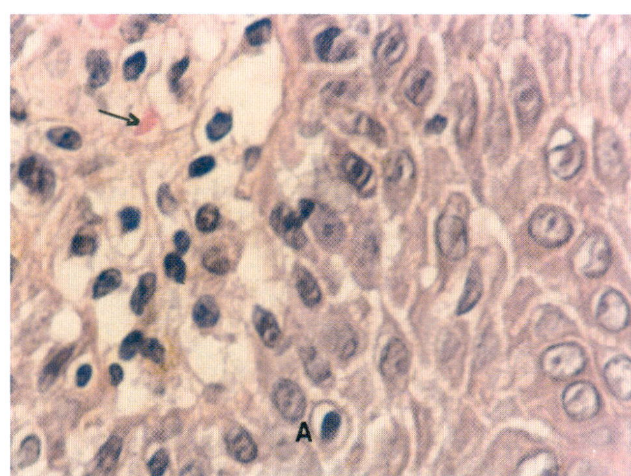

Figure 2.11, H&E x 800

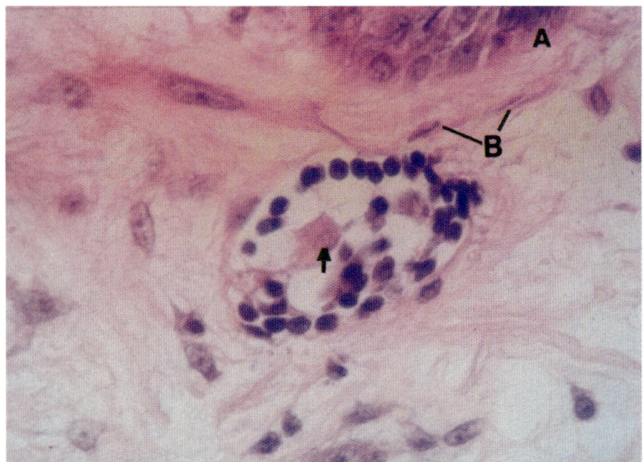

Figure 2.12, H&E x 520

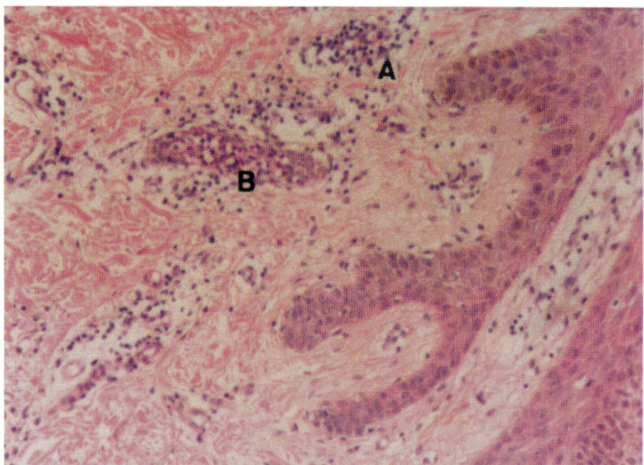

Figure 2.13, H&E x 130

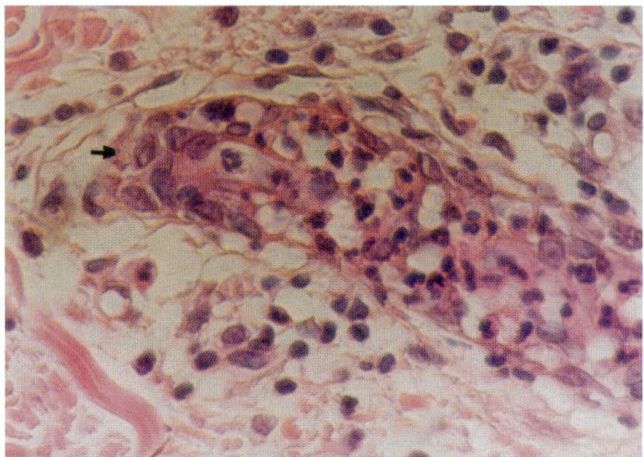

Figure 2.14, H&E x 520

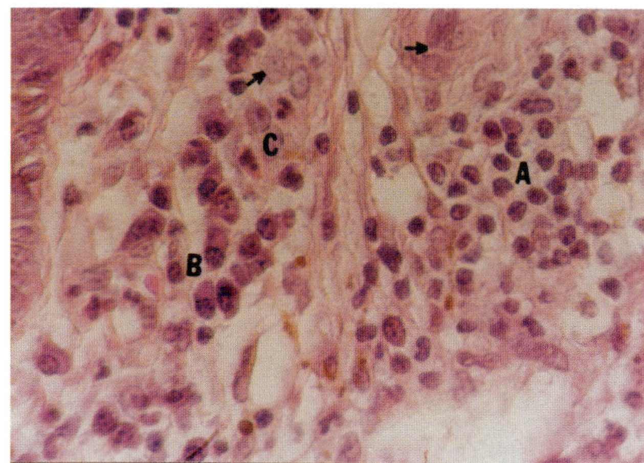

Figure 2.15, H&E x 520

Lymphocytes and lymphatic channel development from papillary dermis (fig. 2.12)

Figure 2.12. Shown here is the development of a lymphatic channel containing plasma and lymphocytes from the upper developing papillary dermis. This figure beautifully demonstrates the undissolved cellular structures of dermal tissue (arrow) within the lumen of a formative lymphatic channel. The presence of local cells resisting lysis and lying within the lymphatic channel clearly demonstrates the local origin of the lymphatic channel and its lymphocytes from dermal tissue. In the lumen, some of the lymphocytes are well formed while others are still adhering to the remains of dermal tissue. (A) denotes the bottom layer of epidermal cells. The loose collagenous fibrils are forming from the lowest layer of these epidermal cells. Two narrow, elongated fibrocyte nuclei (B) are probably arising from the newly developed connective tissue fibrils. (See also figures 8.7 and 8.8 in Chapter 8, Lymph Nodes and Lymphatic Channels.) H&E x 520

Figures 2.13 and 2.14. In figure 2.13, the tips of some isolated rete ridges (A and B) are almost completely replaced by inflammatory cells originating from the epidermis. The remains of epidermal cells in (B) are still recognizable in higher magnification in figure 2.14, where many SN cells (segmented nuclear cells) are arising directly from partially hyalinized and vacuolated epidermal cells. The arrow points to the intercellular bridges of remaining epidermal cells (use a magnifying glass). Figure 2.13 H&E x 130; and figure 2.14 H&E x 520

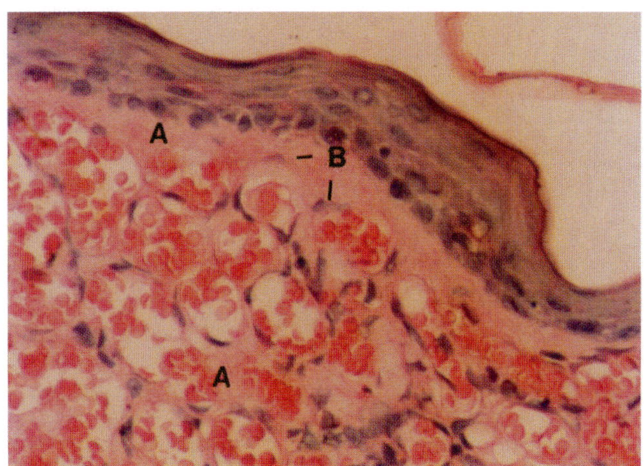

Figure 2.16, H&E x 400

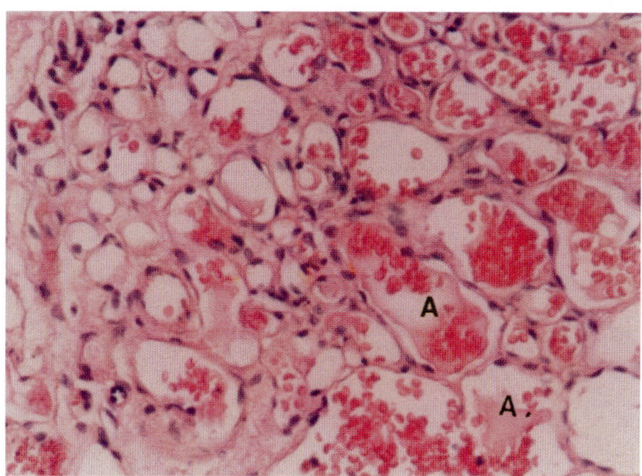

Figure 2.17, H&E x 260

Figure 2.15. Arising from epidermal cells are clusters of lymphocytes (A), plasma cells (B), and a few SN cells (C). The majority of plasma cells are formed in the nonmedullary tissues, especially during chronic inflammation. They may originate as regular sized and shaped cells or as tiny, hyperchromatic bodies (plasma cell bodies) which then mature into regular plasma cells (Patterson, 1986). Arrows point to the remaining faintly visible, but still identifiable, epidermal cells. H&E x 520

2. Capillary Hemangioma (figs. 2.16-2.17)

A capillary hemangioma is a flat, nonelevated benign skin tumor varying in color from pink to dark purplish-red and made up of newly formed vascular units. Local development of blood and blood capillaries, replacing circumscribed subepidermal collagen, in the form of capillary hemangioma is shown here. Figures 2.16 and 2.17 are from a surgically excised capillary hemangioma from the abdomen of a 58-year-old white female.

Blood and blood capillary formation from dermal collagen (figs. 2.16 and 2.17)

Figure 2.16. From dermal collagen, the origin of red blood cells is pointed out by area (A). The endothelial cells of the vascular units (B) are developing from peripheral collagen. H&E x 400

Figure 2.17. In another area of the same tissue section as figure 2.16, several small capillaries on the left have not yet developed red blood cells, and the walls are partially lined with collagen from which endothelial cells are developing. In the lumen of many capillaries (A), red cells can be seen originating from dissolving collagen through the stage of erythrogenic gel. H&E x 260

3. Hemorrhagic Dermatofibroma (figs. 2.18-2.20)

Hemorrhagic dermatofibroma is a benign skin lesion arising from dermal collagen and showing development of red blood cells and hemosiderin pigment; the origin of the latter is commonly thought to be due to hemorrhage. However, I have found that the red blood cells are developing locally, and the hemosiderin pigment is derived as an abortive attempt at red cell development. Figures 2.18-2.20 are from an 8 mm (in greatest diameter) nodular skin lesion surgically removed from a 40-year-old patient diagnosed with cellular hemorrhagic dermatofibroma.

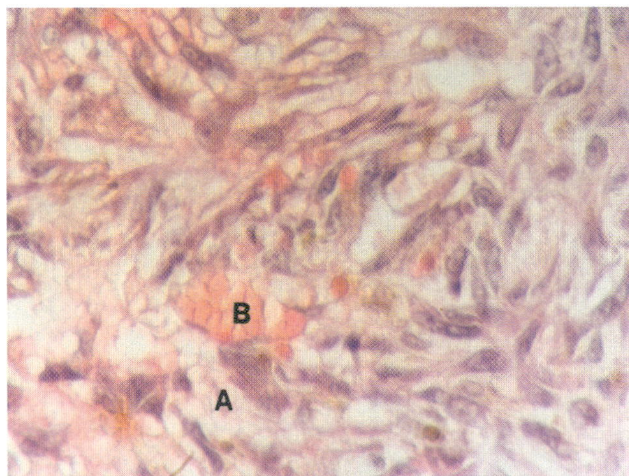

Figure 2.18, H&E x 520

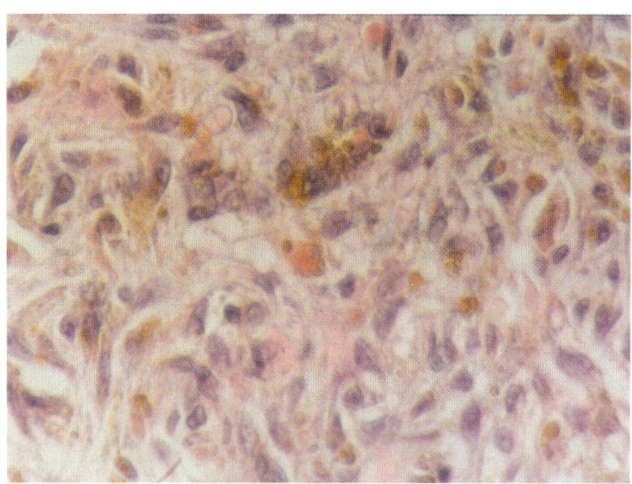

Figure 2.19, H&E x 520

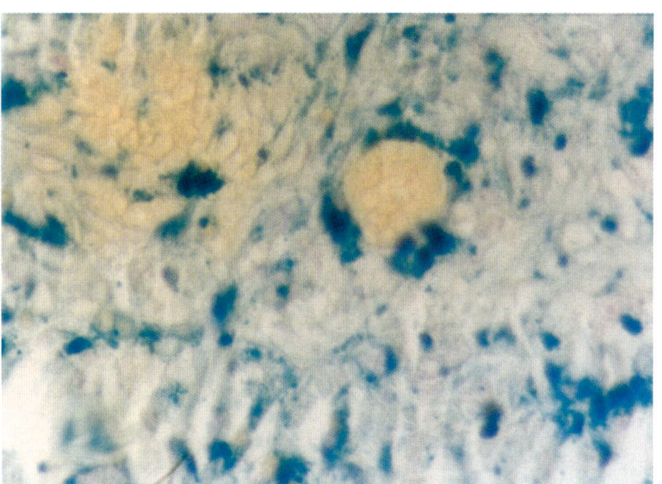

Figure 2.20, Iron Stain x 520

Development of red blood cells and hemo-siderin pigment from fibrocytes (figs. 2.18-2.20)

Figure 2.18. From spindle-shaped fibrocytes there is development of vacuolated net-like structures (A) preceding red cell formation (B). Such red cell sized net-like changes can also be seen in figures 2.19 and 2.20. Also, there is a small amount of brown hemosiderin pigment present. H&E x 520

Figure 2.19. In this figure there is more hemosiderin pigment formation by fibrocytes with very little red cell development. Development of hemosiderin pigment from local tissues as an abortive attempt in red cell development is explained in renal carcinoma in Volume I (McDonald, 1989, page 92). H&E x 520

Figure 2.20. This figure is from the same specimen as figures 2.18 and 2.19. The large amount of blue-stained material is the hemosiderin pigment shown by iron stain. Prussian Blue Stain for Iron x 520

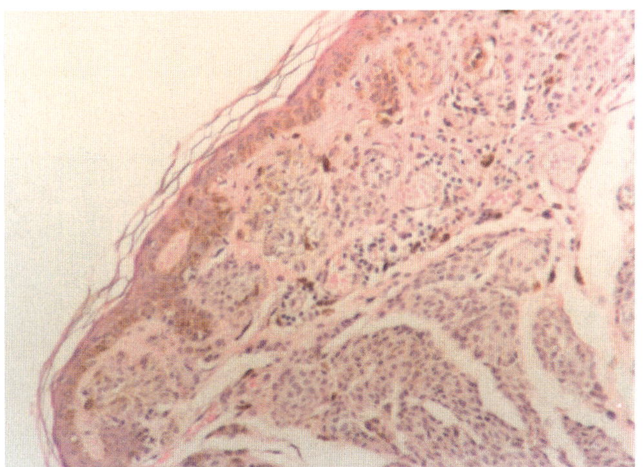

Figure 2.21, H&E x 130

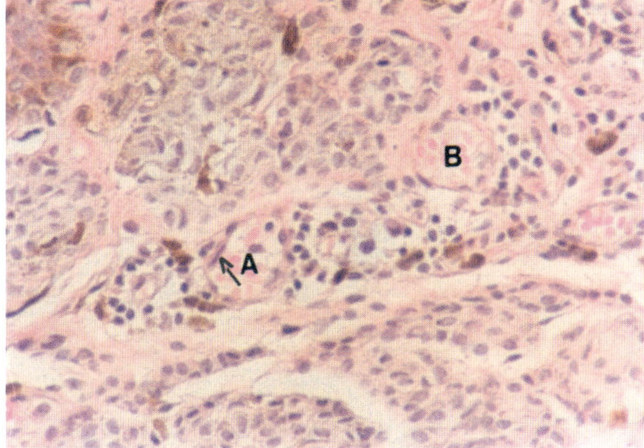

Figure 2.22, H&E x 260

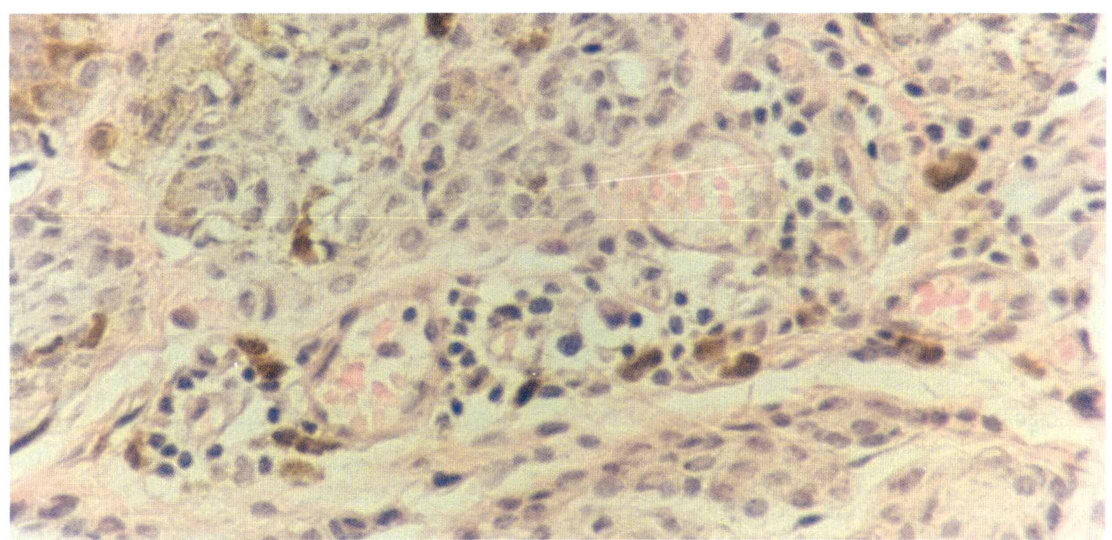

Figure 2.23, H&E x 520

4. Nevus (figs. 2.21-2.23)

The nevus, commonly called *mole*, is a general term for circumscribed new growths of the skin. This usually benign pigmented skin lesion arises from the epithelial cells of the skin. The transformation of nevus cells into red cells, blood capillaries and lymphocytes is shown here. Figures 2.21-2.23 are from a surgically excised nevus from a 44-year-old Caucasian female.

Transformation of nevus cells into red blood cells, blood capillaries and lymphocytes (figs. 2.21-2.23)

Figures 2.21-2.23. Figures 2.21-2.23 are from a moderately pigmented intradermal nevus with a pseudoalveolar pattern. Figure 2.21 is a low-power view of this lesion, and figures 2.22 and 2.23 are magnified views of the central area of figure 2.21. The transformation of the pseudoalveolar structure into developing blood capillaries is better viewed in figures 2.22 (A and B) and 2.23. The arrow (fig. 2.22) points to the endothelial lining cells developing from the nevus cells. In the lumen of the developing vascular units, red cells are forming from nevus cells undergoing vacuolar changes. Note that the nevus cells are transforming into lymphocytes, many of which still contain brownish pigments (best seen in figure 2.23). Figure 2.21 H&E x 130; figure 2.22 H&E x 260; and figure 2.23 H&E x 520

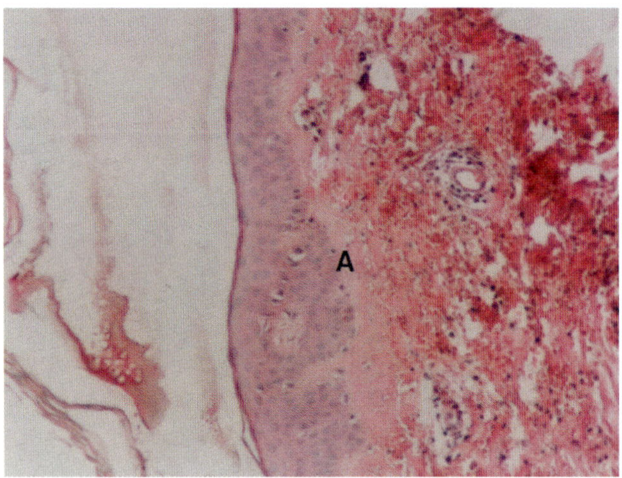

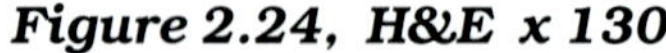

Figure 2.24, H&E x 130

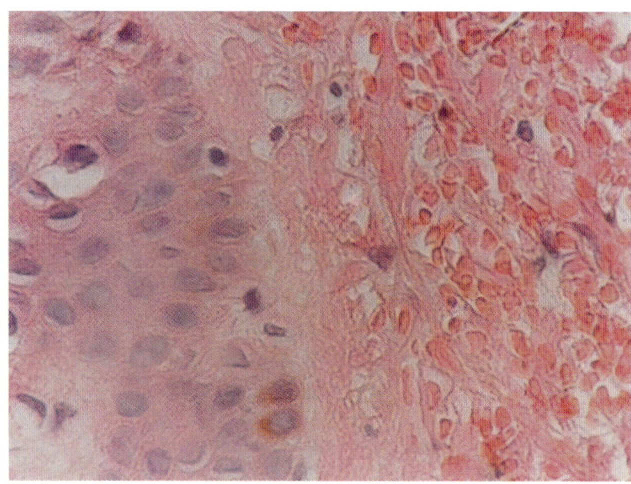

Figure 2.25, H&E x 520

5. Senile Ecchymosis (figs. 2.24-2.25)

Senile ecchymosis is a benign "hemor-rhagic" skin or mucous membrane lesion (without history of trauma) usually seen in elderly persons with thin membranous skin. There is disintegration of collagen bundles in the upper dermis associated with the development of numerous red cells, without the tendency for vessel formation. Figures 2.24 and 2.25 are from a biopsy of an elderly African-American female with unexplained hemorrhagic rashes over both legs and thighs.

Development of red blood cells from collagen bundles of upper dermis (figs. 2.24 and 2.25)

Figure 2.24. In the deeper dermis there is marked disintegration of collagen associated with profuse production of red cells to the extent that is usually designated as hemorrhage. Area (A) is seen in higher magnification in the next figure. H&E x 130

Figure 2.25. At the margin of the lesion, angular-shaped red cells in a mosaic pattern are arising from disintegrating collagen. H&E x 520

Chapter 3

ADIPOSE TISSUE (figs. 3.1-3.16)
(Including Adipose Tissue from Pericolon, Lipoma and Throat)

Normal adipose tissue consists of large cells containing mostly fat. In routine stains, such as H&E, the fat is no longer present, as it has been dissolved in the staining process. Unless it is specially prepared and stained with a fat stain, normal adipose cells (as seen in many figures below) show a very fine and barely recognizable cell membrane, a tiny or absent nucleus, and a cytoplasm (which consists of fat) that becomes a clear space due to routine tissue processing with xylene and alcohol.

Adipose tissue, when disturbed by surgical procedures, by inflammation, or when influenced by the presence of nearby cancer (as well as other causes), shows its capacity to transform into a variety of cellular and tissue structures, including blood, blood vessels, lymphatic channels, and different inflammatory or reactive cells.[1] Cellular changes in malignant adipose tissue are presented in Chapter 23.

All figures are taken from adipose tissue of surgically excised specimens. Figures 3.1, 3.2 and 3.10-3.16 are taken from pericolic adipose tissue of an ulcerative lesion of colon. Figures 3.3 and 3.4 are from an upper leg lipoma. Figures 3.5 and 3.6 are from adipose tissue included in a specimen of total laryngectomy excised for carcinoma of the larynx. Figures 3.7-3.9 are from pericolic adipose tissue near the resection line of a transverse colon with carcinoma.[2] Figure 3.11 is reprinted from Volume III (McDonald, 2001).

Adipose tissue forming blood and blood capillaries (figs. 3.1-3.6)

Figures 3.1 and 3.2. Figure 3.1 shows pericolic fatty tissue resected with a colon ulcer. Area (A) is seen in higher magnification in the next figure. In figure 3.2 red cells are developing in a mosaic pattern from collagenous red fibrillary substance between the adipose cells. Figure 3.1 H&E x 130; and figure 3.2 H&E x 520

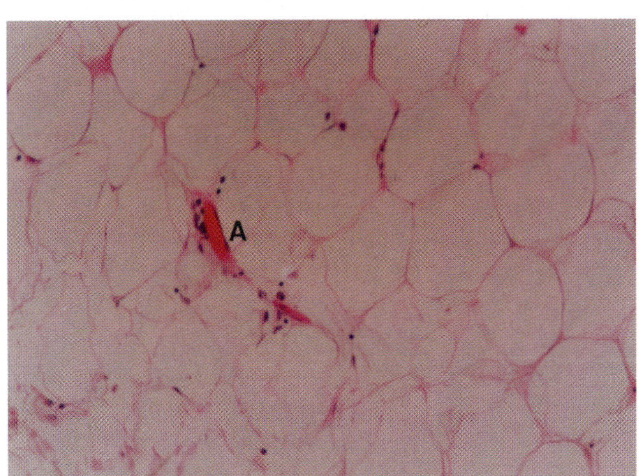

Figure 3.1, H&E x 130

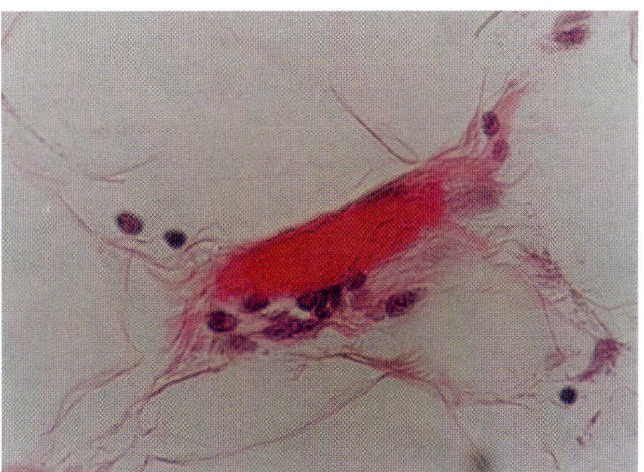

Figure 3.2, H&E x 520

---

1. See figures 3.1-3.16 below, figure 30.12 of this volume, and also figures 48-55 of Volume I (McDonald, 1989). Adipose tissue when stimulated by a near-by cancer, may transform into similar malignant tissue with the same morphologic characteristics as the nearby cancer, thus suggesting an additional mechanism for the local spread of tumors, as shown in adipose tissue near breast cancer in figures 48-51 of Volume I and figures 24.16 and 24.17 of this volume.
2. This colon carcinoma lesion is presented in figures 30.13-30.20.

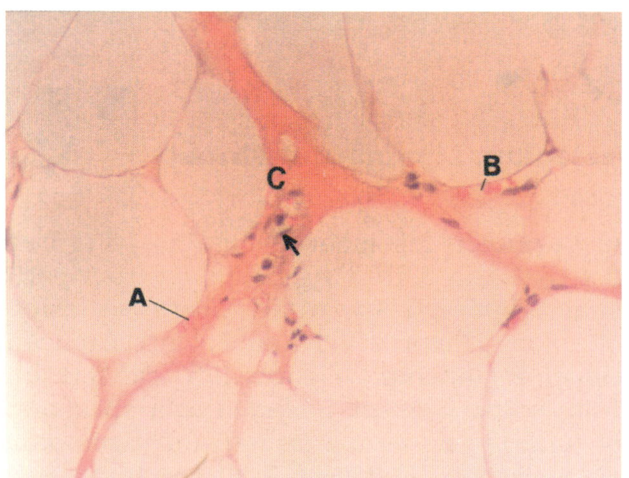

Figure 3.3, H&E x 280

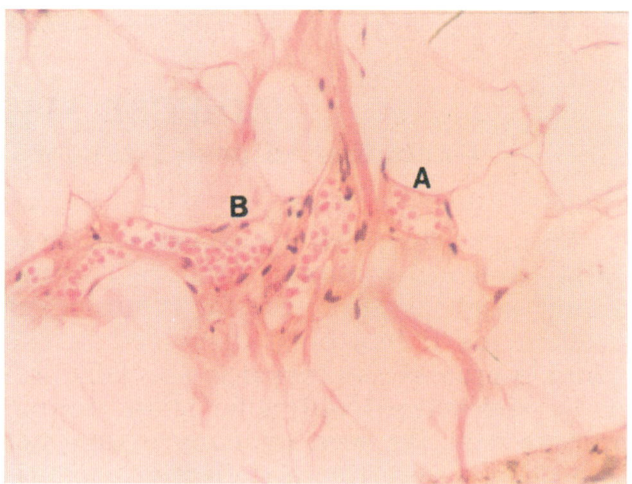

Figure 3.4, H&E x 260

Figure 3.3. A light-pink fibrinoid band with developing blood capillaries has formed in the adipose tissue. A single column of barely visible red cells (A) has appeared. One developing capillary already contains red cells (B) in its lumen. Other formative capillaries (C) are also in the early stages of development. Note the appearance of the endothelial cells (arrow) associated with the canalization of the lumen. H&E x 280

Figure 3.4. The early stages of adipose tissue hyalinization and the appearance of red cells are seen in area (A). Further developmental stages of blood capillaries are present in (B). H&E x 260

Figure 3.5. Broad bands of red hemoglobinized substance (a) appear between the adipose cells. From this red substance, clusters of compressed red cells arise and these areas of development form into wide blood capillaries (b). Tiny endothelial nuclei have appeared around the borders of capillaries (b). In figures 3.5 and 3.6 the linings of the adipose cells are stained dark red; this is probably due to the bloody hemorrhagic laryngectomy procedure performed. H&E x 200

Figure 3.6. This is a higher magnification of a blood vessel similar to the ones seen in figure 3.5. The lumen of this large blood vessel is filled with red hemoglobinized substance which is starting to break down in development of red cells. In area (a) the

red hemoglobinized substance has not yet started to form red cells. The blood vessel wall is formed of a single layer of endothelial cells. H&E x 400

Development of erythrogenic SN cells from adipose tissue, and blood capillary development from the erythrogenic SN cells (figs. 3.7-3.9)

Figure 3.7. In this figure there are many erythrogenic segmented nuclear cells (A) developing from this edematous adipose tissue. Erythrogenic SN cells are seen in the columns of developing red cells (B). SN cells are also surrounding the developing capillary (C). Some endothelium (D) is forming associated with the developing hemoglobin. H&E x 260

Figures 3.8 and 3.9. In figure 3.8 many large blood capillary tubes are forming through erythrogenic SN cells developed from adipose tissue. Area (A) is seen in higher magnification in the next figure. In figure 3.9 erythrogenic SN cells form the outer wall of the large capillaries. As the SN cells transform into red cells, the capillaries enlarge in width and more erythrogenic SN cells develop along the new capillary wall and continue the growth of the capillary. (A) points to a portion of the intact outer wall of this large blood capillary. Notice that the outer wall is made up of erythrogenic SN cells. (B) points to a section of the inner lumen filled with compact

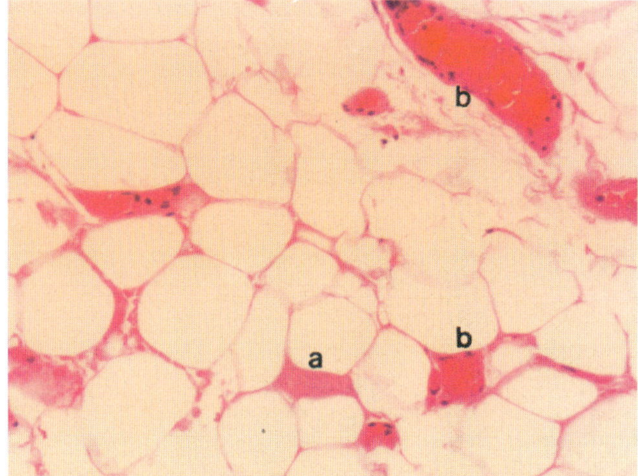

Figure 3.5, H&E x 200

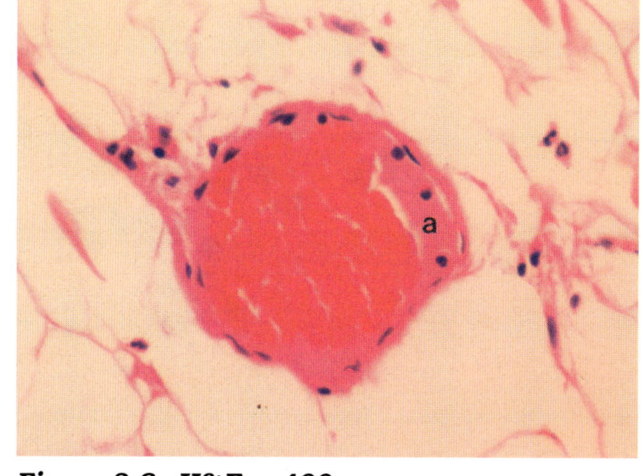

Figure 3.6, H&E x 400

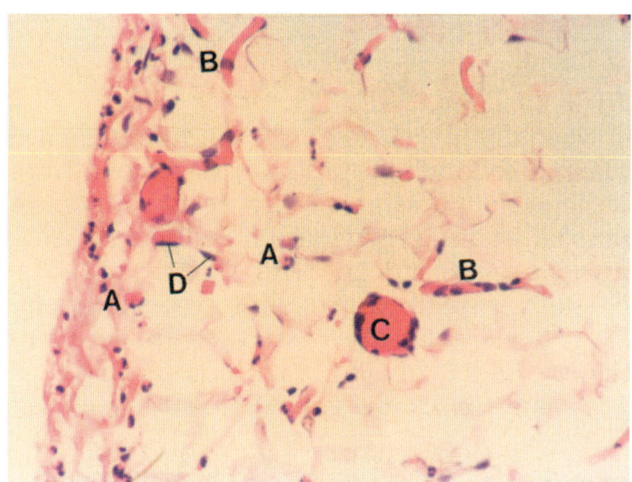

Figure 3.7, H&E x 260

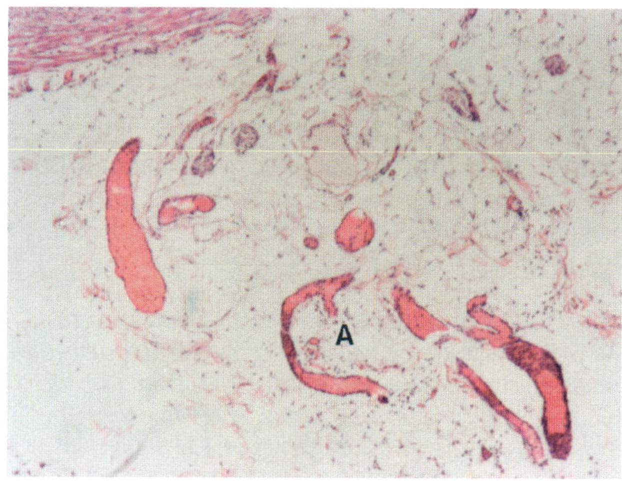

Figure 3.8, H&E x 52

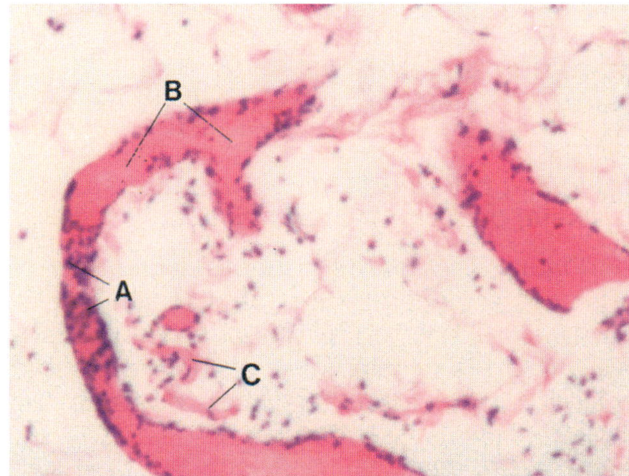

Figure 3.9, H&E x 128

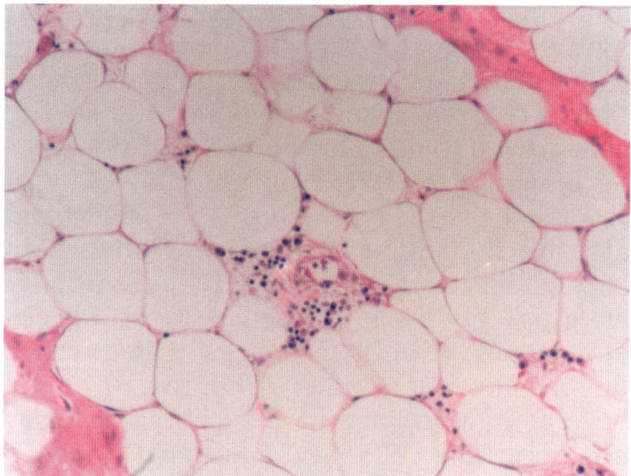

Figure 3.10, H&E x 130

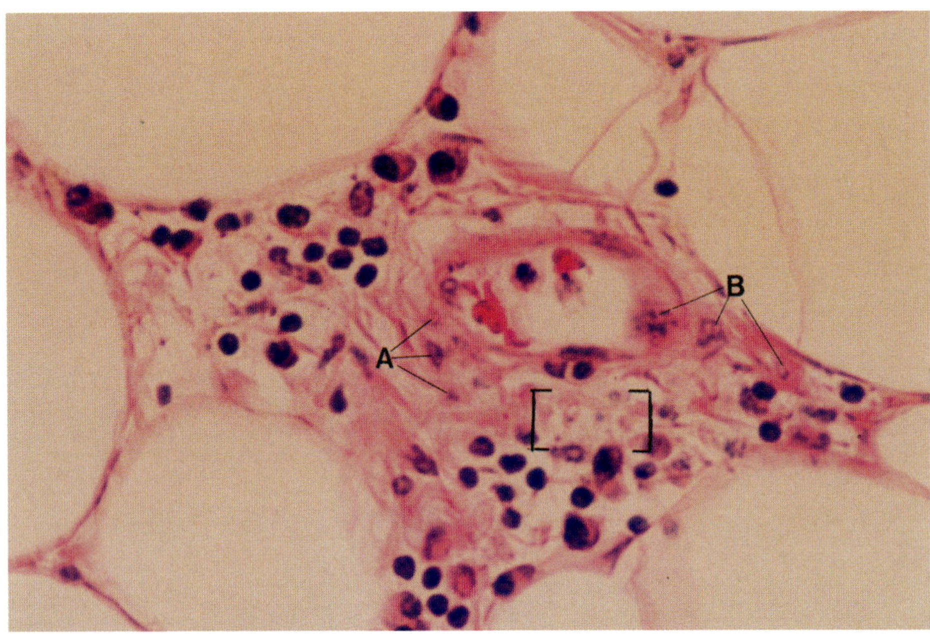

Figure 3.11, H&E x 400

red cells already formed from SN cells. (C) points to segments of small blood capillaries developing from erythrogenic SN cells. Figure 3.8 H&E x 52; and figure 3.9 H&E x 128

Development of RVR cells (Redefined Vasoformative cells of Ranvier) from adipose tissue, and reactive cells and blood capillary development from RVR cells (figs. 3.10 and 3.11)

Figures 3.10 and 3.11. In the center of figure 3.10 there is an area of reactive cell and blood capillary development. This area is seen in higher magnification in the next figure. In figure 3.11, the development of RVR cells (Redefined Vasoformative cells of Ranvier) from adipose tissue mimics the spawning of fish. Also shown is the further development of RVR cells into a variety of reactive cells, which includes red cells, and also into vascular channels. Starting as tiny, barely visible, lightly basophilic, irregular bodies (area between the brackets) the RVR cells progress into further developmental stages (A) and (B). The developing capillary in the center is forming from RVR cells, and the linearly placed RVR cells (B) are participating in the extension of this capillary to the right. Also

shown are erythrogenic lymphocytes and plasma cells which probably developed from RVR cells (similar to the cells shown in figures 21.28-21.32). Such cellular appearance as shown in this figure may be casually termed extramedullary hematopoiesis. (RVR cells are discussed in detail in figures 21.18-21.32.) Figure 3.10 H&E x 130; and figure 3.11 H&E x 400

Fat cells replaced by pools of red hemoglobin-like substance (figs. 3.12-3.14)

Figures 3.12 and 3.13. In this inflamed pericolic adipose tissue there are scattered fat cells replaced by pools of red hemoglobin-like substance. Area (A) is seen in higher magnification in figure 3.13. Figure 3.12 H&E x 52; and figure 3.13 H&E x 260

Plasma cell development from interstitial tissue between fat cells (figs. 3.14-3.16)

Figure 3.14. This is another example of a fat cell replaced by a pool of red hemoglobin-like substance. In the lower right quadrant, there is development of rows of plasma cells from the interstitial area between the fat cells. H&E x 520

Figure 3.15. Many plasma cells are arising from the edematous interstitial spaces

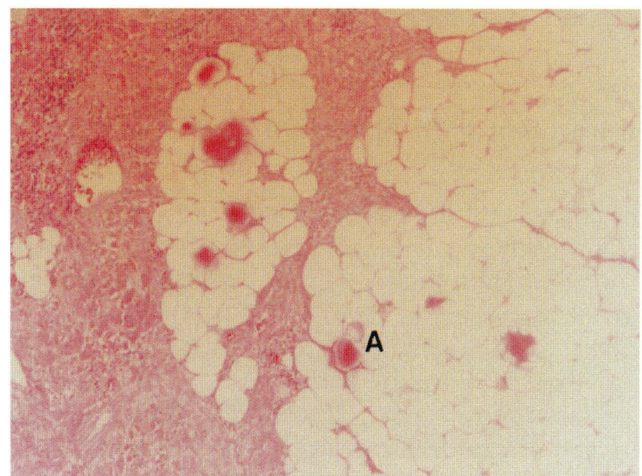

Figure 3.12, H&E x 52

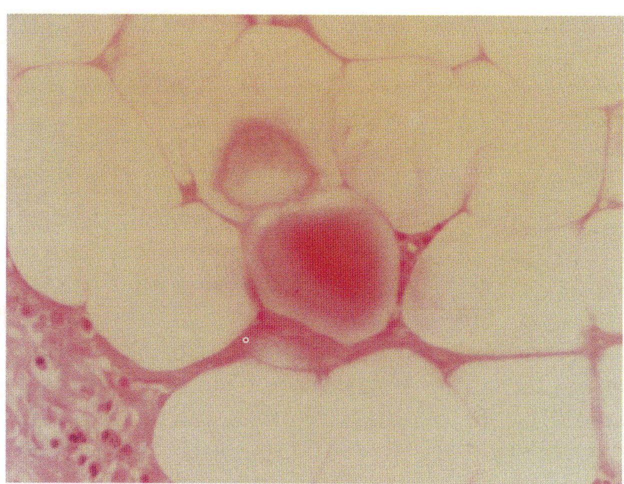

Figure 3.13, H&E x 260

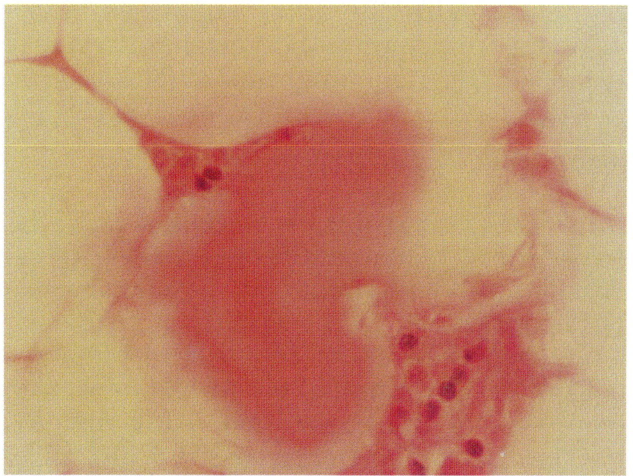

Figure 3.14, H&E x 520

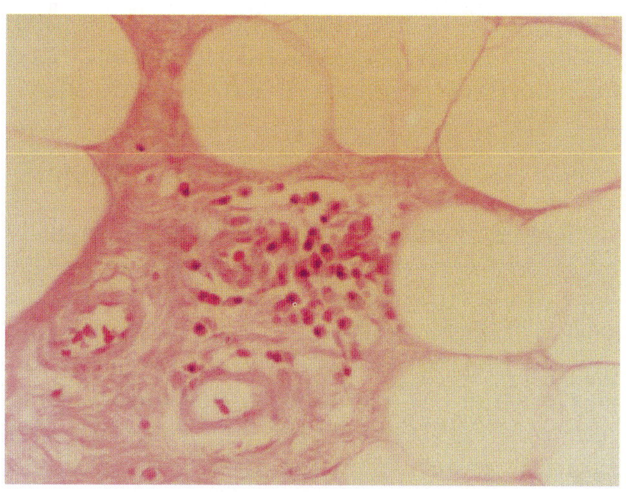

Figure 3.15, H&E x 260

between the fat cells. There are three small blood vessels enclosing red cells. H&E x 260

Figure 3.16. In the center of this figure, multiple faintly stained nuclei are arising from the fine reddish fibrillary substance between the fat cells. H&E x 260

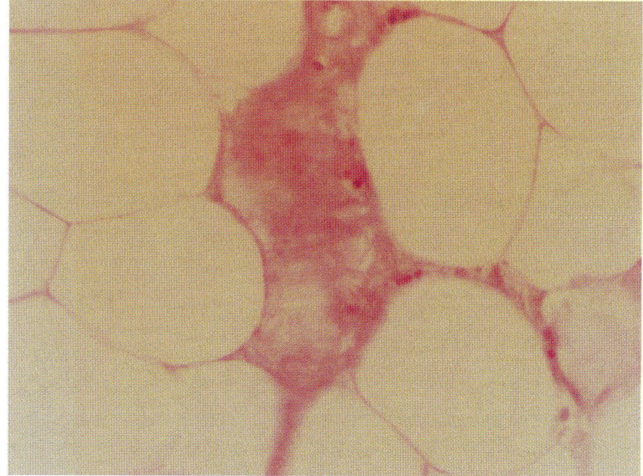

Figure 3.16, H&E x 260

Chapter 4

MUSCLES (figs. 4.1-4.40)
(Including Cardiac, Skeletal and Smooth Muscles)

Muscle tissue is an organ which, by contraction, produces the movements of the body. There are two varieties of muscle: *striated*, including all the muscles in which contraction is voluntary and the heart muscle; and unstriated or *smooth*, including all the involuntary muscles, except the heart, such as the muscles of the intestines, bladder, blood vessels, etc. In the human body there are three different kinds of muscles: 1. Cardiac, which has a structure somewhat similar to skeletal muscle except that the fibers are short and thick and form a dense mesh; 2. Skeletal, in which each muscle fiber consists of sarcoplasm which is composed of alternate light and dark portions (whence the name striated) and is surrounded by a sarcolemma; 3. Smooth, which are composed of elongated spindle shaped nucleated cells arranged parallel to one another and to the long axis of the muscle.

Demonstrated below is the direct origin of red blood cells and/or erythrogenic inflammatory cells from all three types of muscle tissue. This development of red cells and erythrogenic inflammatory cells usually occurs in association with blood vessel formation, with the lining endothelium developing from either the peripheral muscle tissue or the remaining inflammatory cells.

1. Cardiac Muscle [Heart] (figs. 4.1-4.21)

I have studied cardiac muscle extensively in children and adults, in various disease conditions, as well as in embryonic tissues and in tissue cultures. I have published my findings in the second volume of my series entitled *New Concepts in Blood Formation and Cell Generation in Malignant and Benign Tissues, Cardiac Muscle* (McDonald, 1995). A brief synopsis of Volume II is presented in "My Observations and Comments on Volume II" at the end of this book. To save space and expense, only a few of the photomicrographs from Volume II are included in this volume (figs. 4.1-4.15).[1]

Figures 4.1-4.15 presented here demonstrate the development of red cells and blood capillaries directly from cardiac muscle fibers, and the development of red cells and fat from chick embryonic cardiac muscle *in vitro*. In rheumatic fever, the origin of Aschoff bodies from degenerated muscle is demonstrated. In acute myocardial infarction, coagulation necrosis and the subsequent regeneration of red cells, blood vessels, and inflammatory or reactive cells is described.

Also presented in this volume (figures 4.17-4.21) are photomicrographs reprinted from one of my earlier scientific articles published by *The British Journal of Cancer* (Ghosh, see McDonald, 1959a). Figures 4.16-4.21 demonstrate the presence of independently growing metastatic mammary carcinomas within the cardiac lumina in mice. The complete indifference of the mammary carcinoma to invasion of the heart muscle and endocardium is noted in this study. The development of the stroma in a tumor growing in the medium of the circulating blood, without direct contact with the host tissue, contradicts the generally accepted theory that the stroma is supplied by the host tissue only.[2]

Figures 4.1 and 4.2 are from a two hour

---

1. For a detailed discussion of my findings in cardiac muscle see Volume II (McDonald, 1995), figures 1-90.
2. The neoplastic origin of stroma, as earlier suggested (McDonald 1970a), is demonstrated in Volume I (McDonald, 1989) in mammary carcinoma of mice and humans, and basal cell carcinoma and squamous cell carcinoma of humans, and in this volume in Part Two, Malignant Tumors, figures 24.11-24.14, 29.3, 30.19, 30.20, 31.6-31.10, 31.19-31.21, 31.23 and 31.28.

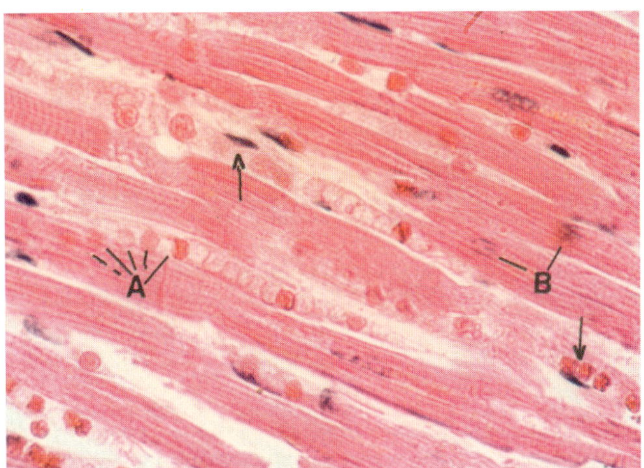

Figure 4.1, H&E x 520

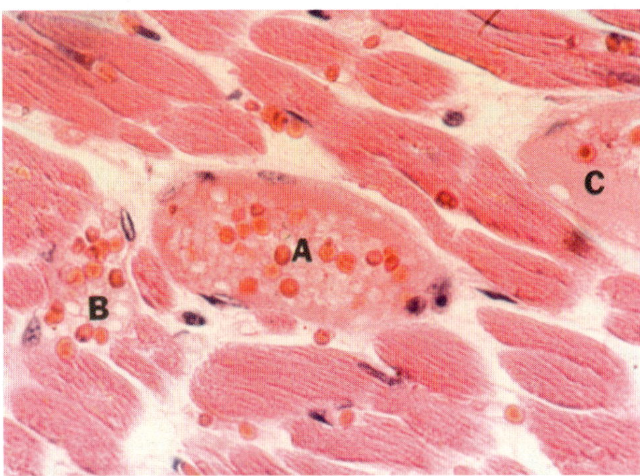

Figure 4.2, H&E x 520

post-mortem specimen of cardiac muscle from a 43-year-old man who died from a widely spreading naso-pharyngeal carcinoma, with terminal pulmonary emboli. Figure 4.3 is taken from the cardiac muscle of a 7-day-old chick embryo grown *in vitro* for 6 days. Figure 4.4 is obtained from the cardiac muscle of a 12-day-old chick embryo grown *in vitro* for 4 days. Figures 4.5-4.9 are from the postmortem specimens of the left ventricular myocardium of patients who died of acute rheumatic fever. Figures 4.5 and 4.8 are from a 9-year-old boy who died after he developed a mild decompensation. Figures 4.6 and 4.9 are taken from a 17-year-old girl who died of acute rheumatic heart disease. Figure 4.7 is from a 4-year-old girl who died from acute rheumatic heart disease. Figures 4.10-4.15 are all autopsy specimens from the ventricular wall of a 57-year-old man with congestive heart failure and arteriosclerotic heart disease with severe coronary arteriosclerosis.

Figures 4.16-4.21 are taken from autopsy sections of the hearts of C3H mice, with either spontaneous or transplanted mammary carcinomas, used in an experimental chemotherapy study. All the figures are from mice with transplanted mammary carcinomas, except figure 4.19 which is from a mouse with a spontaneous mammary carcinoma.

Development of red cells and blood capillaries from cardiac muscle fibers (figs. 4.1 and 4.2)

Figure 4.1. These capillaries are identified by a single column of red cells, some of which are in the ghost red cell stage (red cells with little hemoglobinization). Endothelial nuclei (arrows) are arising from the peripheral remaining myofibrils. The capillary (A) is exhibiting extension in a curved line across the muscle fiber. This occurrence is probably taking place in preparation for anastomosis with other capillaries. There are regular cardiac muscle nuclei (B), which are fading away in the sarcoplasm. H&E x 520

Figure 4.2. In these three large developing blood capillaries (A, B and C) of muscle cell origin, most of the red cells are in the vacuolar stage (ghost red cell stage), but a few are fully hemoglobinized. The right side of the capillary (B) shows a partially developed endothelial membrane with one nucleus, but the left side has not yet completely separated from the muscle fiber. Within the capillary (C), the red cells are developing from the liquefied muscle elements. All these vessels are surrounded by developing endothelium of muscle origin. The cross striations of the vanishing muscle fibers are still visible upon close examination of the blood vessels (A and C). H&E x 520

Development of red cells from 7-day-old chick embryonic cardiac muscle in vitro (fig. 4.3)

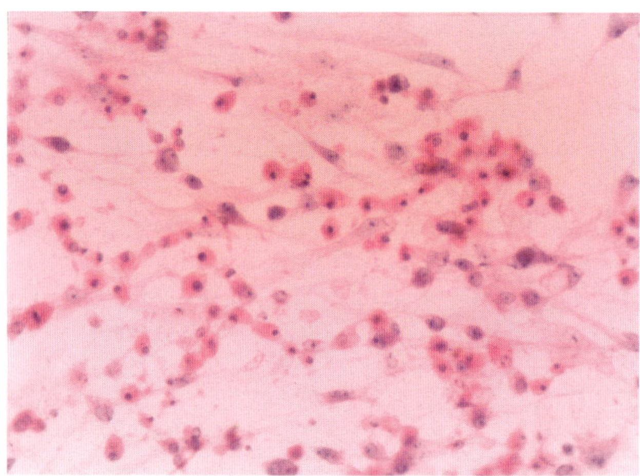

Figure 4.3, H&E x 260

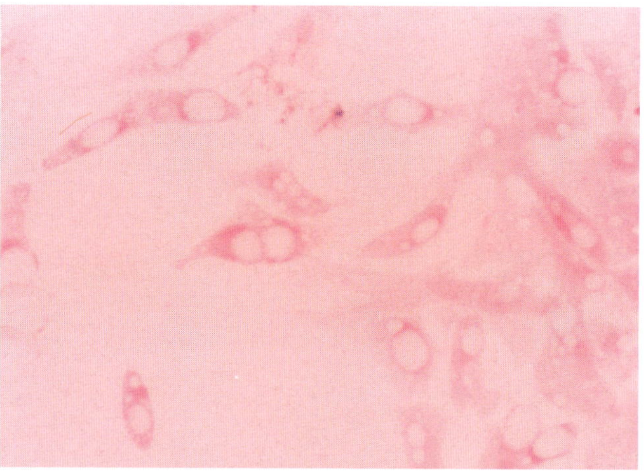

Figure 4.4, H&E x 400

Figure 4.3. From the cultured muscle cells, there arise many small, round, developing red cells with a tiny single nucleus. At this stage these developing red cells are not showing any tendency to acquire a spindle-shape, which is normal for chicken red cells in both adult and embryonic life. (*In vitro* growth of chick embryonic cardiac muscle for 6 days; see also figure 1.9 in Chapter 1, Embryology.) H&E x 260

Capacity for intracellular development of fat by 12-day-old chick embryonic cardiac muscle in vitro (fig. 4.4)

Figure 4.4. This photomicrograph reveals the development of fat in growing cardiac muscle *in vitro.* The fat appears as variable-sized round vacuoles, in H&E stain, because the fat has been removed by alcohol and xylene during the normal staining process. (*In vitro* growth of chick embryonic cardiac muscle for 4 days) H&E x 400

Degenerated muscle element giving rise to fibrinoid material and various types of Aschoff cells (figs. 4.5-4.7); predominantly Type A Aschoff cells (fig. 4.5); Type B Aschoff cells (fig. 4.6); and Type C Aschoff cells (fig. 4.7)

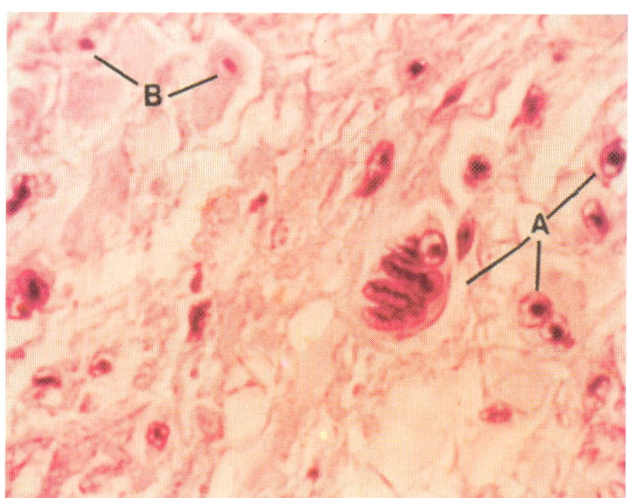

Figure 4.5, H&E x 800

Figure 4.5. There are mixed degenerative changes in this cardiac muscle (including hyaline, fibrillary/fibrinoid degeneration and liquefaction) and the predominant appearance of Anitschkow myocytes and their further development into Type A, owl-eyed Aschoff cells with single, double, and multiple nuclei (A). Note that many nuclei are present in longitudinal as well as transverse planes of Anitschkow myocytes within the large multinucleated Aschoff cell. There are also earlier stages of Anitschkow myocytes in transverse planes (B).[3] H&E x 800

---

3. Figures 4.5-4.9 are color reproductions of black and white photomicrographs which were originally presented in three of my previous articles on rheumatic fever (McDonald, 1963 & 1975, and McDonald and Calkins, 1978). I greatly appreciate the permission for reproduction given by the following editors and publishers: *Texas Medicine,* formerly *Texas State Journal of Medicine* (figs. 4.5 and 4.8); *Experimental Pathology* (figs. 4.6-4.9); and the *Journal of Clinical Pathology* (fig. 4.9).

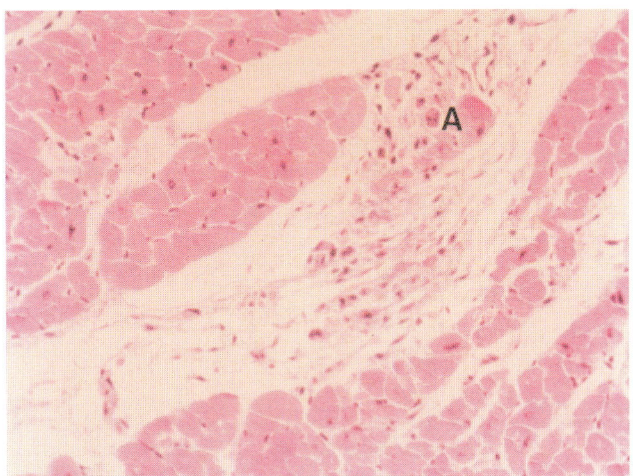

Figure 4.6, H&E x 150

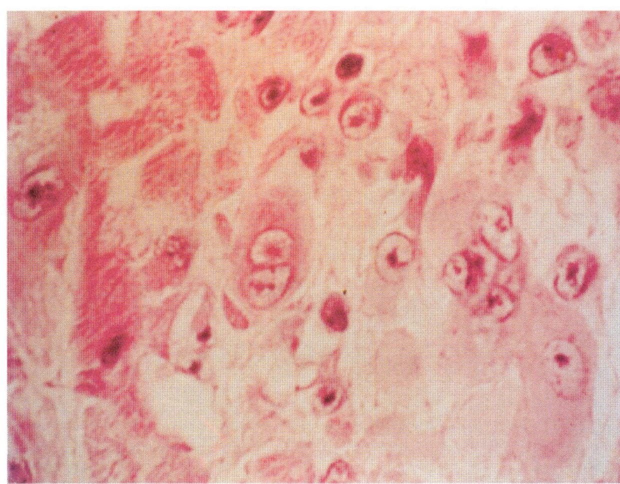

Figure 4.7, H&E x 708

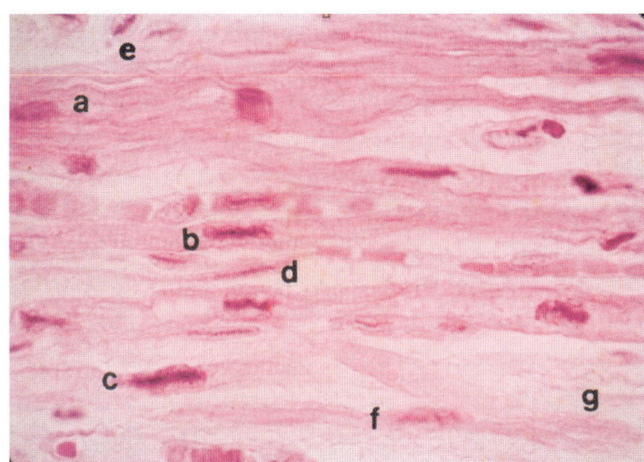

Figure 4.8, H&E x 500

Figure 4.6. Shown here is a spindle-shaped Aschoff body (with Type B Aschoff cells) having shaggy, fibrillary, eosinophilic cytoplasm and hyperchromatic nuclei of variable sizes. (A) points to regenerating (redifferentiating) muscle fiber with a partly fibrinoid cytoplasm derived through Type B cytogenesis. Appropriate staining methods have also shown that in addition to muscle regeneration a small amount of fibrillary material, with or without attachment to the dedifferentiated nuclei, acquires the affinity for collagen stains.[3] H&E x 150

Figure 4.7. Presented here is a preponderance of Type C Aschoff cells with large, hyalinized, usually lightly basophilic cytoplasm and single or multiple large vesicular nuclei. The origin of such Aschoff cells associated with hyalinization is shown in figure 28-1 of Volume II.[3] H&E x 708

Origin and multiplication of Anitschkow myocytes from cardiac muscle fibers (figs. 4.8 and 4.9); cytogenesis of Type A Aschoff cells (fig. 4.8) and Type B Aschoff cells (fig. 4.9)

Figure 4.8. Successive stages of nuclear and sarcoplasmic changes in muscle fibers involving the development of Anitschkow myocytes (a-d), Type A Aschoff cells, are depicted in this figure. Degeneration or alteration of muscle fibers seems to be associated with the origin of Anitschkow myocytes. The presence of the remaining myofibrils at both ends of the formative Anitschkow myocyte (d) appears to be responsible for keeping the long narrow shape of this nucleus. With dissolution of remaining myofibrils, the Anitschkow myocytes assume an oval form (e). Proliferation of Anitschkow myocytes, as if by fragmentation within the muscle fiber, is shown by (f) (use a magnifying glass). (g) points to wavy fibrillary changes in muscle fibers, with the loss of cross striations.[3] H&E x 500

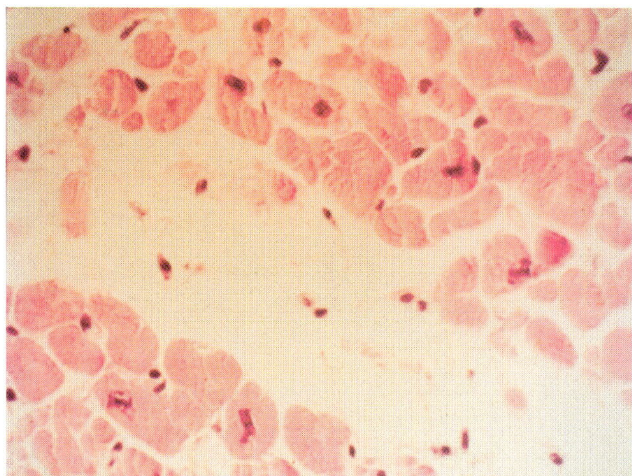

Figure 4.9, H&E x 382

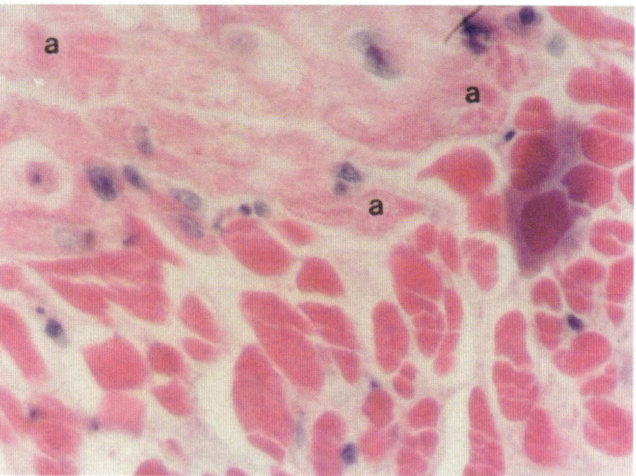

Figure 4.10, H&E x 520

Figure 4.9. In a cleft-like space created by the lysis of cardiac muscle fibers, tiny hyperchromatic dedifferentiated nuclei are noted. These nuclei may have the potential for the development of Type B Aschoff cells, as well as new cardiac muscle fibers and possibly new connective tissue fibers. Some of the nuclei have not yet been cleared of remnants of lysing muscle elements. Note that there are several similar tiny nuclei at the periphery of the partially damaged surrounding muscle fibers.[3] H&E x 382

Acute myocardial infarction causing coagulation necrosis (fig. 4.10); and the subsequent developments (figs. 4.11 and 4.12)

Figure 4.10. The infarcted area is beginning to extend into the adjoining muscle fibers, as demonstrated by the appearance of speckled bright color (a) in the muscle fibers. With the disappearance of a certain amount of fully infarcted myoplasm through dissolution, more gaps appear between the remaining infarcted myofibrils. H&E x 520

Prior to nuclear regeneration, frequently seen changes in color and consistency of bright-red infarcted myofibers into

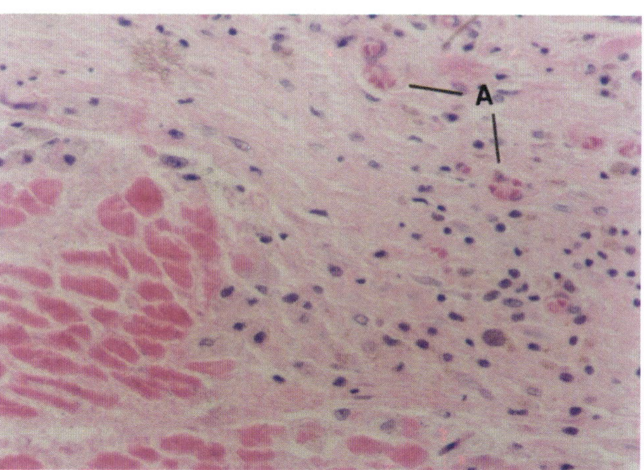

Figure 4.11, H&E x 260

bluish-gray, ground glass-like substance; and subsequent development of lymphocyte-like cells, developing blood capillaries, endothelial cells, and spindle-shaped nuclei (fig. 4.11)

Figure 4.11. There are many developing small, round, hyperchromatic, lymphocyte-like nuclear structures, some spindle-shaped nuclei, and also several developing blood capillaries (A), all arising from this bluish-gray slate-like background substance derived from infarcted muscle. H&E x 260

3.   Figures 4.5-4.9 are color reproductions of black and white photomicrographs which were originally presented in three of my previous articles on rheumatic fever (McDonald, 1963 & 1975, and McDonald and Calkins, 1978). I greatly appreciate the permission for reproduction given by the following editors and publishers: *Texas Medicine,* formerly *Texas State Journal of Medicine* (figs. 4.5 and 4.8); *Experimental Pathology* (figs. 4.6-4.9); and the *Journal of Clinical Pathology* (fig. 4.9).

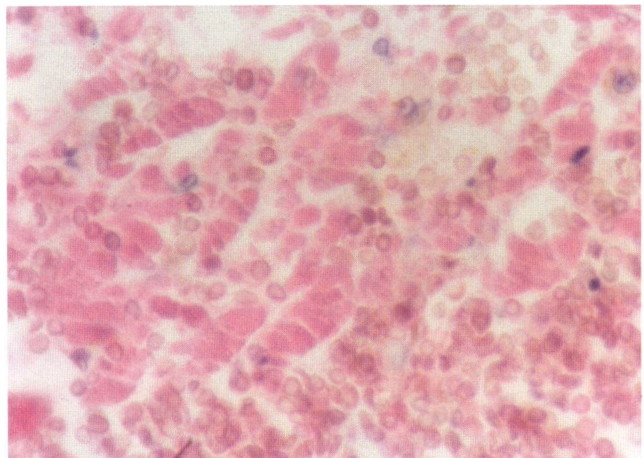

Figure 4.12, H&E x 520

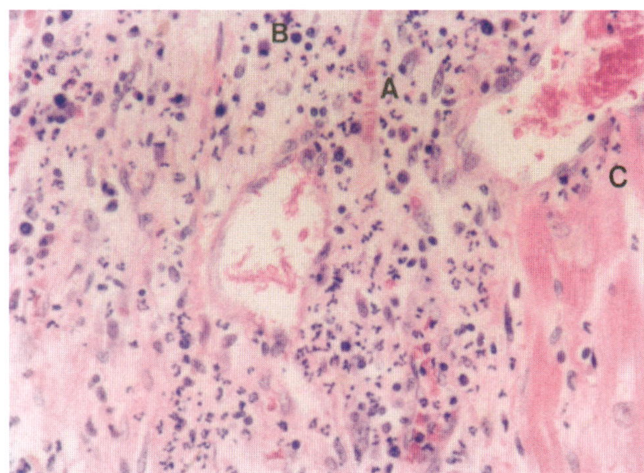

Figure 4.13, H&E x 260

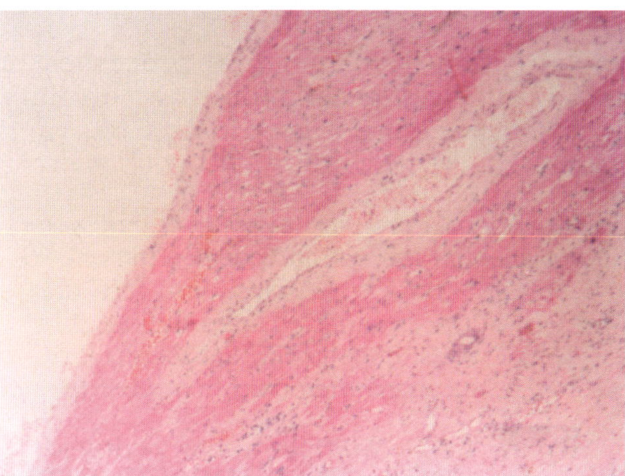

Figure 4.14, H&E x 52

Massive red cell formation directly from fresh infarcted muscle fibers with replacement of the latter (fig. 4.12)

Figure 4.12. This figure shows profusely developing red cells from infarcted myofibers. Many of these red cells are free from muscle elements. H&E x 520

Development of acute inflammatory cells (mainly segmented nuclear cells, SN cells like those of suppurative inflammation) and blood capillaries from disappearing myofibers (fig. 4.13)

Figure 4.13. In this figure there is marked proliferation of acute inflammatory cells, mostly segmented nuclear cells, a few colonies of developing plasma cells, and some unclassified reactive cells arising from and replacing most of the cardiac muscle, except for a segment of myofibers on the right. Also shown in this figure are two large developing blood capillaries arising from liquefied muscle fibers; the shredded remnants of the latter are still visible within the capillary lumen together with developing red cells. Note that there are no segmented neutrophils in the lumen, and the endothelial lining has barely developed. (A) points to a developing capillary with a single column of pressed-biscuit-like red cells which have not yet fully formed. (B) shows a group of plasma cells of various sizes developing from hyalinized muscle fibers. On the right side near (C), SN cells are arising as minute particles from noninfarcted muscle fibers. H&E x 260

Possible transformation of ischemic cardiac muscle (in sub-optimal coagulation necrosis) into blood and a large blood vessel (figs. 4.14 and 4.15)

Figures 4.14 and 4.15. In lower magnification in figure 4.14, upon casual observation this large blood vessel may appear to be well established. However, in higher magnification (figure 4.15), the developing status of both the vascular content and the vessel wall is evident. Within the lumen, red cells are shown in various stages of development, from fine light pink granular undissolved elements to fully formed red cells. Also the endothelium has not yet fully developed, and the apparently thick wall is formed of loose fibrillary structures derived from ischemic muscle. (The mechanism of blood vessel development in ischemic cardiac muscle, as shown here, possibly

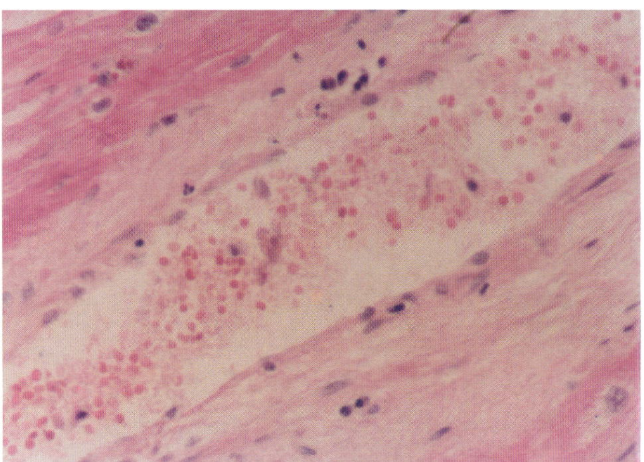

Figure 4.15, H&E x 260

explains the developmental processes of collateral circulation following coronary occlusion.) Figure 4.14 H&E x 52; and figure 4.15 H&E x 260

Independently growing mammary carcinoma emboli within the cardiac cavities (figs. 4.16-4.20) and coronary vessels (figs. 4.16 and 4.21), but not connected with the heart or vessel walls, questioning the theory that stromal tissue originates only from host tissue[4]

Figure 4.16. Multiple tumor nodules are growing inside the right ventricular cavity. Note the presence of delicate cell membranes loosely encapsulating the tumor tissue. Necrosis and dissolution of necrotic tumor tissue are seen in the formation of the cystic cavities. H&E x 45

Figure 4.17. This figure demonstrates metastatic tumor forming a multiloculated cystic papillary pattern inside the right ventricular cavity of a mouse bearing mammary carcinoma. No connection of this tumor with the cardiac wall is seen in serial section, and the cardiac wall is free of tumor tissue. Note the delicate pattern of the tumor tissue with well differentiated stroma. A magnified view of the tumor tissue (A) is shown in the following figure. H&E x 45

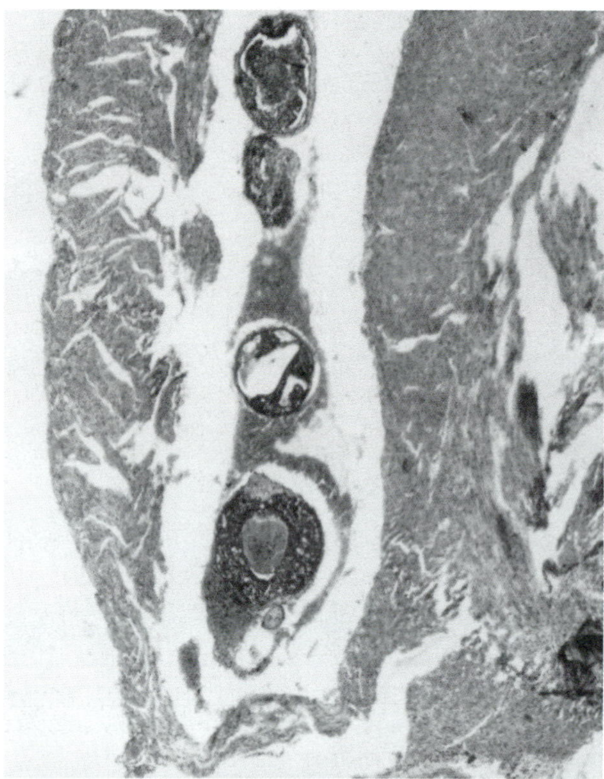

Figure 4.16, H&E x 45

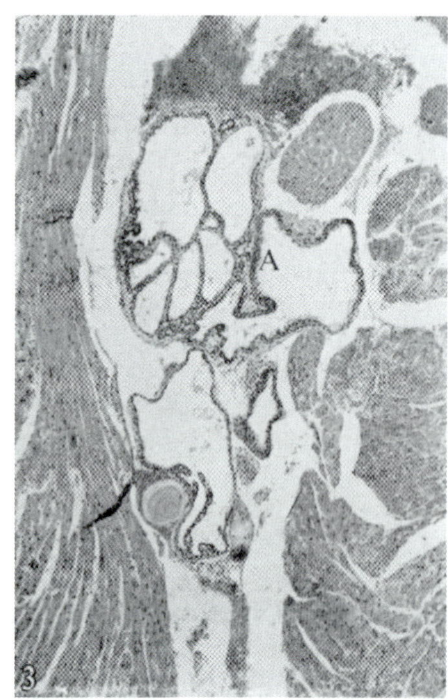

Figure 4.17, H&E x 45

---

4. Figures 4.17-4.21 are photomicrographs reprinted from one of my earlier scientific articles (Ghosh, see McDonald, 1959a). I greatly appreciate the permission for reproduction given by the editor and publisher of *The British Journal of Cancer.*

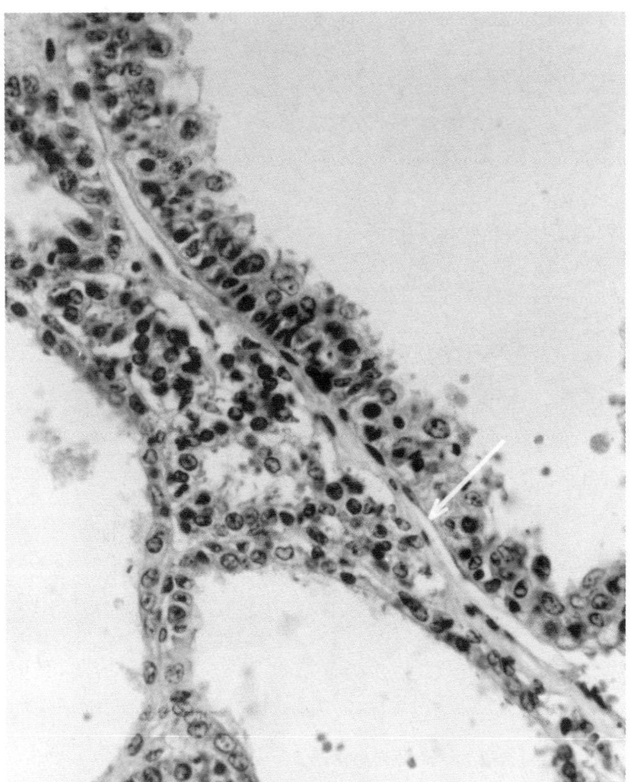

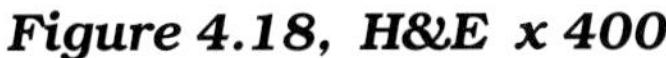

Figure 4.18, H&E x 400

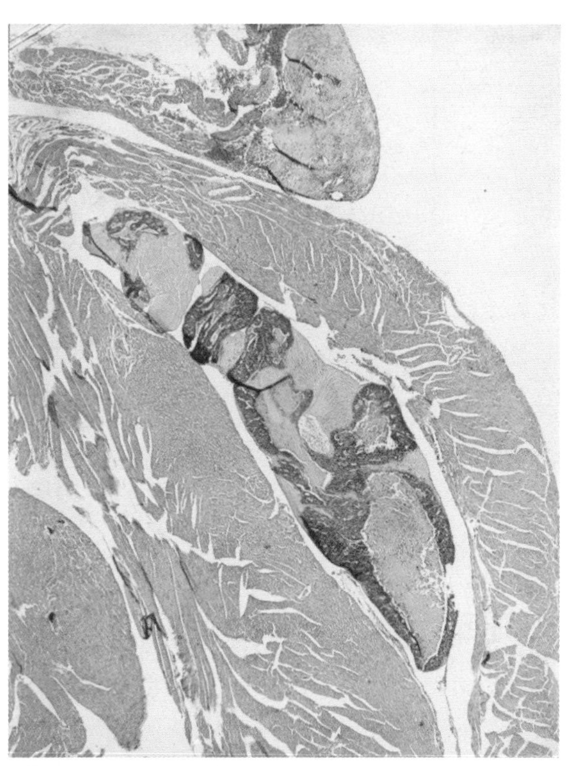

Figure 4.19, H&E x 25

Figure 4.18. This is a magnified view of a portion (A) of figure 4.17. A narrow channel (arrow) lined with endothelial-like flattened cells is shown here. H&E x 400

Figure 4.19. The right ventricular cavity is distended with a tumor growth. Prominent necrosis and hyaline degeneration can be seen in the tumor. The cardiac wall is thickened and shows no invasion by the tumor. H&E x 25

Figure 4.20. The right atrial cavity (A) is distended with a growing tumor embolus (from a well-differentiated breast carcinoma) showing a combination of tumor patterns with delicate stroma and vascular channels. A tumor embolus (B), top of figure, is distending the superior vena cava. A tumor embolus (C) can be seen in the inter-atrial septum distending a coronary vessel. The cardiac wall and coronary vessel wall show no invasion by the tumor upon examination of several serial sections. H&E x 35

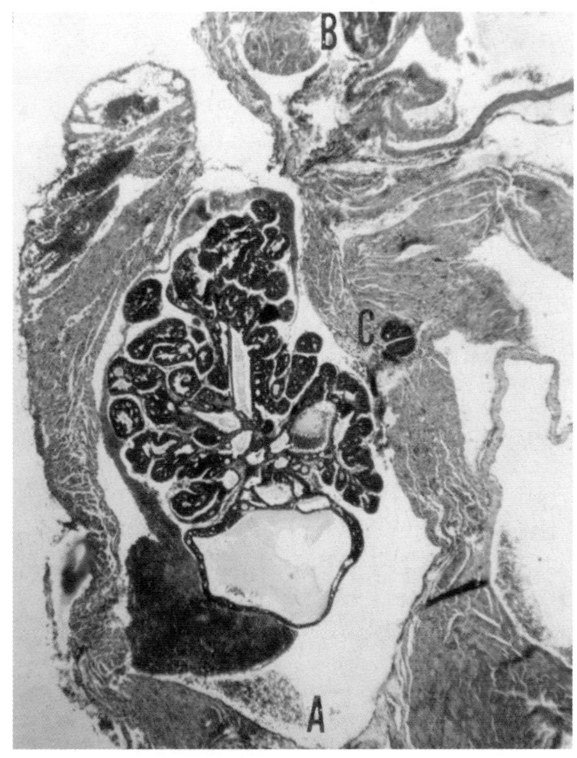

Figure 4.20, H&E x 35

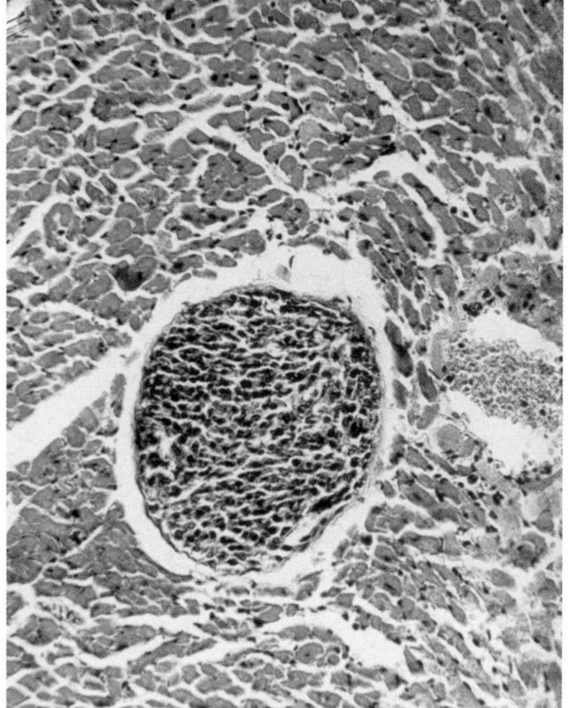

Figure 4.21. In the inter-ventricular septum of a heart, a coronary vessel is filled with solid carcinoma cells. No other tumor metastasis was seen in the heart. Sections of the lungs showed pulmonary metastasis. H&E x 200

Figure 4.21, H&E x 200

2. Skeletal Muscle (figs. 4.22-4.33)

It has been my experience as a pathologist to find many inflammatory (reactive) cells inside and outside both the intact or remains of muscle fibers. As Altschul (1948) pointed out, and also shown by me here in figure 4.32, some muscle fibers may be packed with numerous dedifferentiated cells without any suggestion of mitosis.

The origin of these reactive cells, some of which fall under the category of white blood cells (such as segmented neutrophils, eosinophils, lymphocytes and monocytes/ histiocytes) and some under the category of dedifferentiated cells, can be seen in the following figures. It has been noticed within muscle fibers that some of these reactive cells are devoid of cytoplasm; therefore, they are usually classified according to their nuclear characteristics. In my previous investigations on myocardial changes in acute rheumatic fever, I found that different reactive cells including dedifferentiated cells with the potential for regeneration of cardiac muscle, may originate from altered striated cardiac muscle without any signs of mitosis.[5] The following photomicrographs of skeletal muscle were taken to demonstrate the direct development of red cells as hemoglobin globules, as well as the developmental changes of cells through various inflammatory cells of muscle origin.

Figures 4.22 and 4.33 are from a healthy adult rabbit sacrificed by the administration of ether as a normal control. Figure

5. See Volume II, figures 35, 36 and 40-42.

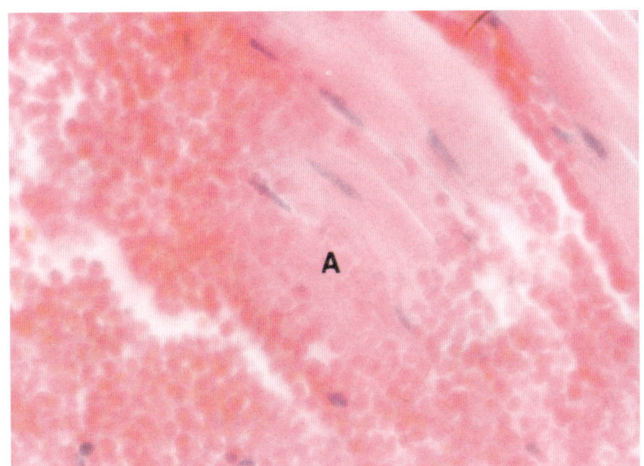

Figure 4.22, H&E x 520

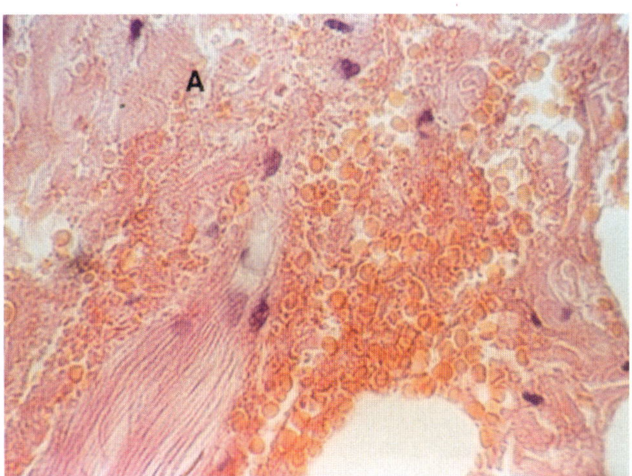

Figure 4.23, H&E x 520

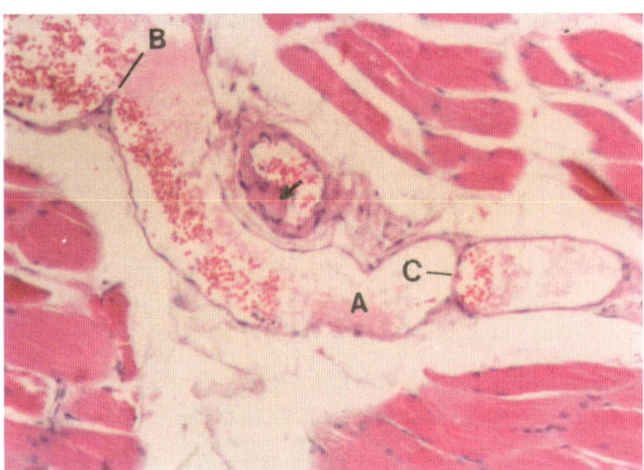

Figure 4.24, H&E x 130

4.23 is from a surgical excision of a subcutaneous nodule of the skin from a 36-year-old male. Figure 4.24 is from an autopsy specimen of skeletal muscle from a 57-year-old man who died of recent and acute myocardial infarction with congested heart failure. Figure 4.25 is from embryonic skeletal muscle taken from a spontaneously aborted human embryo weighing 7.8 grams. Figure 4.26 is from extraorbital skeletal muscle surgically removed with enucleation of eyeball from a clinically blind patient with painful phthysis bulbi. Figures 4.27-4.29 are from skeletal muscle adjacent to a deeply invading squamous cell carcinoma of the tongue that was surgically removed from a 43-year-old male. Figure 4.30 is from skeletal muscle surgically removed near a basal cell carcinoma from the right side of the nose of an 85-year-old male. Figures 4.31 and 4.32 are obtained from surgically removed abdominal skeletal muscle adherent to a colon carcinoma from a 63-year-old male. Figures 4.28, 4.29, 4.31 and 4.32 are reprinted from Volume I (McDonald, 1989).

Direct origin of red cells as hemoglobin globules from skeletal muscle (figs. 4.22-4.25); with developing capillaries (figs. 4.24 and 4.25)

Figure 4.22. From the skeletal muscle, the origin of numerous fully formed red cells as hemoglobin globules is prominently shown here (such findings are commonly known as hemorrhage). The faintly visible pink globules within the muscle fibers (A) show the early stages of red cell development. H&E x 520

Figure 4.23. Along the cross-striation of muscle fibers, the appearance of tiny orange-red granules (A) shows the beginning stage of the development of red cells. Fully developed red cells obtained by the coalescence of these fine granules may be seen mostly on the right of this figure. H&E x 520

Figure 4.24. Prominent dissolution of sarcoplasm with production of clear fluid, usually known as edema fluid, is causing the separation of remaining muscle fibers. In the center, a blood capillary is developing from the product of muscle lysis. The red cells within this capillary were formed

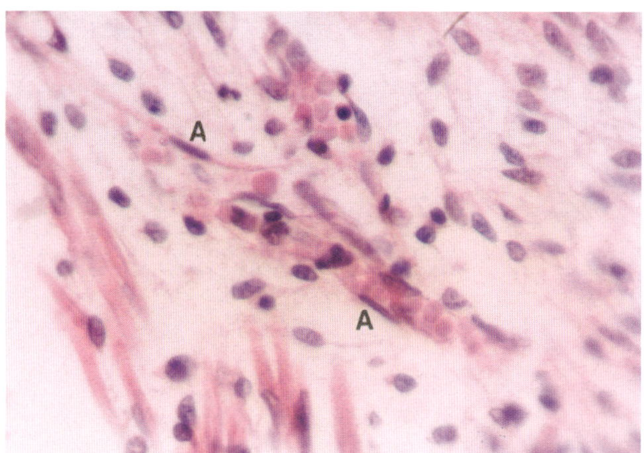

Figure 4.25, H&E x 520

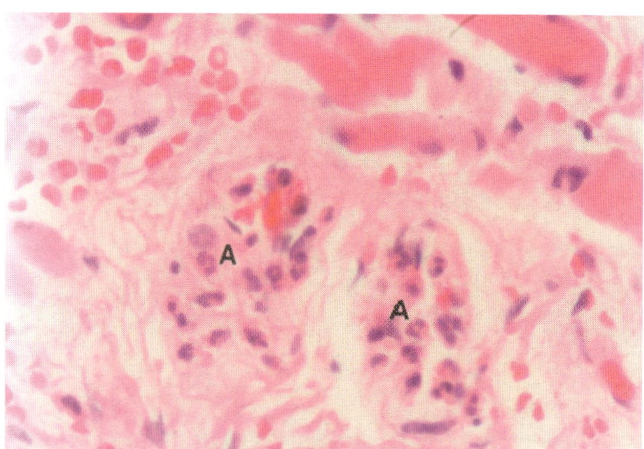

Figure 4.26, H&E x 800

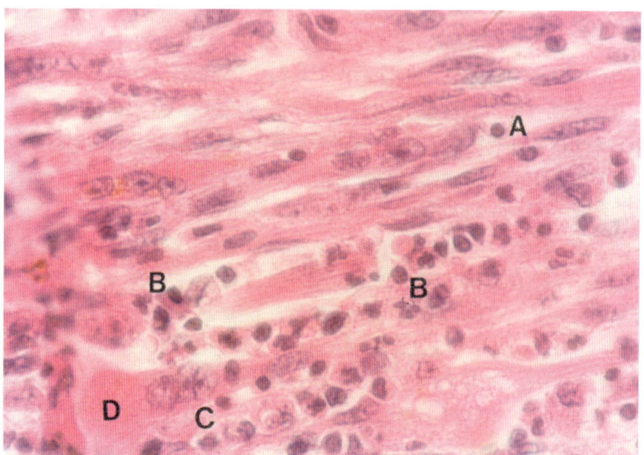

Figure 4.27, H&E x 520

directly from the dissolving skeletal muscle as described in figures 4.22 and 4.23. (A) shows the remnants of dissolving sarcoplasm in the capillary lumen. The irregular girth and the torturous curves of this vessel is caused by end-to-end fusion (B), or impending fusion (C), with adjacent capillaries. Undissolved muscle remnant (arrow) can be seen within a nearby further developed blood vessel in transverse section. H&E x 130

Red blood cells and capillaries developing from embryonic skeletal muscle (fig. 4.25)

Figure 4.25. In this human embryonic skeletal muscle, developing stages of red cells as hemoglobin globules are seen in the formative lumen areas of these capillaries. Note the developing red cells are of different sizes and degrees of hemoglobinization. (A's) are placed near the developing endothelial cells. H&E x 520

Origin of reactive/inflammatory cells from skeletal muscle (figs. 4.26-4.32); erythrogenic SN cells (figs. 4.26 and 4.27); lymphocytes (figs. 4.27 and 4.29); multinucleated giant cells (fig. 4.27); eosinophils (figs.4.27 and 4.29); erythrogenic plasma cells (fig. 4.28); and histiocytes (figs. 4.30 and 4.31)

Figure 4.26. The origin of red cells from erythrogenic SN cells developed from disintegrated skeletal muscle fibers (A) is shown. Also, in the upper left portion of this figure the red cells are probably of erythrogenic SN cell origin because red cells that arise from SN cells often have this characteristic deformity and irregularity in size, shape and in linearity.[6] In the upper right corner, note the disintegration and loss of cross-striations of the remaining skeletal muscle fibers, as well as the origin of a few red cells and SN cells from these muscle fibers. H&E x 800

Figure 4.27. From skeletal muscle, the origin of lymphocytes (A), erythrogenic SN cells and eosinophils (B), and multinucleated giant cells (C) from disintegrated muscles are shown. The myoplasm (D) has become hemoglobinized in preparation for red cell development. H&E x 520

---

6.   As demonstrated in Volume I, figures 46 and 120-1.

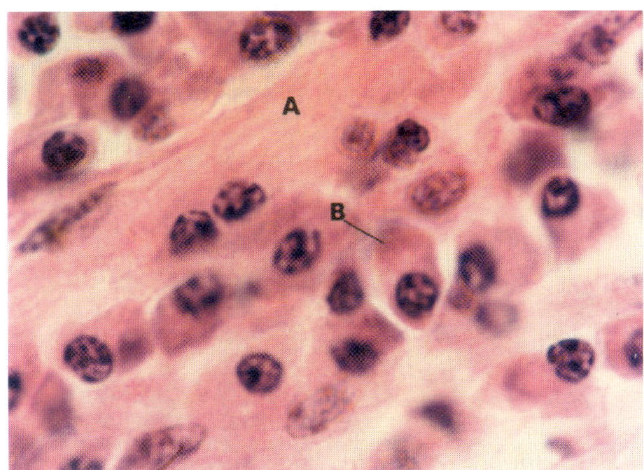

Figure 4.28, H&E x 1300

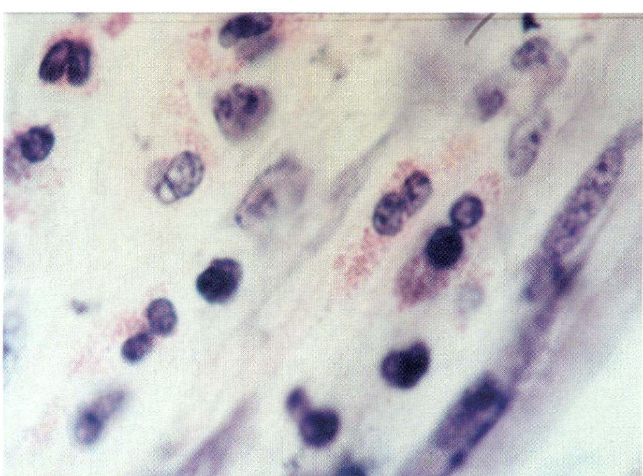

Figure 4.29, Giemsa x 1300

Figure 4.28. There are a large number of plasma cells developing from skeletal muscle, the remains of which can be seen in area (A). The origin of plasma cells started within muscle fibers as lone nuclei. The majority of plasma cells are formed in the nonmedullary tissues, especially during chronic inflammation. They may originate as regular sized and shaped cells or as tiny hyperchromatic bodies (plasma cell bodies) which then mature into regular plasma cells (Patterson, 1986). Schridde (1905) contended that plasma cells can transform into SN cells and eosinophils (see figure 14.1, Chapter 14, Stomach). Changes in the plasma cells toward development of red cells are noted (B). H&E x 1300

Figure 4.29. Within these muscle fibers, eosinophils and a few lymphocytes can be observed to arise. The eosinophilic granules appear over a long distance without any cellular boundaries. Giemsa x 1300

Figure 4.30. The skeletal muscle fibers, stimulated by a nearby basal cell carcinoma, exhibit cellular reaction with marked proliferation of intramuscular reactive cells (typically categorized as histiocytes). These changes are associated with fibrosis. H&E x 520

Without evidence of mitosis, production of various inflammatory cells (including lymphocytes, segmented nuclear cells, fibrocytes and unclassified cells) as a reaction of skeletal muscle fibers against advancing cancer (figs. 4.31 and 4.32)

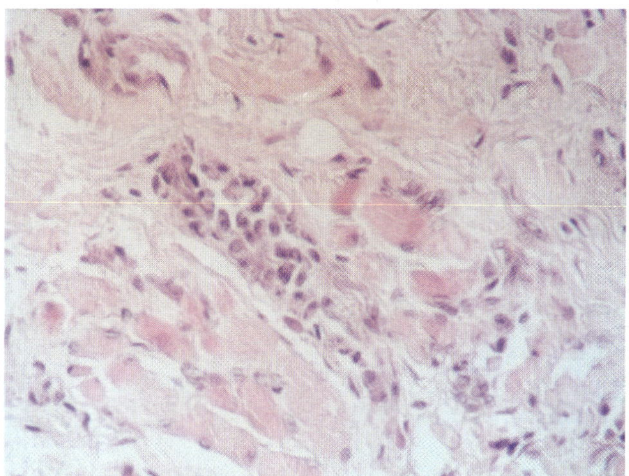

Figure 4.30, H&E x 520

Figure 4.31. The origin and proliferation of dedifferentiated hyperchromatic lymphocyte-like nuclei or small fusiform nuclei within muscle fibers replacing regular sarcoplasm is seen here. Some muscle fibers (A) are filled with such nuclei, nucleosis without evidence of mitosis. Some fibers (B) are partially filled with round or oval vesicular nuclei of moderate chromatism, which are larger than those of lymphocytes. Cells with such nuclei are often categorized as histiocytes or monocytes. Also seen is the proliferation of lymphocyte-like cells from muscle fibers undergoing lysis (C and D). In some muscle fibers no cellular reaction is seen (E). The release of lymphocytes by the liquefaction of mother cells is shown in area (F), where the remaining muscle element is barely identifiable. H&E x 640

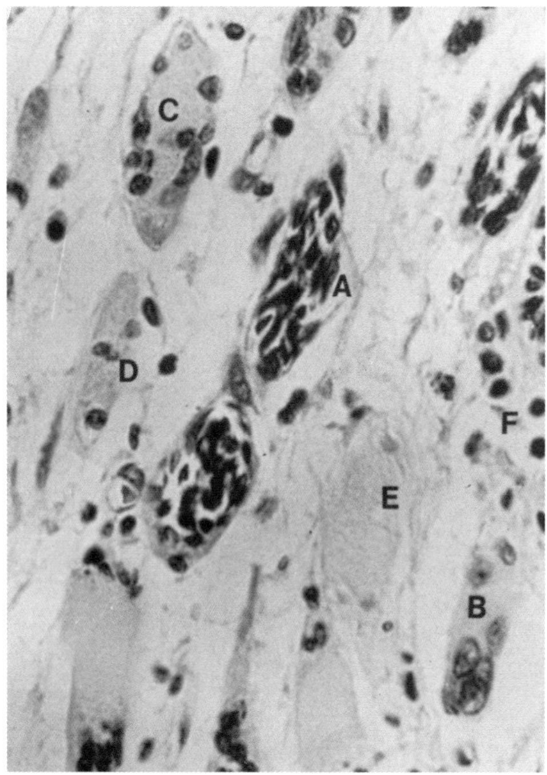

Figure 4.31, H&E x 640

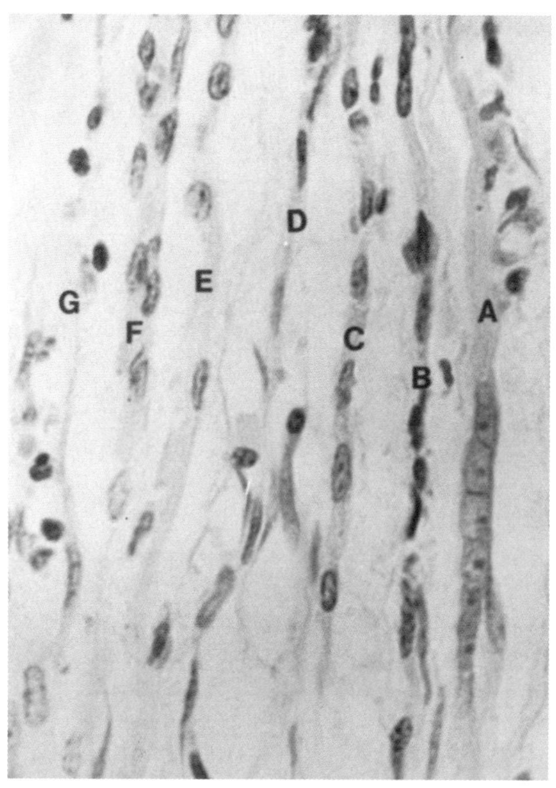

Figure 4.32, H&E x 720

Figure 4.32. Different varieties of nuclear structures, mostly unclassified, originating in separate fascicles of remaining muscle fibers are shown in longitudinal plane (A-G). (A) denotes a single column of cuboidal nuclei with prominent nucleoli. At both ends of this column, tiny nuclear structures are evolving from myoplasm in a creeping manner. In (B) many hyperchromatic, narrow, fusiform or spindle-shaped nuclei are lying in an interrupted linear fashion. In (C) the nuclei are elongated with less chromatin density than those in (B). In (D) nuclei are mainly long, narrow, and spindle-shaped (similar to fibrocyte nuclei). They are attached at both ends to fibrillary material of dissolving myofibers. In the upper part of (E), in the dissolving myoplasm are oval vesicular nuclei. Toward the middle and lower part, the nuclei are narrow and elongated. In (F) are nuclei similar to those seen in (C). (G) represents mostly dissolved myofibrils with a few segmented nuclei of varying degrees of chromatism with no cytoplasmic differentiation. H&E x 720

Dissolution of individual muscle fibers with retention of sarcolemma giving the false appearance of adipose tissue (fig. 4.33)

Figure 4.33. The dissolution of skeletal muscle fibers with the retention of sarcolemma often gives an impression of adipose tissue (A) on casual observation. A similar occurrence has been noticed particularly in subepicardial myocardium of children who suffer from acute rheumatic fever, where individually dissolved cardiac muscle fibers retain their sarcolemma showing the possibility for regeneration (Volume II, figure 48). H&E x 200

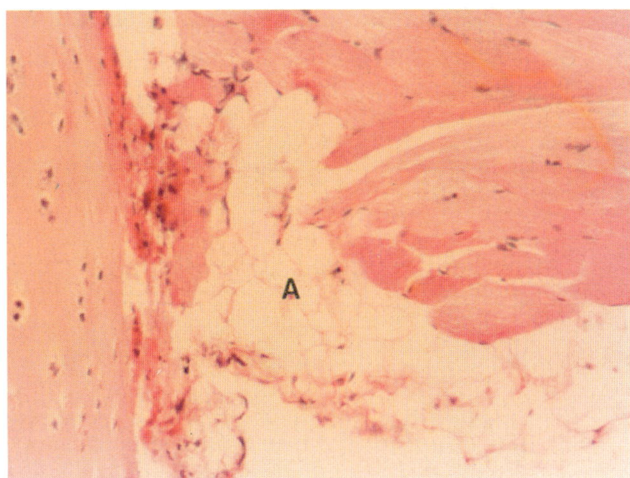

Figure 4.33, H&E x 200

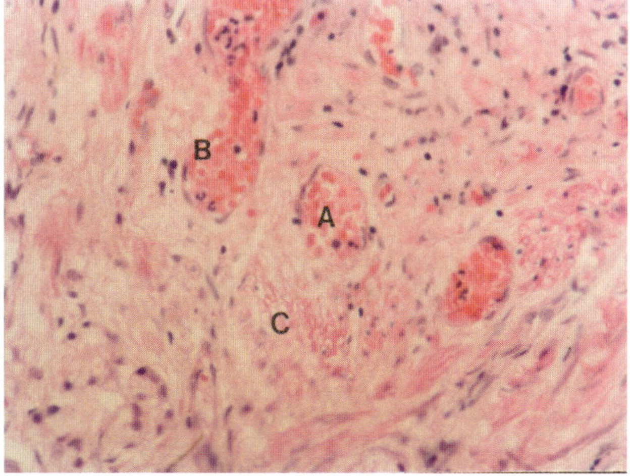

Figure 4.34, H&E x 260

3. Smooth Muscle (figs. 4.34-4.40)
(Including Smooth Muscle from Prostate, Stomach, Colon and Urinary Bladder)

The smooth muscle, which is an involuntary muscle, makes up the contractile portion of the vascular wall, ductal wall, and intestinal wall. Furthermore, some organs such as the uterus, urinary bladder and prostate consist of mainly smooth muscle. Described below is the direct development of red blood cells, erythrogenic inflammatory cells, and blood vessels from smooth muscle tissue.[7]

Figure 4.34 is from the smooth muscle of a prostate that was removed for an adenocarcinoma. Figures 4.35 and 4.36 are from the smooth muscle coat of a gastric wall obtained from a partial gastrectomy due to a carcinoma.[8] Figure 4.37 is from an extensive chronic ulceration of the colon. Figure 4.38 is from the smooth muscle coat of a gastric wall surgically removed due to a chronic gastric ulcer. Figures 4.39 and 4.40 are from the smooth muscle wall of a urinary bladder removed for transitional cell carcinoma. Figure 4.38 is reprinted from Volume I (McDonald, 1989).

*Blood and blood capillary development
directly from smooth muscle
(figs. 4.34 and 4.35)*

Figure 4.34. Blood and blood capillaries are developing from the smooth muscle of a prostate with carcinoma. Many of the red cells are not fully developed, and the endothelial lining is not completely formed. In the early stages of development irregular hemoglobinization of red cells had occurred within the blood capillaries (A and B). In area (C) there is irregular hemoglobinization of smooth muscle fibers. (See also figure 14.13 in Chapter 14, Stomach.) H&E x 260

Figure 4.35. A large blood capillary is developing from smooth muscle that has undergone fibrinoid changes. A portion of the undissolved fibrinoid material (A) is still present inside the capillary lumen. Initially, the red cells were highly deformed and difficult to separate from the undissolved remains of the mother cell substance. H&E x 260

---

7.   The *in vitro* direct development of red cells from smooth muscle is demonstrated in figures 1.44 and 1.45 of Volume III.
8.   Blood and blood vessel development from locally developed inflammatory cells is also presented in figures 30.1-30.4.

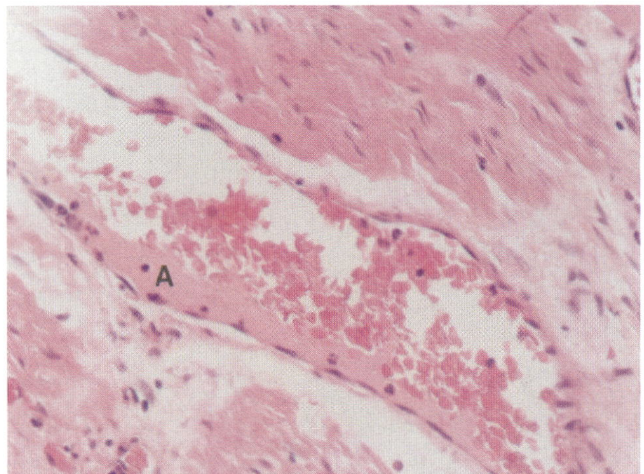

Figure 4.35, H&E x 260

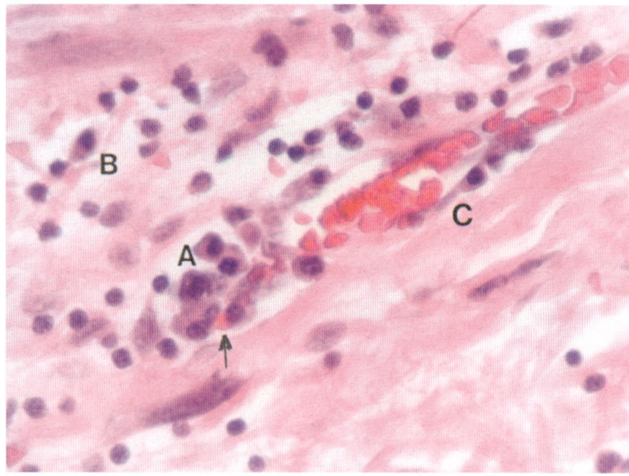

Figure 4.36, H&E x 520

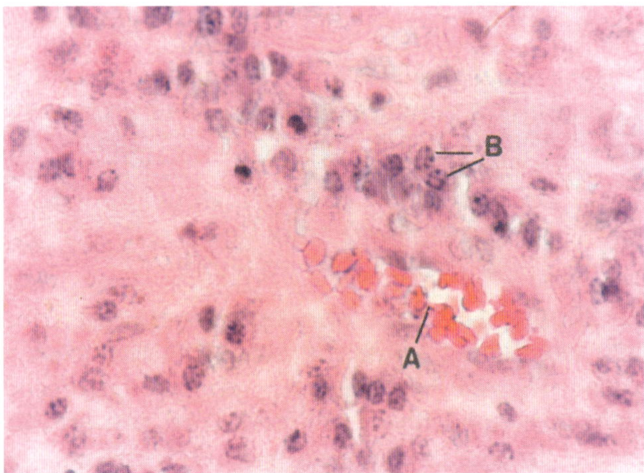

Figure 4.37, H&E x 520

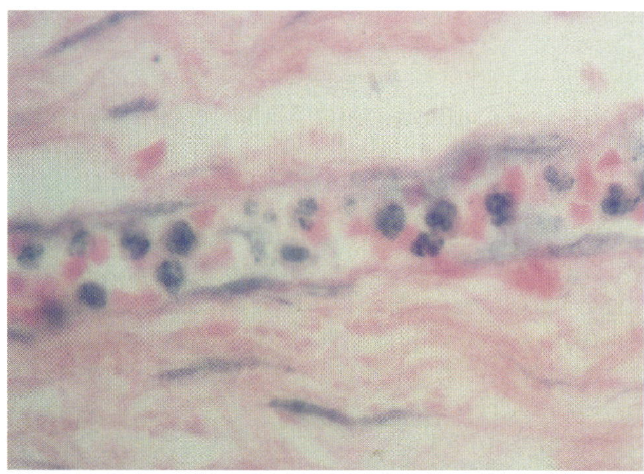

Figure 4.38, Giemsa x 800

Blood and capillary development from reactive plasma cells of smooth muscle origin (figs. 4.36 and 4.37)

Figure 4.36. Lymphocytes are arising from smooth muscle and are transforming into plasma cells (A and B). The plasma cells then give rise to red cells as well as endothelium (arrow and C). As the plasma cells are transformed into red cells, the peripheral plasma cells transform into endothelium around the developing capillary (C). H&E x 520

Figure 4.37. Many plasma cells are disappearing into the background substance from which a new blood capillary is developing. Notice that some of the red cells (A) still contain minute chromatin dots as shown in plasma cells (B). H&E x 520

Development of erythrogenic SN cells and endothelium from smooth muscle (fig. 4.38)

Figure 4.38. A column of erythrogenic SN cells (segmented nuclear cells) within a formative vascular channel lined by developing endothelium is originating from the gastric muscle coat. With the development of red cells, the remainder of the SN cell dissolves and this material is added to the plasma. Giemsa x 800

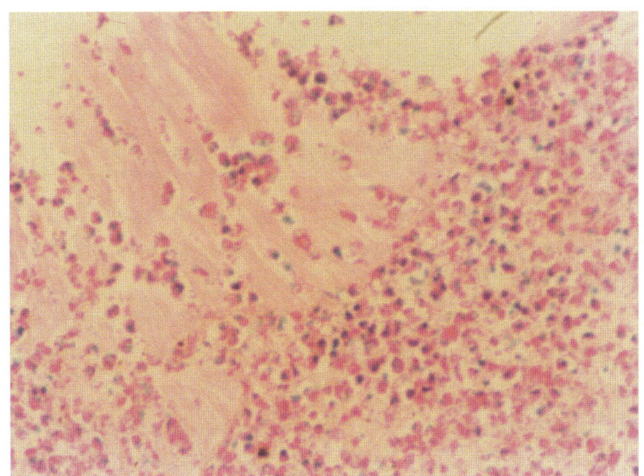

Figure 4.39, Giemsa x 260

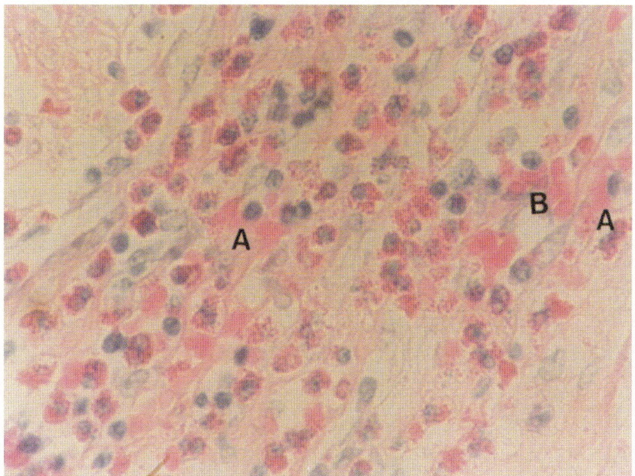

Figure 4.40, Giemsa x 520

Erythrogenic eosinophil development from smooth muscle (figs. 4.39 and 4.40)

Figure 4.39. At the upper left of this figure, eosinophils are arising from the hypertropic smooth muscle of the urinary bladder and then transforming into red cells. This is better seen in figure 4.40. In the lower right, the muscle fibers have been mostly replaced by the developing eosinophils. Giemsa x 260

Figure 4.40. This photomicrograph is taken from the same tissue section as figure 4.39. On the left side of this figure many of the eosinophils are rounded. In other areas, the eosinophil granules are coalescing in the formation of hemoglobin (A), followed by the development of red cells (B). Note that the eosinophil granules may appear in the muscle fiber with or without the association of an eosinophil nucleus. Giemsa x 520

Chapter 5

ENDOCRINE GLANDS   (figs. 5.1-5.50)
(Including Thyroid, Parathyroid, Adrenal, Pituitary and Pancreas)

Many of the body's major functions are controlled by the *endocrine glands*. By producing and secreting chemicals called *hormones* into the blood stream, endocrine glands are able to relay chemical messages to various organs and tissues and stimulate them to carry out such critical processes as growth and reproduction. The glands largely responsible for making and releasing most of the body's hormones are a collection of *ductless* or endocrine glands, so-called because they discharge their products directly into the blood stream and not via a tube or duct, as do the exocrine glands. The endocrine glands discussed in this chapter include thyroid, parathyroid, adrenal, pituitary and pancreas; the latter is actually a mixed gland having both endocrine and exocrine systems. Cellular changes in normal salivary gland tissue, as a reaction to nearby poorly differentiated papillary adenocarcinoma, are presented in Chapter 26.

All endocrine glands are composed primarily of specialized epithelium. In these endocrine glands, I will demonstrate the development of red blood cells and blood vessels directly from the glandular epithelium, as well as the direct transformation of the glandular epithelium into endothelium, hyalinized stromal tissue, and erythrogenic reactive cells such as segmented nuclear cells (SN cells), lymphocytes, plasma cells, etc. As the glandular epithelium transforms into blood and blood vessels, their secretory products (hormones, etc.) are automatically released into the blood stream and carried throughout the body.

1. Thyroid (figs. 5.1-5.24)

The thyroid is believed to be connected in some way to the maintenance of normal erythroid values. The fact that a thyroidectomy, or hypothyroidism, can produce anemia is evidence that it may be associated with red cell development. In fact, in 1947 Duran-Jorda first observed red cell development in thyroid secretions. Below, I will demonstrate the various ways of blood and blood vessel formation directly from the thyroid glandular tissue and from the colloid of thyroid follicles.

The thyroid, an endocrine (ductless) gland, consists of glandular acini or *follicles* which are lined by cuboidal or flattened epithelium. As seen with the H & E stain, the follicles usually contain a red fluid *colloid* that ranges in color from light pink to scarlet red. By using certain stains it can be seen that the nature of the fluid (especially its biochemical composition) may vary considerably from one follicle to the next, or even within the same follicle. This colloid as well as the surrounding epithelium can develop into blood, blood vessels, fibrous tissue and various other types of cells. Development of hemoglobinized colloid from dissolving thyroid cancer tissue is presented in Chapter 25, figures 25.1 and 25.2. Hemoglobinized colloid developing from benign thyroid tissue, as a reaction to nearby adenocarcinoma, is demonstrated in figure 25.3.

The photomicrographs presented here reveal the inherent capacity of thyroid tissue to produce various morphological structures such as red blood cells, blood vessels, lymphocytes, lymphatic tissue, fibrous tissue, hemosiderin, hemosiderin-pigmented cells and other cells. As the glandular epithelial cells transform into blood and blood vessels, their secretory

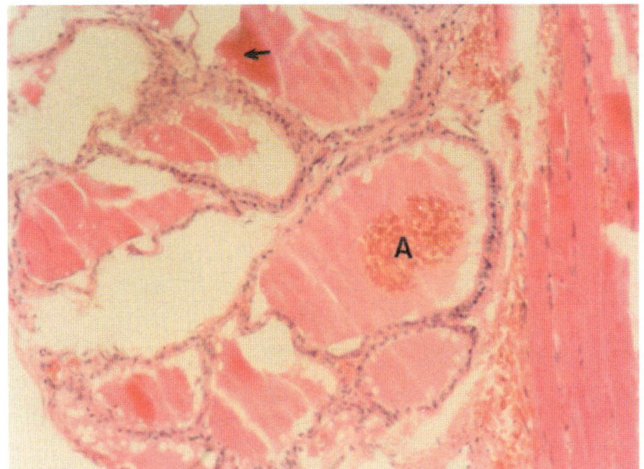

Figure 5.1, H&E x 130

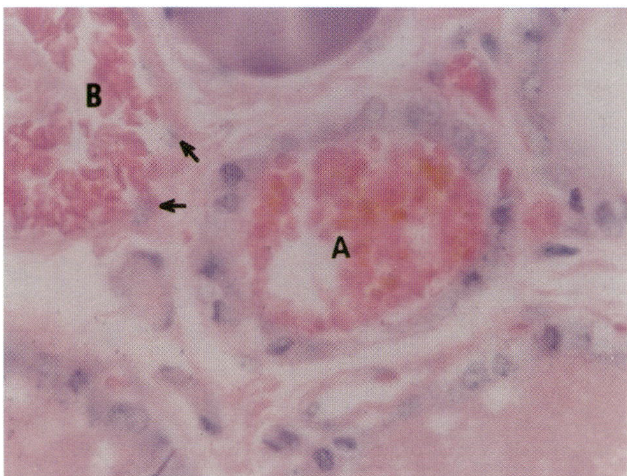

Figure 5.2, Giemsa x 520

products are automatically released into the blood stream and are carried throughout the body.

Figures 5.1, 5.4, 5.5-5.7, 5.8, 5.10-5.13, and 5.22-5.24 are taken from six different human thyroids that were surgically excised for benign lesions. Figure 5.2, figure 5.3, and figures 5.9 and 5.14-5.21 are from three different autopsy specimens. Figures 5.5-5.7 and figure 5.8 are taken from two hyperplastic thyroids. And figures 5.22-5.24 are taken from a case of Hashimoto's disease.

Red cell development directly as hemoglobin globules from the colloid of thyroid follicles (figs. 5.1-5.9)

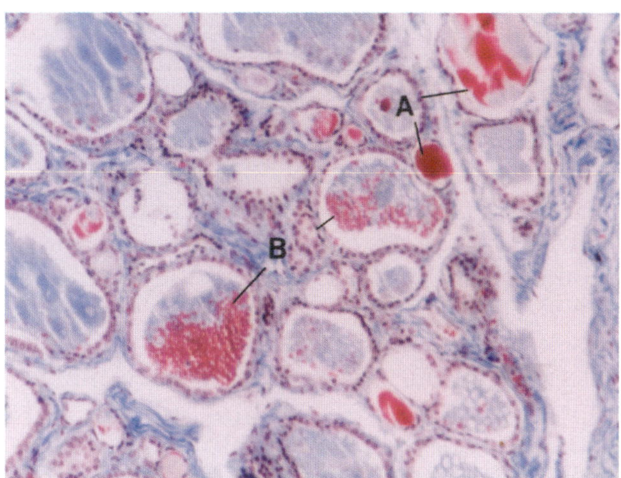

Figure 5.3, Chromotrope x 130

Figure 5.1. Several follicles are filled with colloid. In one follicle (A), red cells are developing from the colloid. In the majority of the follicles the colloid has retracted from the surrounding epithelium. Some parts of the colloid (arrow) are becoming denser and hemoglobin-like (deep red) in color. H&E x 130

Figure 5.2. The colloid in the follicles (A and B) is transforming into clusters of red cells. The majority of the epithelial cells of follicle (A) have lost their characteristics. Hardly any recognizable epithelial cells remain in follicle (B), which is now filled with

developing red cells. Only remnants of the fading epithelial cells (arrows) remain. As these follicles become blood vessels the secretory contents of the dissolved epithelial cells will be added to the blood stream and carried throughout the body. Giemsa x 520

Figure 5.3. The colloid (A) has changed from a light blue to a hemoglobin-like red with a denser consistency. Groups of red cells (B) are already in the later stages of development from the blue colored colloid. Chromotrope Aniline Blue x 130

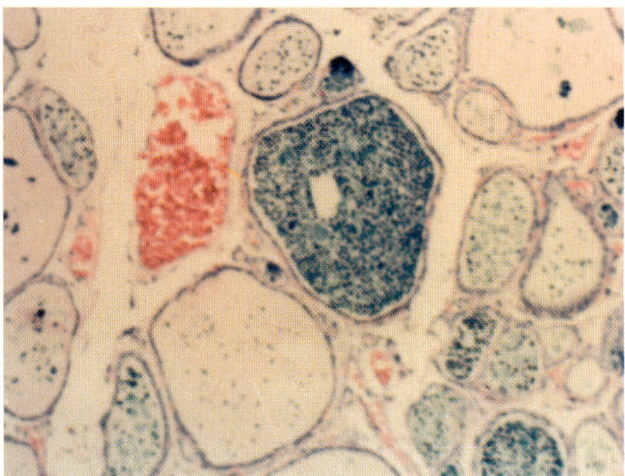

Figure 5.4, Giemsa x 260

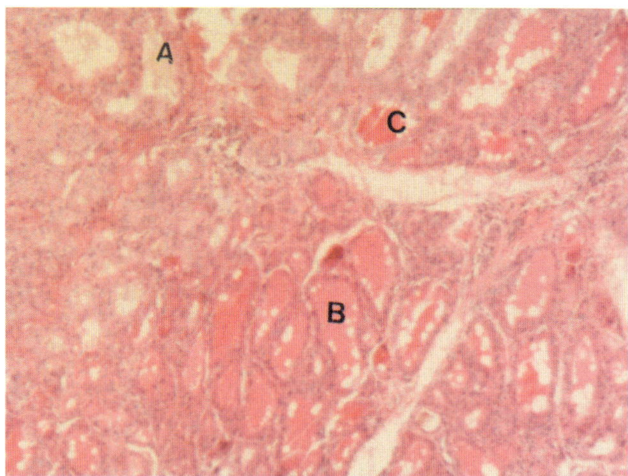

Figure 5.5, H&E x 104

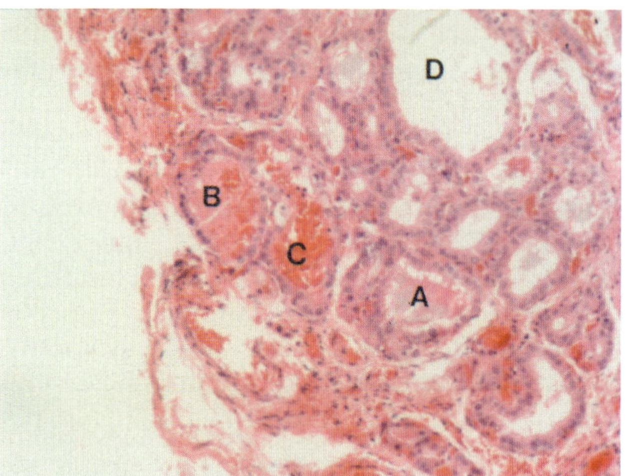

Figure 5.6, H&E x 260

*Blood vessel formation by blood-forming
follicles with endothelium replacing
the disappearing epithelium
(figs. 5.4 and 5.7-5.9)*

Figure 5.4. The dissimilar contents of
the various follicles with flattened or membranous epithelium are prominent in this
figure. The large center follicle appears to
be mostly filled with basophilic tiny particles. To the left of this follicle is another
follicle with a denuded epithelium and filled
with developing red cells that have not yet
rounded. In place of epithelium, this
follicle has a delicate membrane with tiny
endothelial nuclei. Giemsa x 260

Figure 5.5. These follicles have developed three different shades of colloid. Follicle (A) contains a grayish colloid with some
vacuolar changes. In follicle (B) a light red
colloid has formed. Finally, red cells are
developing from the hemoglobin-red colloid
developed within follicle (C). H&E x 104

Figure 5.6. In the follicle (A), the colloid
shows a transition from gray to light pink.
The vacuolar changes at the periphery of
the colloid are characteristic of thyroid toxi-
cosis. Red cells are forming from the red
colloid in follicle (B). There is further red
cell formation almost filling up the central
follicle (C). This follicle is irregular and
oblong, apparently due to the coalescence
of two adjoining follicles. Follicle (D), which
appears empty, is lined by cuboidal or
columnar epithelium. H&E x 260

Figure 5.7. Three blood-forming adjoining follicles are on the verge of coalescence
for the purpose of making a blood vessel.
There are a few remaining epithelial cells
(A). The endothelial lining (B) has begun
its formation. As this blood vessel fully
develops, the secretory contents of the
glandular epithelial cells will be added to
the original plasma and carried into the
blood stream. H&E x 200

Figure 5.8. An irregular blood vessel (A)
is developing from the coalescence of red
cell forming follicles of different sizes. Three
small scarlet red follicles (B), near the
upper edge of the figure, with denuded epithelium are on the verge of producing red
cells (which is hard to see at this magnifi-

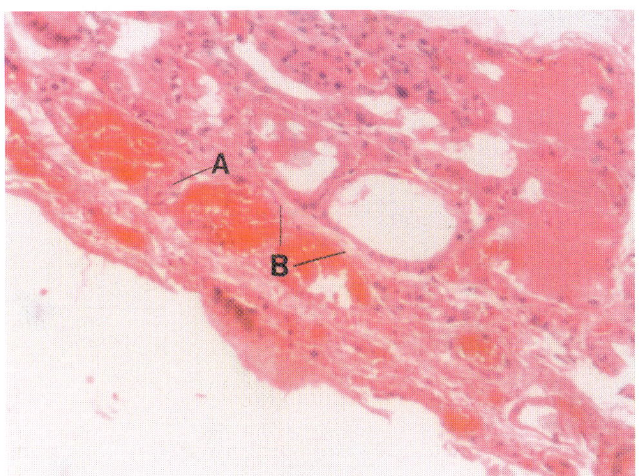

Figure 5.7, H&E x 200

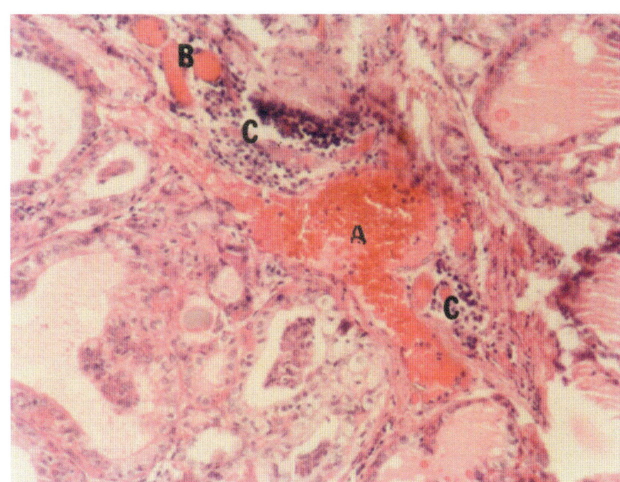

Figure 5.8, H&E x 130

cation). Many lymphocytes are arising from and replacing the follicles (C). H&E x 130

Figure 5.9. Regressive thyroid glandular tissue is associated with the development of blood and blood vessels. The colloid (A) is transforming into red cells, and the vanishing epithelial lining [arrow in follicle (A)] is changing into endothelium that is surrounded by a narrow band of connective tissue-like substance. The collapsed follicle (B) is developing into a narrow blood capillary. The arrow near (B) points to an endothelial cell. There are two large follicles (one on the left and one near the upper right border) which contain hemoglobin-like substance from which red cells have not yet developed. This hemoglobinized colloid is produced by the liquefaction of thyroid tissue (see next figure). H&E 400

Multiplying glandular epithelial cells dissolving and forming colloid and red cells, and development of blood capillaries from the interfollicular space (figs. 5.10 and 5.11)

Figure 5.10. Numerous thyroid epithelial cells are being formed and dissolved (liquefied) in the production of colloid. Red cell development from colloid is presented in the next figure. A developing red cell is situated within the core of the formative capillary (A) developing in the interfollicular region. Giemsa x 800

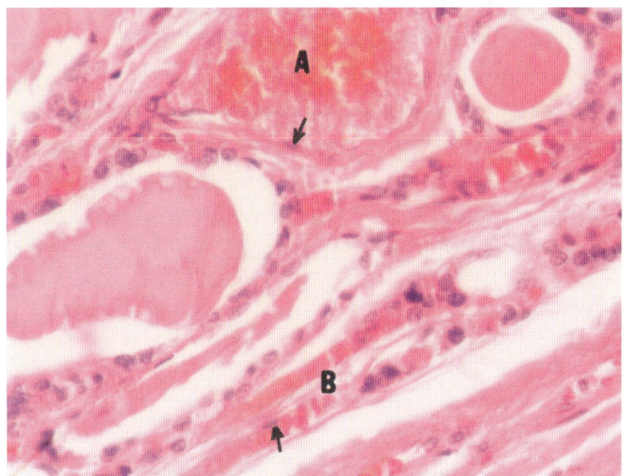

Figure 5.9, H&E x 400

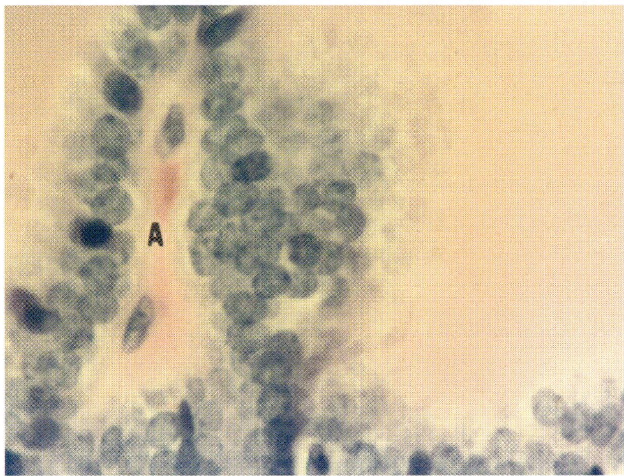

Figure 5.10, Giemsa x 800

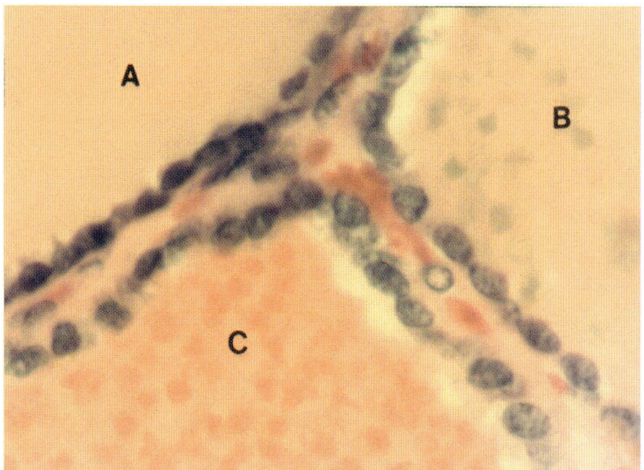

Figure 5.11, Giemsa x 800

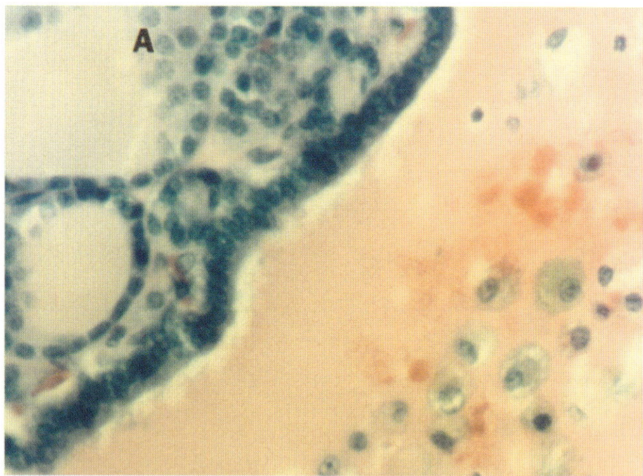

Figure 5.12, Giemsa x 400

Figure 5.11. The progressive stages of red cell formation from colloid are presented here in three large adjacent colloid follicles. (A) is homogenous hyalinized colloid. Very light basophilic particles are appearing in the colloid (B). Note the irregularly shaped red cells in (C) in the process of developing from the colloid. Also, note the developing red cells and possible endothelial cells in the interfollicular region as a blood capillary further develops [the earlier stage is observed in (A) of figure 5.10]. Giemsa x 800

Origin of red blood cells and erythrogenic nucleated cells from colloid of follicles (figs. 5.12 and 5.13)

Figure 5.12. In the lower right of this figure, cells are originating from the colloid. Some only have small nuclei while other enlarging cells contain various amounts of cytoplasm which is finely and uniformly vacuolated. The uniform vacuolation is often preliminary to red cell formation. These large cells are commonly classified as phagocytes or erythrophagocytes, but I prefer the term *floating erythrogenic cells* because the red cells are developing within these cells (see also figure 5.13). The fragmented acellular reddish substance adjacent to these cells is forming red cells. The peripheral epithelial cells of these follicles have retained their shape, but there is dissolution of the internal cells within the follicle (A). Giemsa x 400

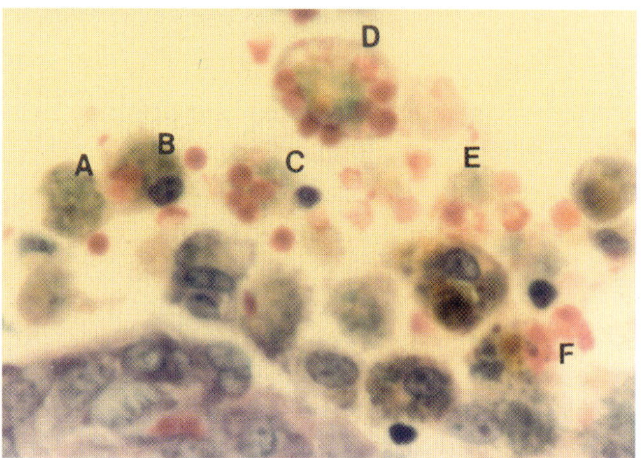

Figure 5.13, Giemsa x 800

Figure 5.13. Different developmental stages (A, B, C, D, E and F) of colloid originated cells are present. Cell (A) does not show red cell development as yet, but it is filled with tiny indistinct vacuoles. In contrast, the cells (B to E) are developing red cells. The large cell (D) contains many red cell sized vacuoles, the majority of which are filled with developing red cells. Remnants of a mother cell (E) can be seen among the extruded red cells. The cell (F), with extruding red cells on the right, contains a greenish-brown pigment on the left as evidence of a failed attempt at red cell formation in that area. These blood forming cells (B to F) are commonly and erroneously termed erythrophagocytes. The cell (A), because it is not forming red cells at this stage, is simply called a phagocyte. Giemsa x 800

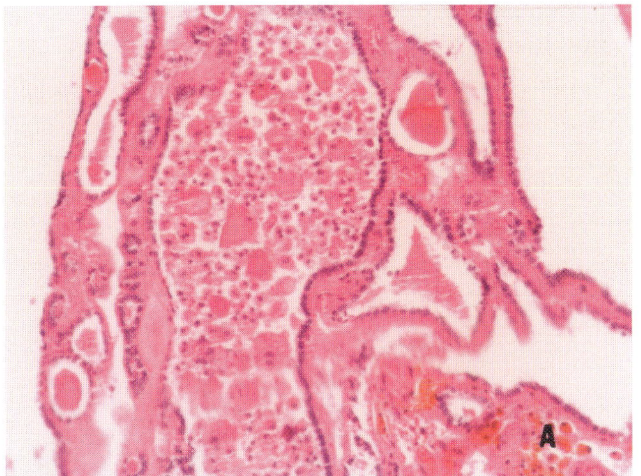

Figure 5.14, H&E x 130

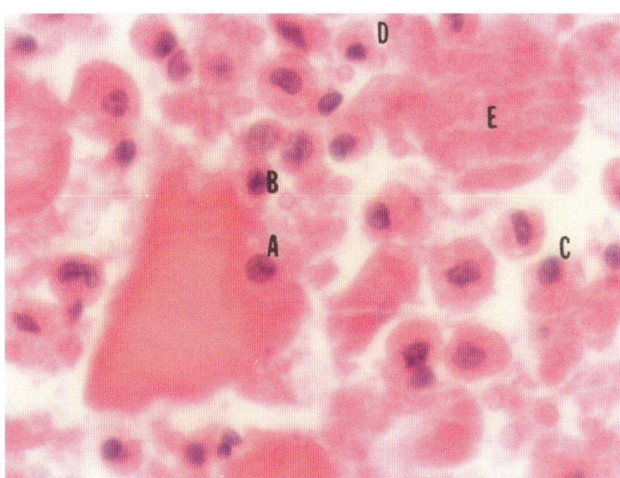

Figure 5.15, H&E x 520

Development of hemosiderin pigmented cells from colloid of thyroid follicles (figs. 5.14 and 5.15); and development of blood capillaries from hemosiderin pigmented cells of colloid origin (fig. 5.16)

Figure 5.14. So-called histiocytes are arising from the fragmented thickened colloid within the large follicle in the center (better seen in magnified view in figure 5.15). Golden brown pigmented hemosiderin in its early stage can be seen in the so-called histiocytes in area (A). H&E x 130

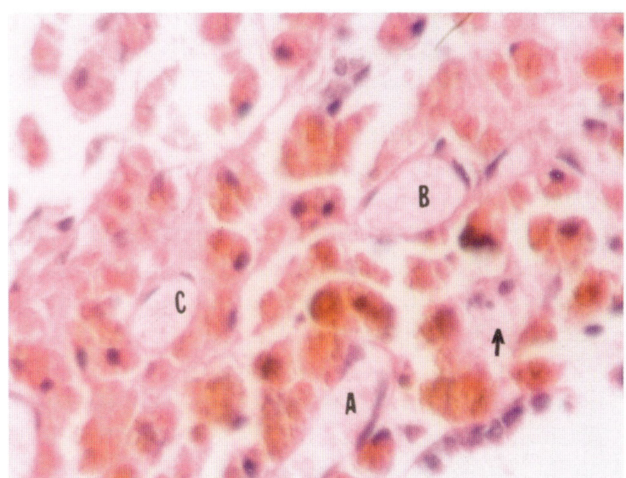

Figure 5.16, H&E x 520

Figure 5.15. This figure is a magnified view of the center of figure 5.14. A nucleus (A) is arising in the hyalinized thickened colloid which is liquefying around this nucleus. The colloid surrounding this nucleus is changing into minute seeds or globules which can only be seen with a magnifying glass. Cell (B) is separating from the thick mother-colloid while forming its own cytoplasm. In this field there are many similarly developed, colloid-generated, blood forming cells. The cells and intercellular substances (C and D) are forming clusters of minute red cells (use hand lens). (E) points to indistinct rows of developing red cells from colloid. H&E x 520

Figure 5.16. Capillaries (A, B and C) are developing from hemosiderin pigmented cells; however, the hemosiderin has vanished during red cell formation. Red cells in capillary (B) are in a vacuolated mesh form. Note that the endothelial lining is developing from the peripheral pigmented cells. The endothelium to the right of (A) has developed from the peripheral substance; on the left, the pigmented cells are forming a boundary. Small faintly visible vacuolar changes (arrow) in the hemosiderin containing cells (commonly known as phagocytes) may give rise to red cells and blood capillaries. H&E x 520

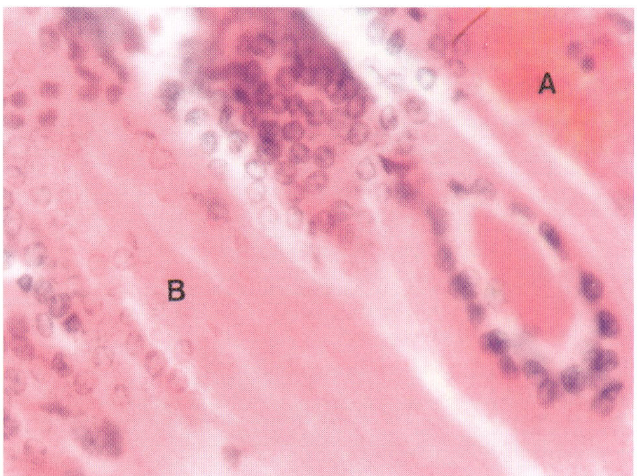

Figure 5.17, H&E x 520

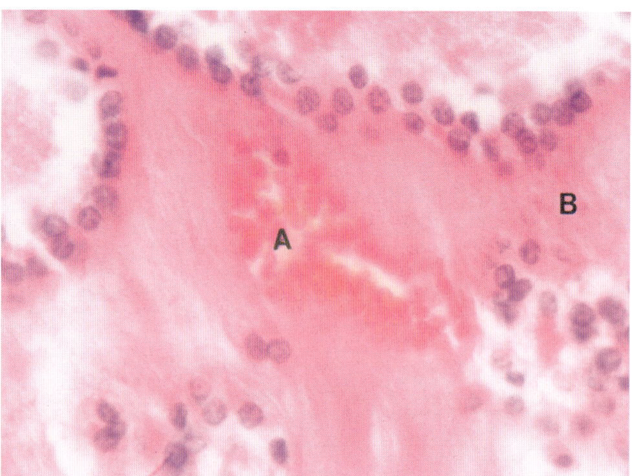

Figure 5.18, H&E x 520

Transformation of colloid and glandular epithelium into collagen (figs. 5.17-5.21); and development of blood and blood vessels from this collagen (figs. 5.18-5.20)

Figures 5.17 and 5.18. Colloid and many epithelial cells (seen more prominently in figure 5.17) are transforming into what appears to be an early stage of extensive collagen formation. In both figures, large clusters of red cells are developing from transformed collagen [area (A) of figure 5.17 is in an earlier stage than (A) of figure 5.18]. Disappearing epithelial cells [(B) in both figures] can be seen in the developing collagen. H&E x 520

Figure 5.19. Two small blood capillaries (A and B) are developing from a precollagenous state of colloid. Red cells are still forming, and the endothelial lining is not yet complete. A portion of a large developing capillary (C) is not yet lined by endothelium and contains red cells that are incompletely separated and in a granular state (use a magnifying glass). The colloid (D) appears to be forming red cells, and the epithelial cells (E) are vanishing in the transforming colloid. Once again, as these capillaries mature the remaining colloid and secretory products of the dissolved epithelial cells will be added to the blood stream. H&E x 260

Figure 5.20. The blood present in this photomicrograph is not due to hemorrhage.

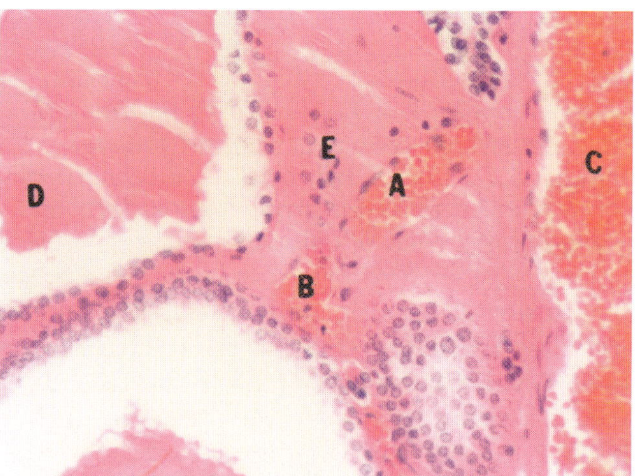

Figure 5.19, H&E x 260

The red cells are developing from numerous irregular areas of the fibrosed thyroid tissue. H&E x 130

Figure 5.21. Collagenization of the colloid is more advanced here with the development of collagen fibers having spindle-shaped nuclei (arrow). Remnants of the colloid and the colloid containing cells, so-called phagocytes, are dispersed over the center and lower right of this figure. H&E x 260

Origin of lymphocytes and lymphatic tissue from thyroid tissue in Hashimoto's disease (figs. 5.22-5.24)

Figure 5.22. The replacement of the thyroid tissue by lymphatic tissue with a promi-

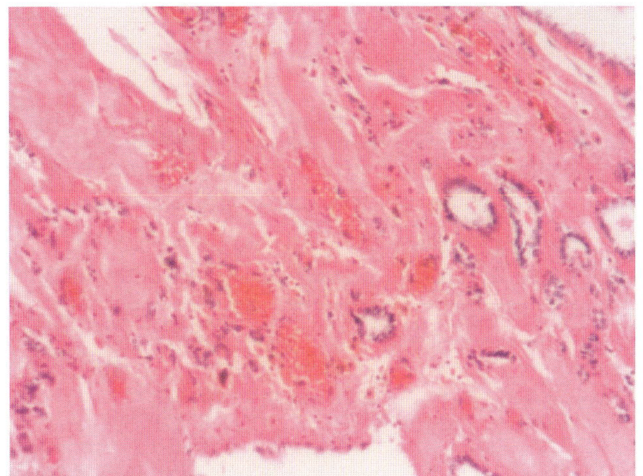

Figure 5.20, H&E x 130

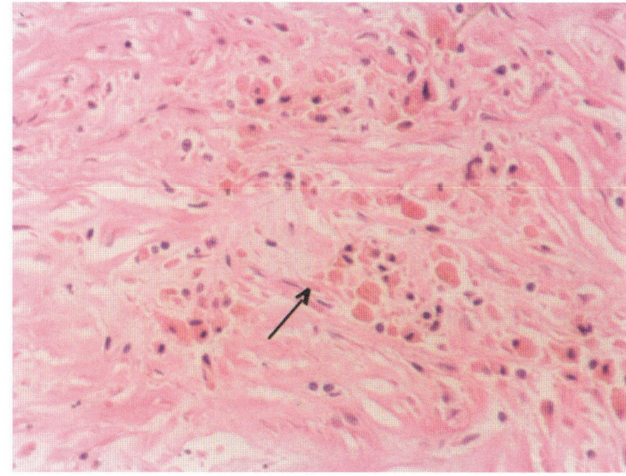

Figure 5.21, H&E x 260

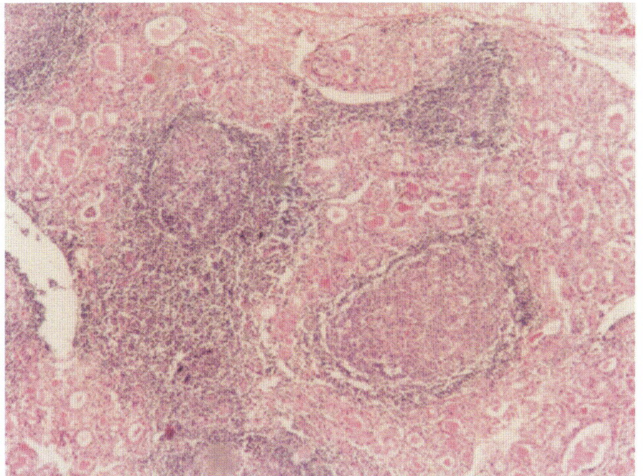

Figure 5.22, H&E x 52

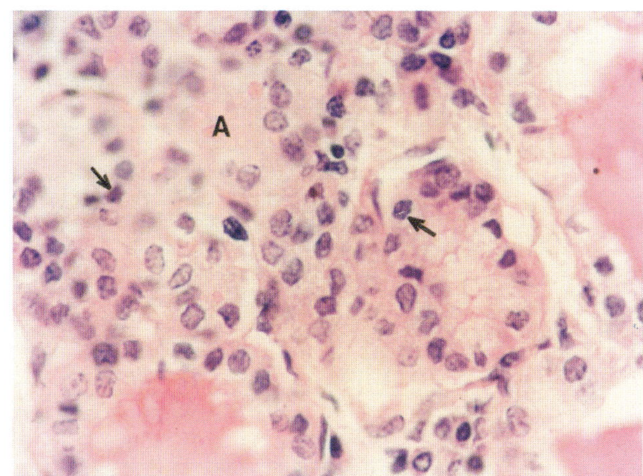

Figure 5.23, H&E x 520

nent germinal center is displayed in this figure. H&E x 52

Figure 5.23. The disappearance of glandular tissue begins with hyalinization of the colloid (A). Then the cuboidal epithelial cells transform (transdifferentiate) into lymphocyte-like cells (arrows), and new barely visible lymphocytes arise from the hyalinized colloid and glandular epithelium. H&E x 520

Figure 5.24. Lymphocytes (A) originate from the colloid and glandular epithelium. The original cuboidal cells (B) of the transforming acinus are still present. So far about half of the thyroid tissue in this photomicrograph has been replaced by lymphocytes in production of Hashimoto's disease. H&E x 260

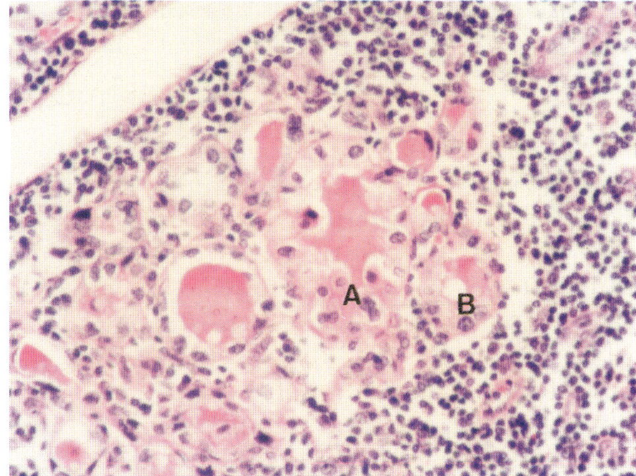

Figure 5.24, H&E x 260

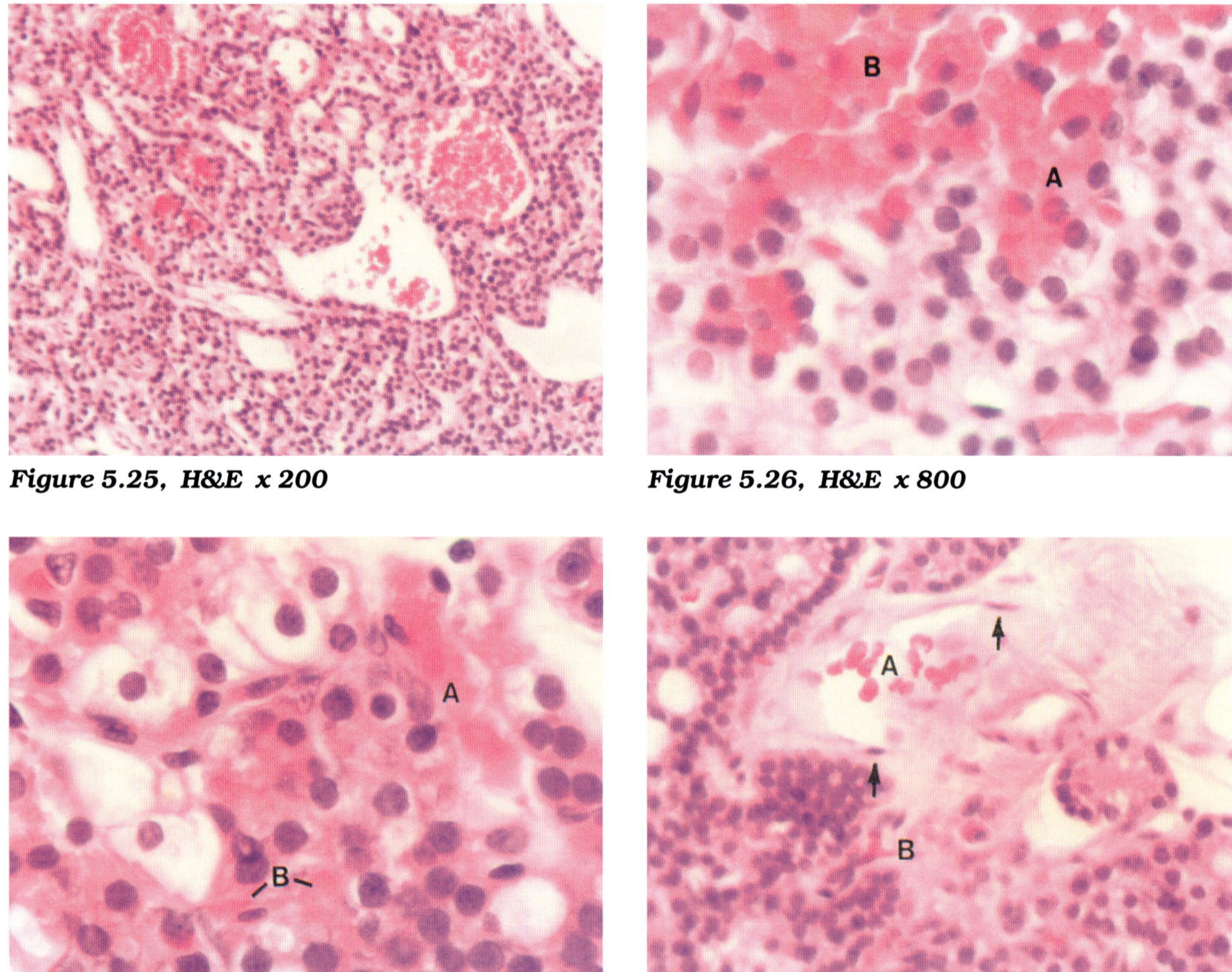

Figure 5.25, H&E x 200

Figure 5.26, H&E x 800

Figure 5.27, H&E x 800

Figure 5.28, H&E x 400

2. Parathyroid Gland (figs. 5.25-5.29)

The parathyroids are usually four tiny glands situated behind the thyroid. These glands are so small that they are often difficult to find. They play an important part in controlling the levels of calcium in the body. The parathyroid is an endocrine gland that consists of densely-packed groups of specialized epithelial cells. The parenchyma of the parathyroid is made of two types of cells, *clear* cells and *oxyphilic* cells (see figure 5.27). The glandular arrangement of these cells is shown in figure 5.25.

The development of red blood cells and blood vessels from the glandular epithelial cells and stromal tissue of the parathyroids is demonstrated below. As the glandular epithelium transforms into blood and blood vessels, its secretory products are automatically released into the blood stream and are carried throughout the body.

Figures 5.25, 5.26, 5.28 and 5.29 are all from a parathyroid surgically excised for adenoma. Figure 5.27 is normal parathyroid tissue incidentally removed during a total laryngectomy for squamous cell carcinoma.

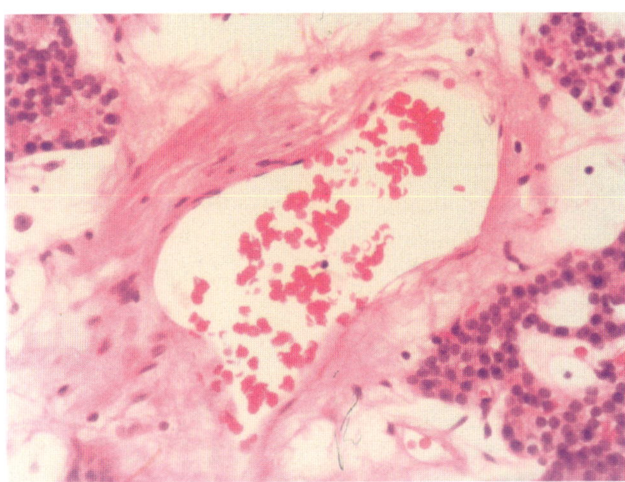

Figure 5.29, H&E x 260

Development of red blood cells and blood vessels from glandular parathyroid tissue (figs. 5.25-5.27)

Figure 5.25. In scattered areas, there are collections of developing blood in which the red cells have not fully formed as individual separated and rounded hemoglobin globules. On casual observation, such a finding would usually be categorized as a hemorrhage or a blood cyst. H&E x 200

Figure 5.26. Prominently shown in this figure are many epithelial cells transforming into red cells. This appearance is usually considered hemorrhage instead of red cell generation. The reddish-purplish tinge in the nuclei near (A) signifies the early stages of the hemoglobinization of epithelial cells. The reddish-purplish area near (B) points to further hemoglobinization of parathyroid cells compared to (A). In most of the areas, the developing red cells are

not yet separated and are smudged together. This dissolving and transformation of glandular epithelium into red cells releases the parathyroid hormone into the blood system. H&E x 800

Figure 5.27. In an irregular column, epithelial cells (A) are disintegrating in association with the early stages of hemoglobinization. There are other focal areas (B) where the hemoglobinization is leading to red cells. A formative red cell is present in the upper left corner. Both clear cells and oxyphilic cells can be easily identified at this magnification. H&E x 800

Development of blood and blood capillaries from hyalinized stromal tissue developed from glandular epithelium (figs. 5.28 and 5.29)

Figure 5.28. Here, red cells in rouleau form (A) are developing from the dissolving glandular epithelium preceded by hyalinization. Vessel formation has begun, as indicated by the developing endothelial cells (arrows) at the periphery of the liquefied product. The fibrous stromal tissue is, in turn, developing from the vanishing epithelial tissue (B). The further stages of collagenization of stromal tissue derived from glandular tissue are present in figure 5.29. H&E x 400

Figure 5.29. A large blood vessel is developing from the hyalinized fibrous stromal tissue. Note that developing red cells are not individual round bodies, but are smudged together as hemoglobinized particles of various sizes. The endothelium of the vessel wall has developed from the inner surface of the fibrous wall. H&E x 260

3. Adrenal Gland (figs. 5.30-5.37)

The adrenal glands (commonly called the adrenals) are located immediately above the kidneys. This important endocrine gland consists of two anatomically and physiologically distinct components, the *cortex* (outer covering) and the inner *medulla*, that differ in embryological origin and development. The adrenal cortex produces hormones essential for life and consists of three zones (the glomerulosa, fasciculata and

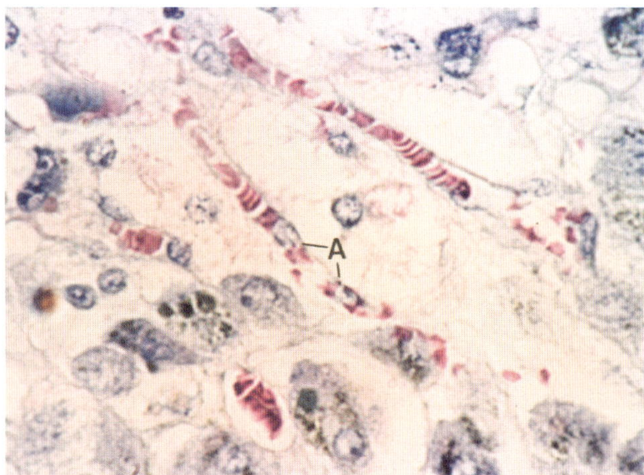

Figure 5.30, Giemsa x 520

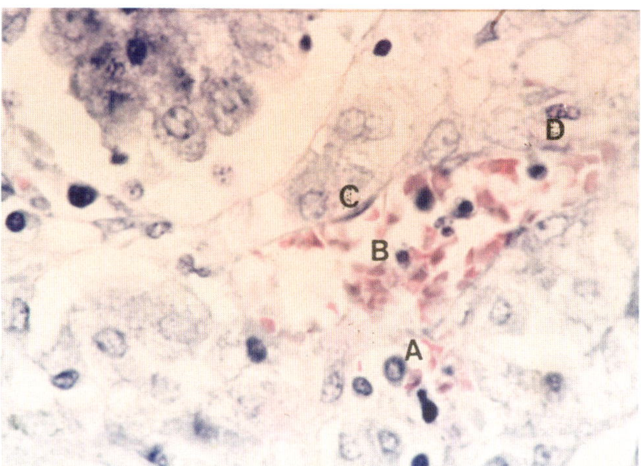

Figure 5.31, Giemsa x 520

reticularis). The medulla, or core, is the part of the gland that secretes adrenalin and noradrenalin.

Demonstrated below is the development of blood and blood vessels from the glandular epithelial cells of the adrenal cortex. In the development of the blood vessels, the glandular epithelium dissolves and the secretory contents of the cells are released directly into the blood stream. Also, the development of lymphocytes, erythrogenic SN cells and plasma cells is described. The last figure demonstrates the development of sickle cells from adrenal tissue, in a case of sickle cell anemia. For cellular changes in malignant adrenal tissue see Chapter 32.

Figures 5.30-5.34 and 5.36 are taken from the zona reticularis of the adrenal cortex from a one and one-half hour post-mortem human specimen. Figure 5.35 is taken from a normal adrenal cortex consisting of the zona glomerulosa of a healthy rabbit. Figure 5.37, from the zona fasciculata, is taken from the autopsy of an adult who died of complications of sickle cell disease. Figures 5.32 and 5.33 are reprinted from Volume I (McDonald, 1989), and figure 5.37 is reprinted from Volume III (McDonald, 2001).

Blood and blood vessel formation from adrenal tissue (figs. 5.30-5.36)

Figure 5.30. A narrow capillary is developing with a single column of red cells. Note

that the red cells are in the shape of pressed biscuits. The endothelial wall is only partially formed, with the appearance of a fine membrane derived from the local protoplasmic substances. Some of the nuclear structures (A) may be forming endothelium. Giemsa x 520

Lymphocyte development from adrenal cells (figs. 5.31, 5.33 and 5.35)

Figure 5.31. Note the progressive developing stages of lymphocytes from the surrounding adrenal cells (A). In the formative blood vessel, in addition to the developing red cells in the lumen, there are also a few lymphocytes (B) of various sizes within the lumen. On the upper wall of this developing blood vessel, there is a well developed endothelial cell (C) with a rounded margin. There is a less developed endothelial cell (D) further along the upper wall. Most of the lower wall's lumen remains ragged, as endothelium has not yet formed from the lysing surrounding adrenal cells. Giemsa x 520

Development of erythrogenic SN cells from adrenal tissue (fig. 5.32)

Figure 5.32. A formative vascular channel is arising diagonally as a column of erythrogenic SN cells (segmented nuclear cells) develop from the adrenal glandular tissue. Some of the SN cells (arrow) are beginning to show erythrogenic capacity with the production of hemoglobin. In the

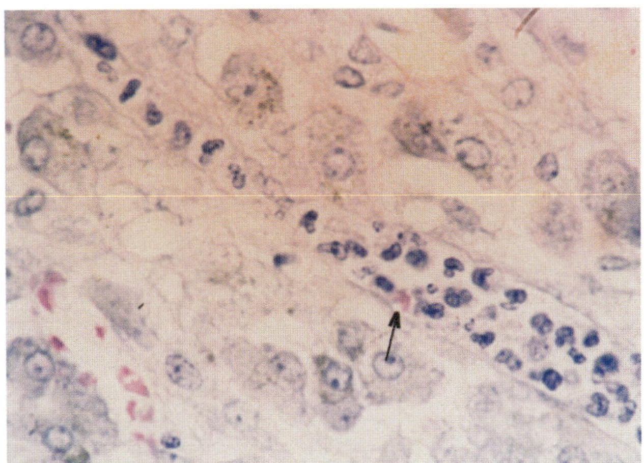

Figure 5.32, Giemsa x 520

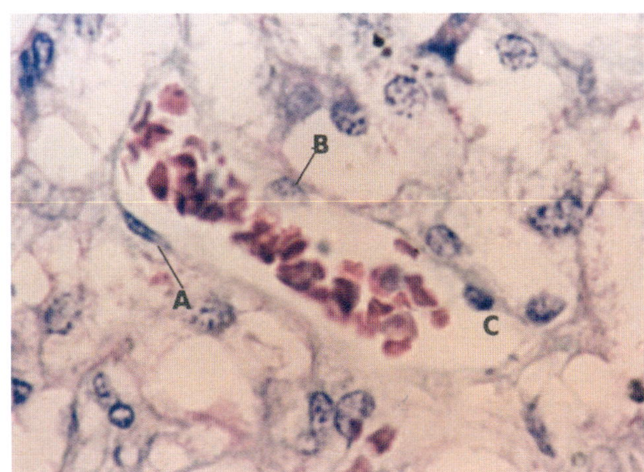

Figure 5.33, Giemsa x 800

lower left corner, red cells are arising directly from the glandular epithelial tissue which is undergoing canalization for vascular channels. Giemsa x 520

Figure 5.33. This formative blood vessel shows only one fully developed endothelial cell (A). The epithelial cells (B) on the opposite wall are in the process of dissolving and releasing their hormonal contents directly into this developing blood capillary. Within the lumen there are fragments of the original cellular substance (adrenal cells) mixed in with the irregularly shaped red cells. (C) is a possible developing stage of a lymphocyte from a peripheral remaining epithelial cell. Giemsa x 800

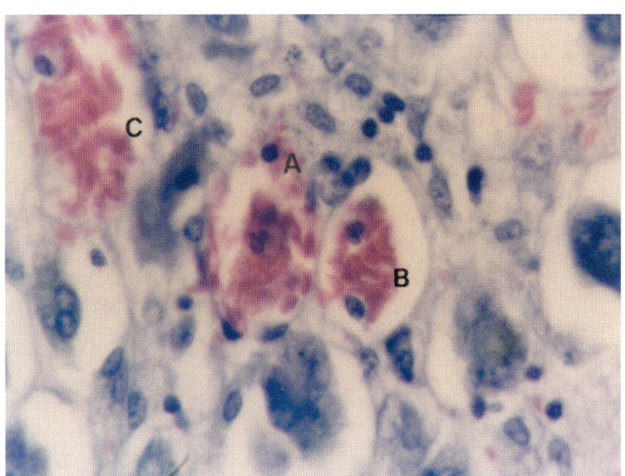

Figure 5.34, Giemsa x 520

Figure 5.34. Three blood capillaries (A-C) are developing from the adrenal tissue. Red cells that have not yet rounded are formed by the adrenal cells (A). Note the intimate connection between the red cells and the adrenal cells. Giemsa x 520

Figure 5.35. In this figure, red cells are directly arising as hemoglobin globules from the lysing parenchyma cells. Remnants of the lysing adrenal cells (A) are still present in the formative vascular lumens. These remnants will dissolve and release their hormonal content into the blood stream. (B) points to developing lymphocytes from adrenal cortical cells. Endothelial cells (C) are developing from the lining epithelial cells at the edge of the formative capillaries. H&E x 520

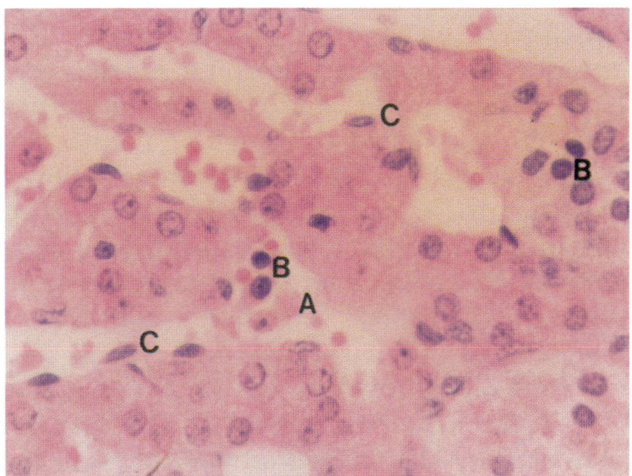

Figure 5.35, H&E x 520

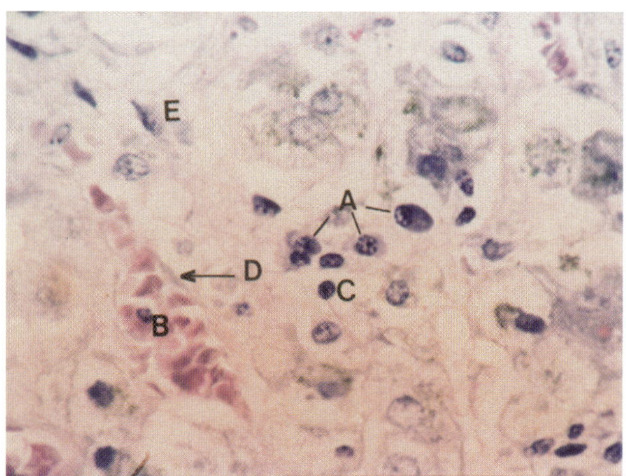

Figure 5.36, Giemsa x 520

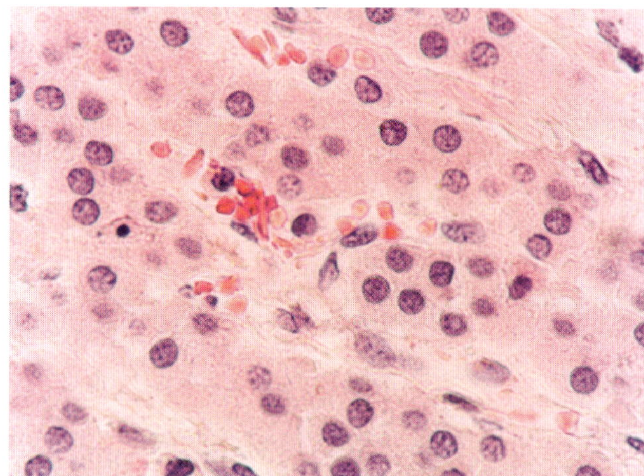

Figure 5.37, H&E x 520

Transformation of adrenal epithelium into plasma cells (fig. 5.36)

Figure 5.36. The stages shown in (A) portray the transformation (trans-differentiation) of adrenal epithelium into fully formed plasma cells. The developing red cells in vessel (B) show extreme irregularity in shape and size. A lymphocyte (C) is present in the adrenal tissue. A barely stained endothelial cell (D) has begun formation around the lumen of the capillary. A small endothelial cell (E) has formed from vanishing nuclei of adrenal cells in anticipation of blood vessel formation. Giemsa x 520

Sickle cell development from adrenal cells (fig. 5.37)

Figure 5.37. This section of an adrenal gland shows sickle cell formation from adrenal cells. H&E x 520

4. Pituitary [Hypophysis Cerebri] (figs. 5.38-5.43)

The pituitary is the master gland of the body. It not only produces its own hormones, but also influences the hormonal production of the other glands. This organ consists of two main parts or lobes. The anterior lobe is called the *pars distalis* and is formed by groups and columns of epithelial cells supported by reticular connective tissue. These epithelial cells consist mainly of oxyphil (chromophil) cells and chromophobe cells. The posterior lobe consists of the *pars intermedia* and the *pars nervosa*. The pars intermedia consists of basophilic epithelial cells, which blend into the cells of the anterior and posterior lobes. The pars nervosa consists mostly of fusiform and irregularly shaped cells. Discussed below is the development of red blood cells and blood vessels from the oxyphil and chromophobe cells of the anterior lobe. As the glandular epithelium transforms into blood and blood vessels, the hormonal products are automatically released into the blood stream and are carried throughout the body. Figures 5.38-5.43 are from the autopsy of a 57-year-old male who died of a recent cardiovascular accident.

Three zones of the pituitary (figs. 5.38 and 5.39)

Figure 5.38. Three zones of the pituitary, the anterior lobe with mainly oxyphil cells (A), the pars intermedia with mostly basophilic cells (B), and the posterior lobe con-

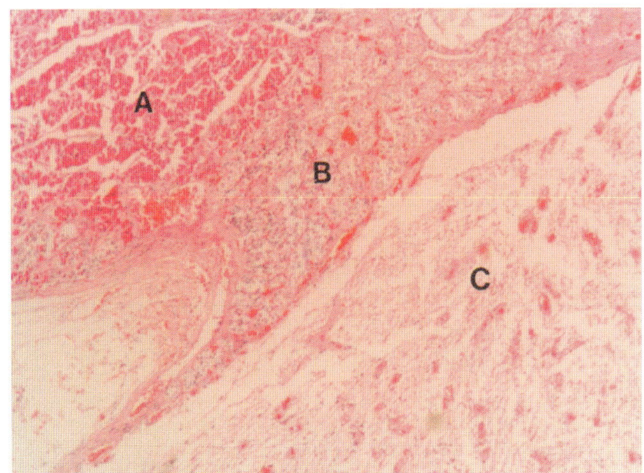

Figure 5.38, H&E x 52

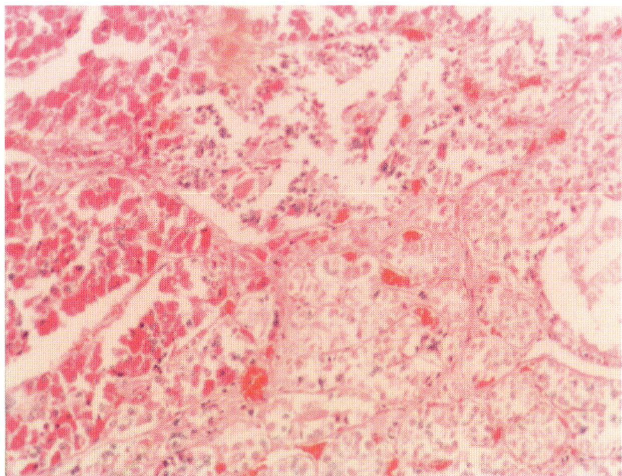

Figure 5.39, H&E x 130

sisting of fusiform and irregularly shaped cells (C), are shown from upper-left to lower-right. The pars intermedia in humans are quite small and its boundaries are not always sharply demarcated. H&E x 52

Figure 5.39. In this higher magnification of figure 5.38, oxyphil and chromophobe cells are seen from upper-left to lower-right. H&E x 130

Oxyphil cells developing red blood cells and blood capillaries (figs. 5.40-5.43); and chromophobe cells developing red blood cells (figs. 5.40 and 5.42)

Figure 5.40. Oxyphil cells are producing hemoglobin in clumps (A), which later form clusters of individual red cells (B). In the lower right corner of this figure, chromophobe cells are also forming red cells. H&E x 520

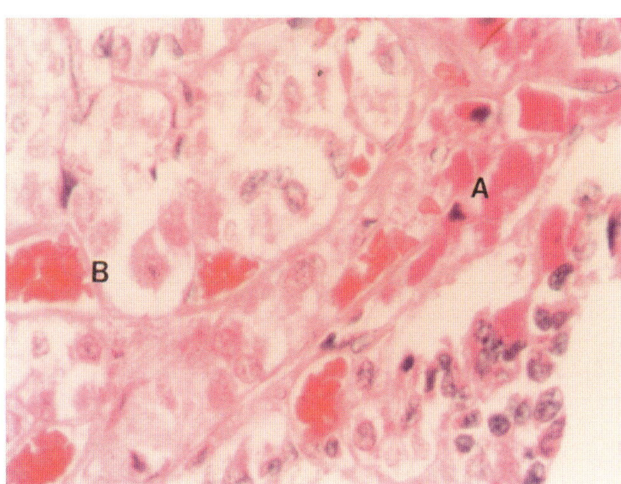

Figure 5.40, H&E x 520

Figure 5.41. Within engorged developing blood capillaries, there are oxyphils producing red cells which are not completely rounded in many areas (use a magnifying glass). And there are undissolved local cellular substances, light-purplish areas, within the lumen (A). As these cellular substances dissolve, their secretory products are added to the plasma and carried throughout the body. Despite prominent vascularization, the endothelium is still deficient in most places. H&E x 200

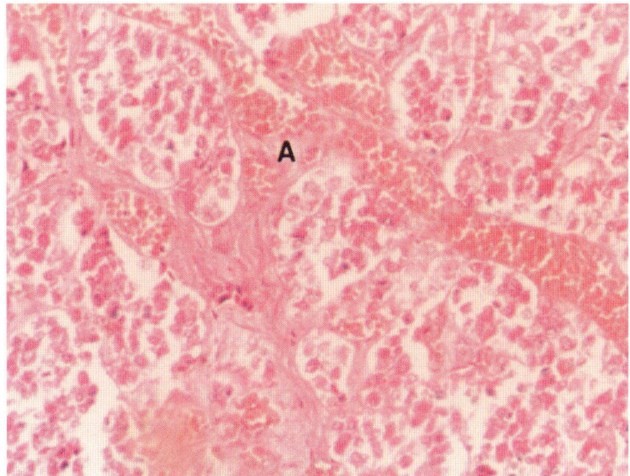

Figure 5.41, H&E x 200

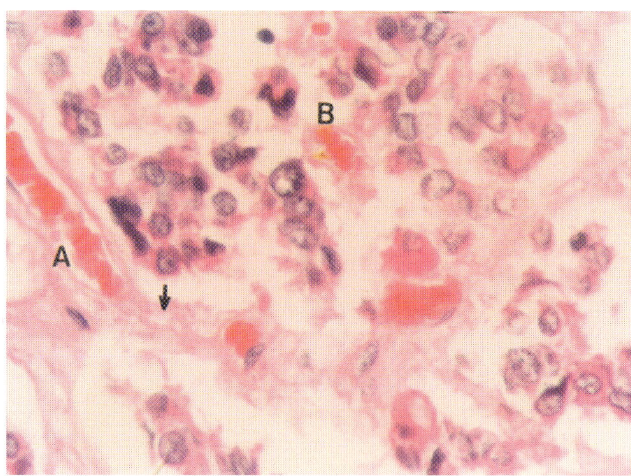

Figure 5.42, H&E x 520

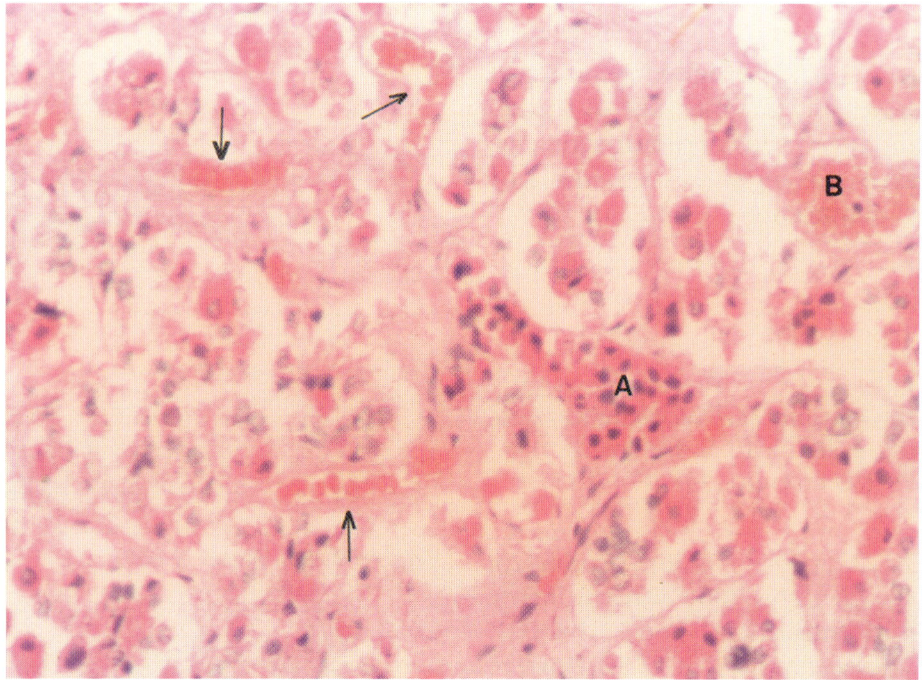

Figure 5.43, H&E x 400

Figure 5.42. In this figure, chromophobe cells predominate; these cells are transforming (transdifferentiating) into oxyphil cells towards the development of blood and blood vessels. Blood capillaries are formed first by hyalinization of the glandular tissues in the septa area (arrow). And then, between the remaining cell lobules, the blood capillaries arise (A and B). The red cells shown in this figure are still immature and are not yet rounded. Capillary (A) is shown in longitudinal plane, and capillary (B) is shown in the transverse plane. H&E x 520

Figure 5.43. In this figure oxyphil cells predominate. There are several narrow developing blood capillaries (arrows); however, the endothelial development is lagging behind. The oxyphils in the triangular area (A) are compact and the nuclei are small and hyperchromatic. With the disappearance of these nuclei the red cytoplasm may transform into red cells, as shown in area (B). H&E x 400

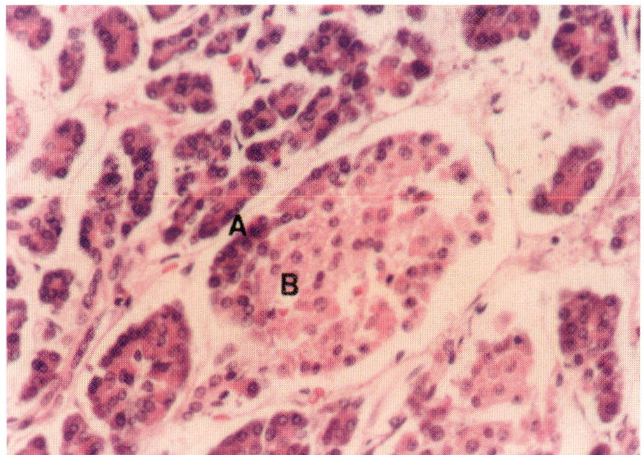

Figure 5.44, H&E x 260

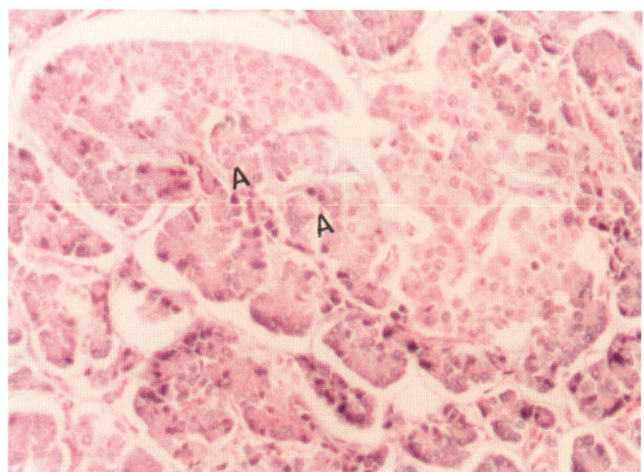

Figure 5.45, H&E x 200

5. Pancreas (figs. 5.44-5.50)

The pancreas consists of an endocrine portion composed of the Islets of Langerhans which is devoid of acinar and ductal systems. Insulin, the important product of the Islet of Langerhans, is released directly into the circulatory system and plays an important role in controlling carbohydrate metabolism. The major portion of the pancreas consists of an exocrine gland, which is composed of acinar structures and elaborate digestive juices which are carried by the ductal system into the digestive canal.

The following photomicrographs demonstrate the transformation of acinar structures in development of Islet of Langerhans. Subsequently, the islet cells may develop into blood and blood capillaries through which the liquefied insulin-containing products go directly into the general blood circulation. A portion of the acinar product is also carried into the circulating blood by the acinar tissue itself becoming blood and blood capillaries. Cellular changes in benign pancreatic tissue near adenocarcinoma are presented in Chapter 28.

Figure 5.44 is taken from the autopsy of a 68-year-old male with advanced arteriosclerosis, and who died of a cerebral vascular accident. Figures 5.45 and 5.48-5.50 are taken from the autopsy of a 58-year-old male with generalized arteriosclerosis, who also died of a cerebral vascular accident. Figures 5.46 and 5.47 are taken from the autopsy of a 51-year-old male with chronic pancreatitis who died of malignant hypertension and uremia.

Transformation of acinar structures into Islets of Langerhans (figs. 5.44 and 5.45)

Figure 5.44. This figure shows the continuous transformation of acinar structures (A) into islet cells (B) in the Islets of Langerhans. H&E x 260

Figure 5.45. In this figure, the lower left section of the islets (consisting of well-preserved acinar structures) are in the process of becoming islet cells (A). H&E x 200

Formation of red blood cells and blood capillaries from Islets of Langerhans (figs. 5.46 and 5.47); and from acinar glandular structures (fig. 5.48)

Figures 5.46 and 5.47. These figures demonstrate the Islet of Langerhans transforming into large blood capillaries. As the islet cells dissolve and develop into blood and blood vessels, their secretory products are added to the blood plasma and carried throughout the body. In figure 5.47, within the developing blood capillary there is development of red cells from well-preserved and partially vanished islet cells (A). These developing red cells have not completely separated and are not uniformly rounded. The development of surrounding endothelial lining from peripheral tissue is more advanced in figure 5.47. H&E x 520

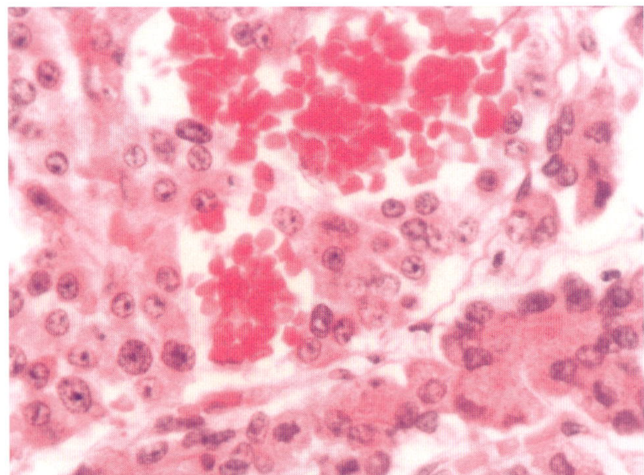

Figure 5.46, H&E x 520

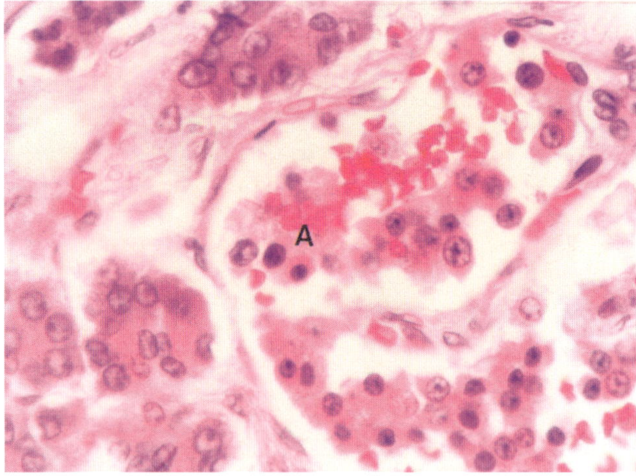

Figure 5.47, H&E x 520

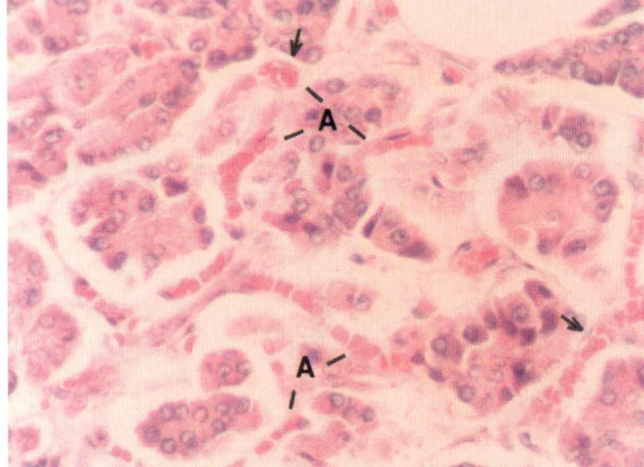

Figure 5.48, H&E x 400

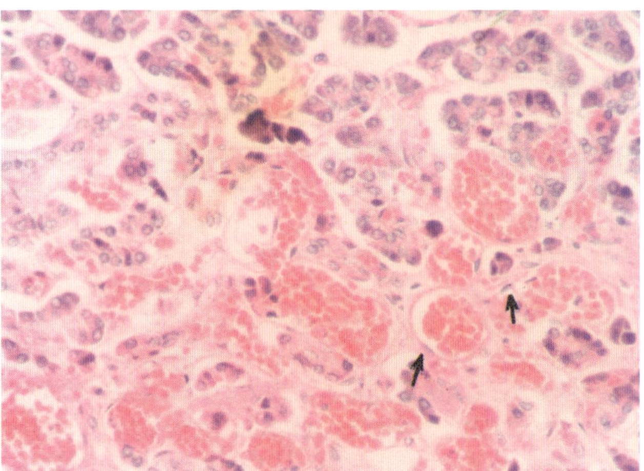

Figure 5.49, H&E x 260

Figure 5.48. The disappearance of acinar structures (A) takes part in the development of narrow blood capillaries. Note the origin of columns of red cells from disappearing acinar structures resulting in the development of narrow blood capillaries which are encircling the glandular tissues. In a few areas, the beginning of endothelial lining (arrows) can be seen encircling the developing capillaries. H&E x 400

Transformation of glandular tissue into large blood capillaries (figs. 5.49 and 5.50)

Figures 5.49 and 5.50. Mainly seen in transverse sections in figure 5.49 and longitudinal sections in figure 5.50, these two different areas of the same tissue section reveal the transformation of glandular tissue into wide blood capillaries. In figure

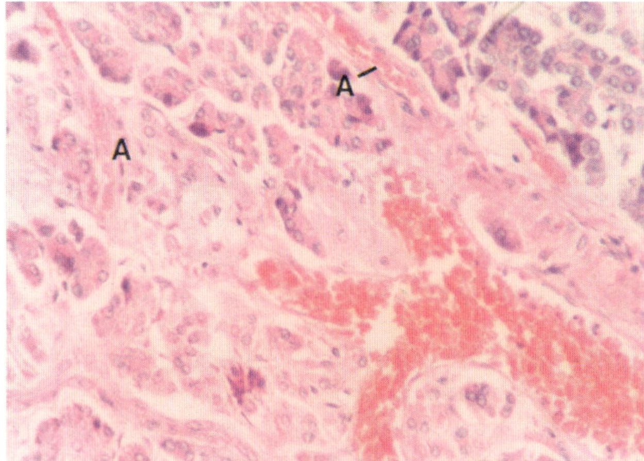

Figure 5.50, H&E x 260

5.49, the endothelial lining formation (arrows) has barely started. In figure 5.50, (A) points to two narrow blood capillaries developing from parenchyma tissue. H&E x 260

Chapter 6

CARTILAGE AND BONE (figs. 6.1-6.5)
(Blood and Bone Marrow Development from Cartilage and Bone)

The origin of cartilage from mesenchymal tissue and its transformation into bone coincides with the origin of undifferentiated tiny irregular nuclei, with the surrounding protoplasm formed by the dissolution of local cartilage cells or bone. These dissolved cartilage cells, as well as already developed bone, eventually form the bone marrow. Confluence of such marrow forming cartilage cells forms irregular sinuses, and the intervening cartilaginous substance may transform into bone through what are known as osteoid stages.

In addition to the participation of undifferentiated cells,[1] the cartilage and already formed bone may also directly form hemoglobin globules, sometimes in rouleau formation.[2] Following the dissolution of bony and cartilaginous tissue, there arises initial marrow plasma. Additional marrow fluid is also formed by other means, such as absorption from surrounding tissue, as well as through blood circulation. This newly formed marrow plasma then mainly becomes the original mother substance of red cells, white blood cells, platelets and megakaryocytes.

In early embryos, bone marrow originates from cartilage and bone with the latter structures dissolving in pre-determined areas towards the development of bone marrow. Surrounded by bony or cartilaginous tissue, the initial marrow spaces are filled with fluid (marrow plasma) and undifferentiated cellular structures. Arising from bone and/or cartilage, the marrow spaces enlarge by dissolution of surrounding bony or cartilaginous tissue. Before the cartilage becomes bone, one can see the "osteoid changes" that take place.

My findings in cartilage and bone are demonstrated in Volume III (McDonald, 2001). To save space and expense, only a few of the photomicrographs from Volume III are included in this volume. For a detailed discussion of my findings in cartilage and bone see Chapter 2, Volume III (figs. 2.1-2.12). For cellular changes in cancers of the bone see Chapter 35 of this volume.

Figures 6.1-6.5 are taken from the vertebra of a healthy adult rabbit which was sacrificed by the inhalation of ether for the purpose of detailed microscopic examination of various fresh tissues and organs. Figures 6.1-6.5 are reprinted from Volume III (McDonald, 2001).

Original bone marrow cells developing from lysing cartilage cells producing sinuses filled with original marrow cells and marrow fluid (figs. 6.1-6.3)

Figures 6.1-6.3. These figures demonstrate original marrow cells developing from lysing cartilage cells, (A) in figures 6.1 and

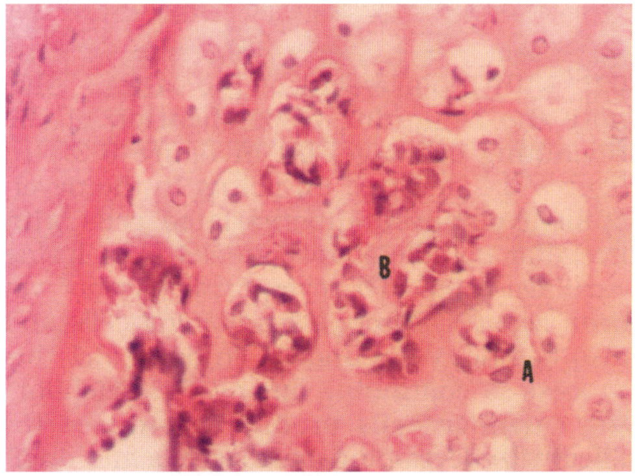

Figure 6.1, H&E x 400

---

1. Red cell development through cells somewhat similar to erythropoietic series cells in bone marrow developed from chick embryonic rib cartilage is demonstrated in figures 1.10-1.12 of Volume III (McDonald, 2001).
2. Red cell development directly as hemoglobin globules from periosteal fibroblastic tissue is demonstrated in figures 2.9 and 2.10 of Volume III, and red cell development from dissolving human cortical bone is demonstrated in figures 2.11 and 2.12 of Volume III.

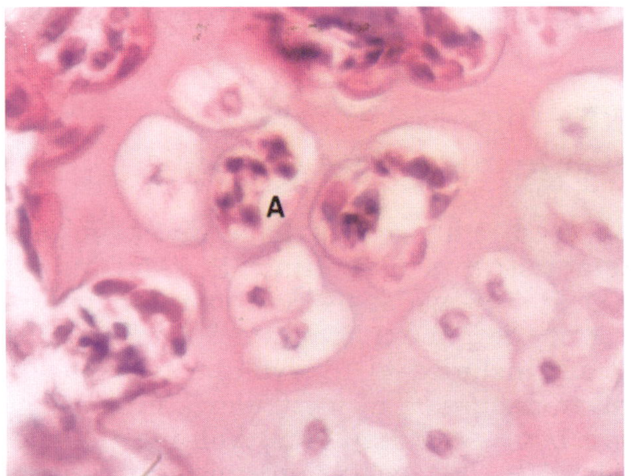

Figure 6.2, H&E x 520

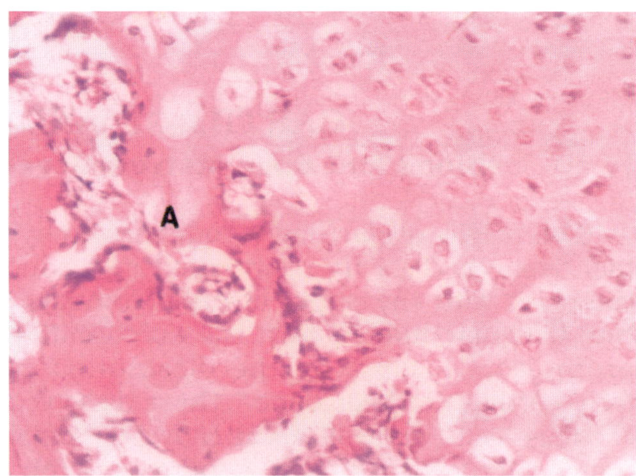

Figure 6.3, H&E x 260

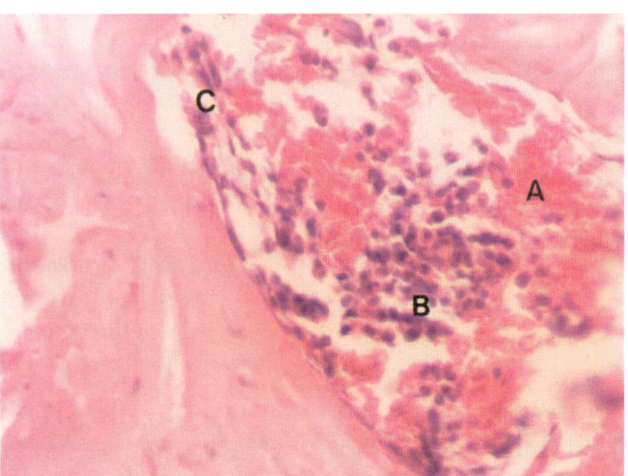

Figure 6.4, H&E x 400

6.2, with replacement of the latter. Also note the beginning of ossification of the remaining cartilage, (A) in figure 6.3. In figures 6.1 and 6.2, (A) points to the development of tiny undifferentiated cells with irregular spindle-shaped nuclei from lysing cartilage cells. The cytoplasm has not yet been separated from the adjoining cells and also has not yet rounded. In figure 6.1 this process is continuing among the neighboring cartilage cells producing sinuses (B) filled with developing marrow cells and original marrow fluid. In figure 6.3, calcification and the beginning of bone formation from original cartilage between marrow spaces is pointed out by (A). Figure 6.1 H&E x 400; figure 6.2 H&E x 520; and figure 6.3 H&E x 260

Red cells developing directly as hemoglobin globules, often in rouleau formation; and tiny undifferentiated nucleated cells arising from dissolved cartilage cells in marrow cavities (fig. 6.4)

Figure 6.4. Clumps of developing red cells in rouleau formation (A) adhere to each other and occupy a considerable amount of space in a marrow cavity surrounded by bone. In the center, many nucleated cells (B) with fuzzy borders are not yet rounded. (C) points to cells with narrow spindle-shaped nuclei arising from recently dissolved bone. (See also figures 2.5-2.7 of Volume III.) H&E x 400

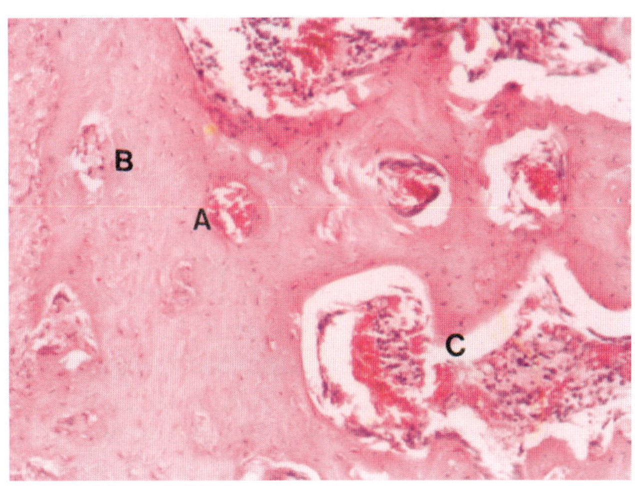

Figure 6.5, H&E x 130

Dissolution of bone in the development of blood and vascular channels (fig. 6.5)

Figure 6.5. New vascular channels are developing in cortical bone with localized lysis of bony substance in the formation of the lumen and developing blood; (B) is an earlier stage and (A) is a later stage. Continued enlargement of marrow spaces by further dissolution of the surrounding peripheral bone is shown by (C). H&E x 130

Chapter 7

BLOOD AND BONE MARROW (figs. 7.1-7.20)

My findings on blood and bone marrow are demonstrated in detail in Volume III of this series, *New Discoveries in Hematology* (McDonald, 2001). A discussion of my findings and a brief synopsis of the chapters in Volume III is presented in "My Observations and Comments on Volume III" at the end of this book. To save space and expense, only a few selected photomicrographs from Volume III are included in this volume (figs. 7.1-7.20). For a detailed discussion of my findings in blood and bone marrow see Volume III.

The main original mother substance of all cellular elements of bone marrow is found to be of marrow plasma origin. The vitality of bone marrow is amazing. The following figures provide brief demonstrations of the development of red cells directly from bone marrow plasma, nucleated cells developing from various coagulation products of marrow plasma, and platelets developing from red blood cells and bone marrow plasma. The blackish reaction of red cells in bone marrow giving rise to nucleated cells is also demonstrated. In both peripheral blood and bone marrow the participation of blue fibers in development of red cells and nucleated cells is presented. Reticulocyte development from clumps of reticulum and surrounding blood plasma is shown, and platelet development from red cells and blood plasma is also demonstrated. Finally, the nonparticipation of megakaryocytes in development of platelets is explained.

Figures 7.1 to 7.12 are taken from the following patients with various hematological diseases: figure 7.1 - a 76-year-old female with essential thrombocythemia and myelofibrosis resulting in marked anemia; figure 7.2 - a 62-year-old male with hyperplasia with dyserythropoiesis; figure 7.3 - an adult patient with no hematological problems; figures 7.4 and 7.6 - a 73-year-old male with sideroblastic anemia; figure 7.5 - a 45-year-old female with iron deficiency anemia; figures 7.7, 7.8 and 7.11 - a 71-year-old male with idiopathic thrombocytopenic purpura; figures 7.9, 7.10 and 7.12 are taken from peripheral blood smears from three different patients with no particular hematological disorders.

Figures 7.13 to 7.20 are taken from the following patients: figure 7.13 - a 74-year-old female with increased granulopoiesis diagnosed with microcytic hypochromic anemia associated with iron deficiency; figure 7.14 - an 82-year-old male with iron deficiency anemia, mildly hypercellular marrow with an increased number of megakaryocytes and large platelet clusters, and polycythemia vera; figure 7.15 - a patient with marked anemia (Hgb- 4.4 gram%) due to marked gastrointestinal bleeding resulting from duodenal ulcer and gastric erosion; figures 7.16 and 7.20 - a 66-year-old female with autoimmune hemolytic anemia; figure 7.17 - a 72-year-old female with essential thrombocythemia, platelet count 2,112,000/ cu mm and Hgb of 10.9 gram%; figure 7.18 - a 76-year-old female with essential thrombocythemia and myelofibrosis resulting in marked anemia, platelet count - 1,446,000/ cu mm and Hgb - 7.7 gm%; and figure 7.19 - a 64-year-old male with thalassemia minor.

Red cells developing directly as hemoglobin globules from erythrogenic mesh (fig. 7.1); and crystallization of erythrogenic plasma gel (fig. 7.2)

Figure 7.1. This figure is taken from a patient with thrombocythemia, and it demonstrates red cell development from erythrogenic marrow fluid. There are different stages of red cell development from an erythrogenic mesh, starting as tiny hemo-

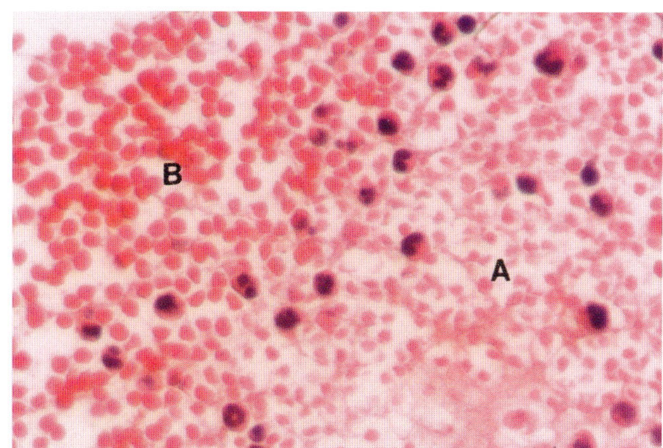

Figure 7.1, H&E x 520

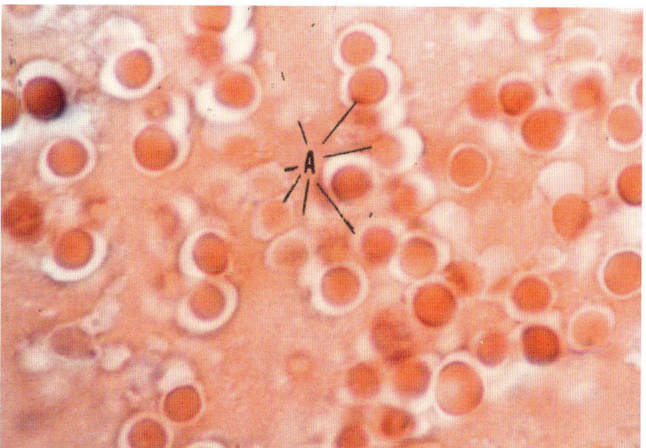

Figure 7.2, H&E x 1300

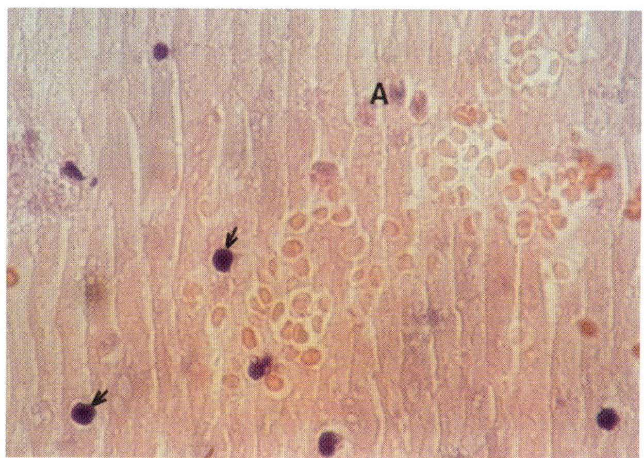

Figure 7.3, H&E x 520

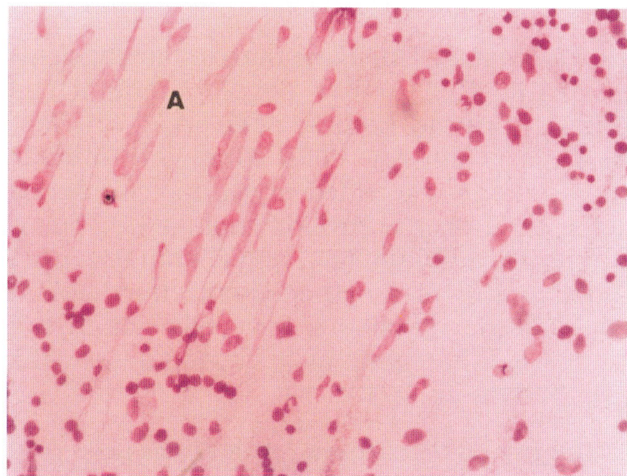

Figure 7.4, Wright x 130

globinized red cells (A) which develop into fully-formed red cells (B). Bone marrow clot section, H&E x 520

Figure 7.2. In this bone marrow clot section, red cells are developing from erythrogenic gel by crystallization. (A) points to the progressive stages of red cell development from vacuole formation to fully developed red cells. These progressive stages are shown in the following order: 10:00 o'clock position, 6:30, 8:00, 7:30, 3:00, 5:00 and 1:00. H&E x 1300

Red cells and nucleated cells developing from clotted blood plasma (fig. 7.3)

Figure 7.3. Red cells and a few nucleated cells are developing in a supernatant plasma clot section from a patient with no hematological problems. Fresh whole blood was drawn from a patient and quickly separated by centrifugation. The plasma was drawn by capillary pipette and allowed to clot. Plasma clot sections were stained with hematoxylin and eosin. (A) points to developing nucleated cells that are not yet differentiated. Circulating blood plasma is also capable of producing nucleated cells as evidenced in this figure (arrows). H&E x 520

Smeared fibrinoid bodies developing from coagulated marrow plasma (fig. 7.4); and transforming into naked nuclear bodies (fig. 7.5); and then developing into so-called hematopoietic nucleated marrow cells (fig. 7.6)

Figure 7.4. Round fibrinoid bodies are arising separately over a large area from coagulated marrow plasma particles of this bone marrow aspiration smear. In the up-

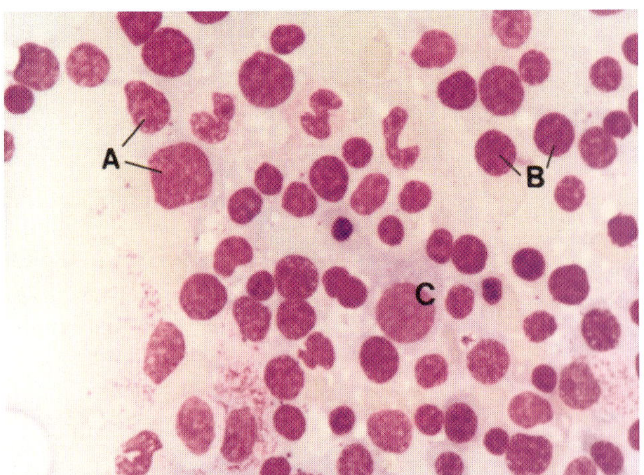

Figure 7.5, Wright x 520

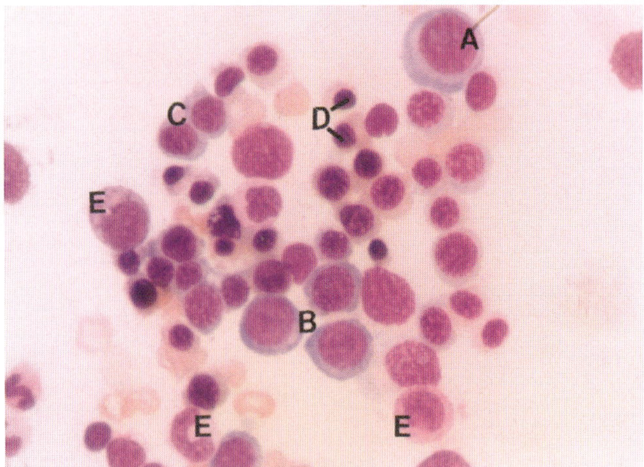

Figure 7.6, Wright x 520

per left quadrant of this figure (A), the plasma has not had time to solidify to form fibrinoid bodies. Therefore, the fibrin was smeared and drawn out in long strands while making the smear. Wright x 130

Figure 7.5. Seen in this bone marrow smear is the transformation of the fibrinoid bodies shown in figure 7.4 into many naked nuclear bodies (A and B). (A) is an early stage and (B) is a later stage with darker nuclear chromatism. (C) points to what may be called blast cell development, showing a large nuclear body with a faintly visible nucleolus in the center. Wright x 520

Figure 7.6. This bone marrow smear demonstrates the further development of the round fibrinoid bodies, seen in figure 7.4, into many marrow cells which are similar to commonly classified erythropoietic series cells, including a pronormoblast (A), basophilic normoblast (B), polychromatic normoblast (C), and orthochromic normoblast (D). There are also a few cells which fall under the category of granulopoietic cells (E). Wright x 520

Large fibrinoid particles forming red cells and segmented nuclear cells (figs. 7.7 and 7.8)

Figures 7.7 and 7.8. These two figures are from the same bone marrow aspiration smear. In figure 7.7, the peripheral part of two fibrin plaques is giving rise to a large number of red cells in rouleau formation. The disappearing fibrin (A) is shown between

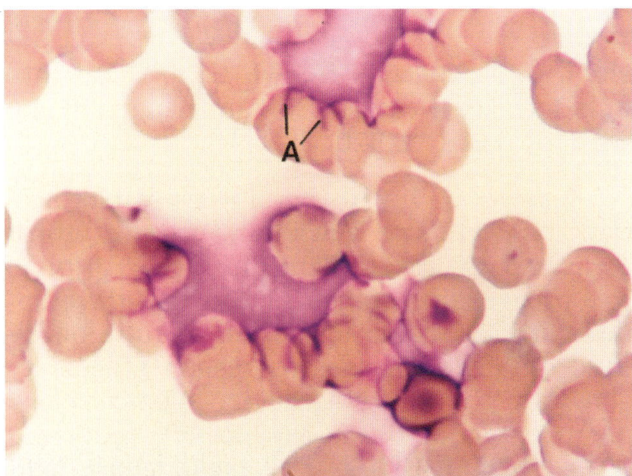

Figure 7.7, Wright x 1300

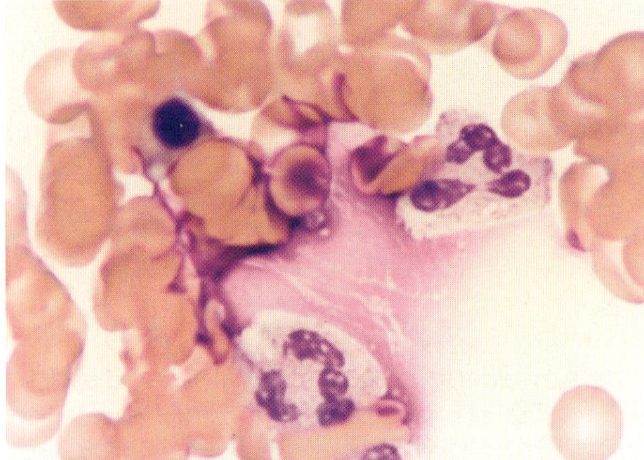

Figure 7.8, Wright x 1300

the red cells. In figure 7.8, both red cells and SN cells (with neutrophilic granules) are developing from a large fibrin plaque. Wright x 1300

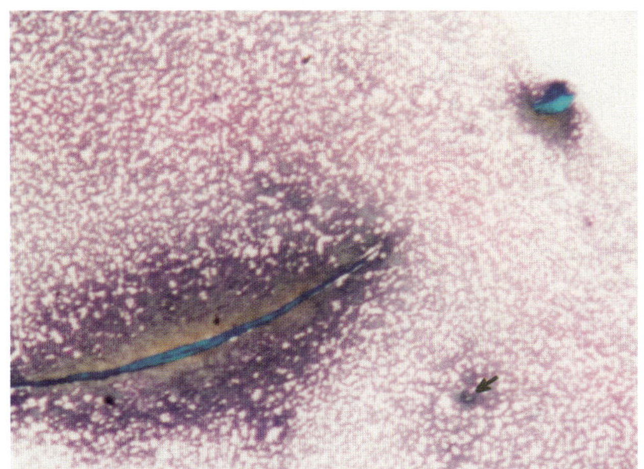

Figure 7.9, Wright x 80

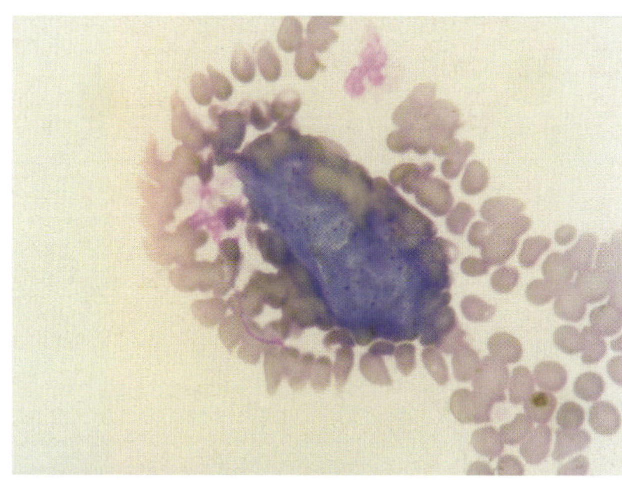

Figure 7.10, Wright x 520

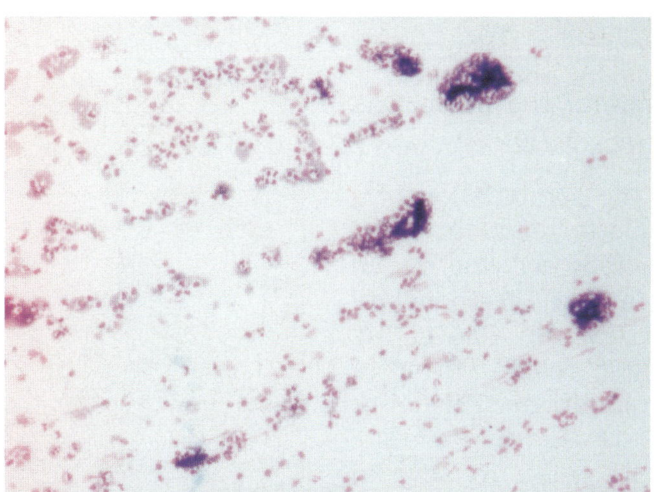

Figure 7.11, Wright x 200

Participation of blue fibers in the development of red cells (figs. 7.9-7.11)

Figure 7.9. This is a routine peripheral blood smear taken from the author's blood. Surrounded by dark-purple crowded red cells, a long deep-blue fiber with tapering ends stretches diagonally across half of the field. In the upper right, one can see another similarly-colored blue fragment, again surrounded by darkly-stained crowded red cells. The arrow points to a nearly-vanished light-blue fragment surrounded by less crowded purplish red cells. Wright x 80

Figure 7.10. Red cells as non-uniform hemoglobin globules are developing in compressed form from the outer part of a blue fiber. There is a fine, blue, dust-like material (the significance of which is not clear) in the blue fiber from which the red cells are arising. Peripheral blood smear, Wright x 520

Participation of blue fibers in the development of white blood cells, mainly segmented nuclear cells (figs. 7.11 and 7.12)

Figure 7.11. Fragments of blue fibers generating red and white blood cells are seen in a very low magnification of this bone marrow smear. In many peripheral blood smears, particularly those with a high white blood cell count, many similar blue fiber fragments can be seen actively generating red and white blood cells, especially SN cells, at the tapering end of the blood smear. Wright x 200

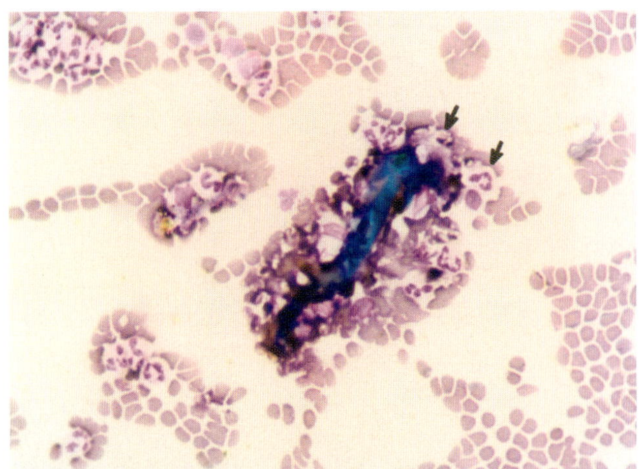

Figure 7.12, Wright x 260

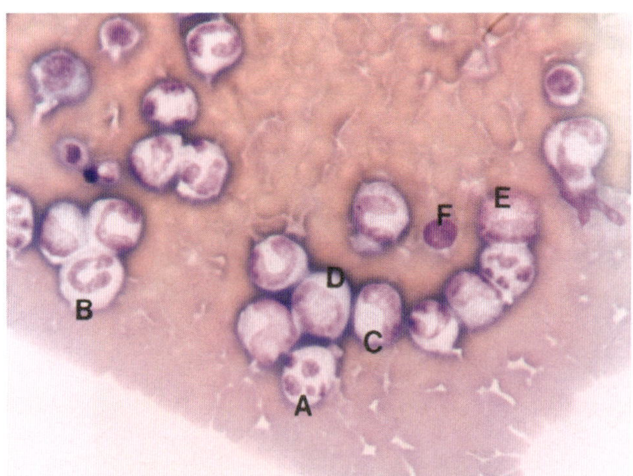

Figure 7.13, Wright x 520

Combination of blue fibers and blackish reaction of red cells in development of white blood cells (fig. 7.12)

Figure 7.12. Here, the combination of a blue fiber and the blackish reaction of red cells is seen. The blue fiber is giving rise to red cells and SN cells, and the blackish reaction of red cells (arrows) is also giving rise to many granulocytic cells. This process is beautifully demonstrated in figure 7.13. On the left side of this figure, note the development of SN cells, in several groups, from the blackish reaction of red cells. (See figures 7.13 and 7.14 for more examples of blackish reaction.) Peripheral blood smear, Wright x 260

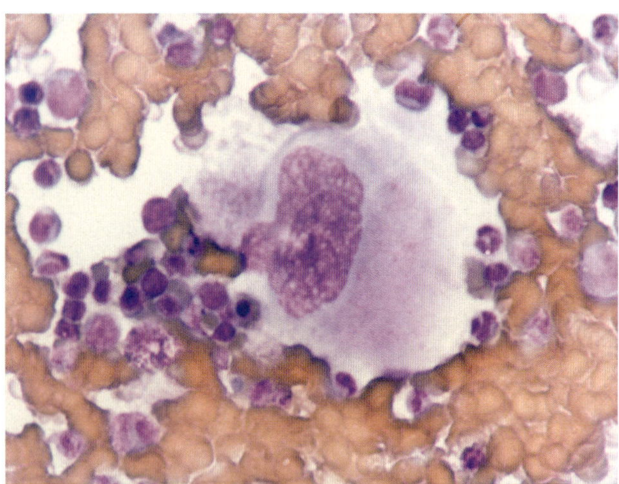

Figure 7.14, Wright x 520

Blackish reaction of red cells in bone marrow giving rise to commonly known erythropoietic and granulopoietic series cells including segmented nuclear cells, eosinophils and lymphocytes (fig. 7.13)

Figure 7.13. Within the spaces created by the blackish reaction and disappearance of red cells, many single granulocytic cells have appeared. These granulocytic cells can be categorized as follows: SN cell (A), band cell (B), metamyelocyte (C), myelocyte (D), eosinophil (E) and lymphocyte (F). A black rim made by the blackish reaction of red cells surrounds each of these newly-formed cells. Bone marrow smear, Wright x 520

Blackish reaction of red cells giving rise to megakaryocytes (fig. 7.14)

Figure 7.14. The large irregular space seen in this bone marrow smear was created by the blackish reaction and dissolution of red cells. The margin of the cavity undergoing the blackish reaction is producing nucleated cells (including SN cells). Occupying the central area is a megakaryocyte whose nucleus and cytoplasm are both produced as a result of the blackish reaction of red cells. Towards the left, there are clusters of small nucleated cells originating via the blackish reaction of red cells. Wright x 520

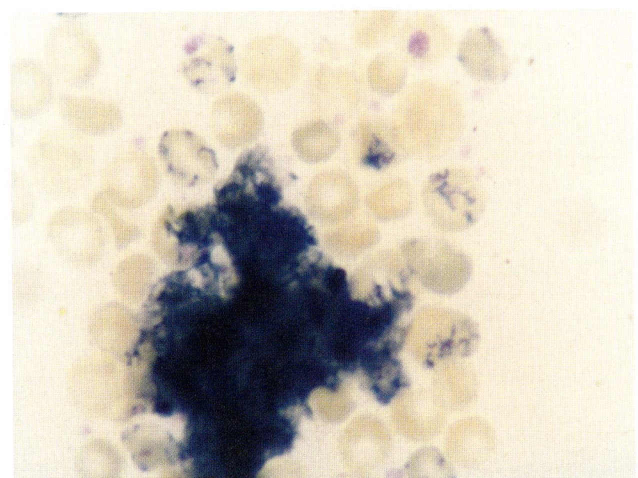

Figure 7.15, Reticulocyte Stain x 1300

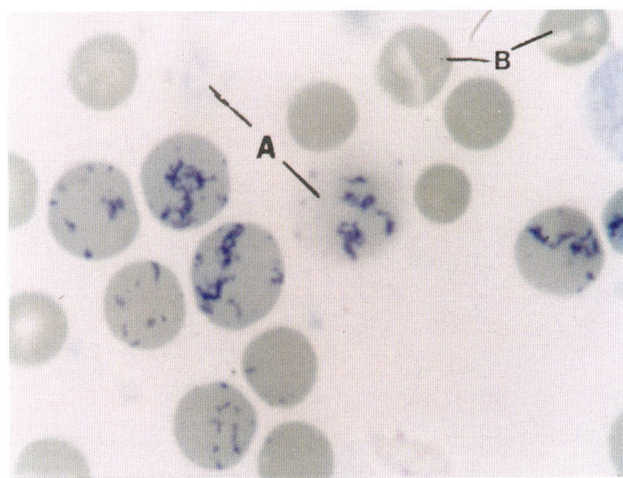

Figure 7.16, Reticulocyte Stain x 1300

Reticulocytes developing from minute clumps of reticulum of finely-beaded fibrils (SGF, the name given by Griffith, et al., 1970) of unknown origin and surrounding blood plasma (fig. 7.15)

Figure 7.15. In this peripheral blood smear prepared for reticulocyte study, the process of red cell development associated with reticulum is suggested. Reticulocytes are first seen to contain thick blue reticulum fibrils in clumps, which gradually disappear. As hemoglobin further develops from the surrounding plasma the dissociated reticulum disappears. New Methylene Blue x 1300

Colorless hemolysis as a means of disappearance of red cells (fig. 7.16)

Figure 7.16. In this peripheral blood smear prepared for reticulocyte study, the hemolytic process is shown by the fading away of the reticulocytes (A). In addition, two red cells (B) show irregular bar-shaped colorless hemolysis with almost-ruptured cell membranes. New Methylene Blue Stain x 1300

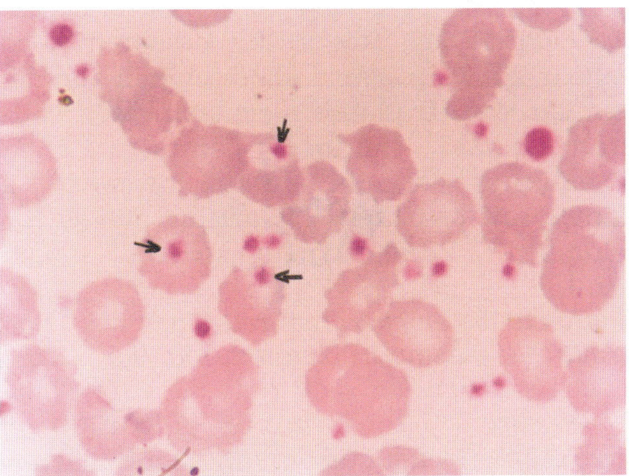

Figure 7.17, Wright x 1300

Platelets developing from red blood cells and plasma in peripheral blood (figs. 7.17 and 7.18)

Figure 7.17. In this EDTA mixed venous blood smear, three red cells (arrows) show the intracellular development of platelets, each surrounded by a clear zone. From two of these red cells, platelets are about to emerge by breaking through the cell membrane.[1] Wright x 1300

---

1. For a detailed discussion of the historical consideration of platelet development see Volume III, page 119.

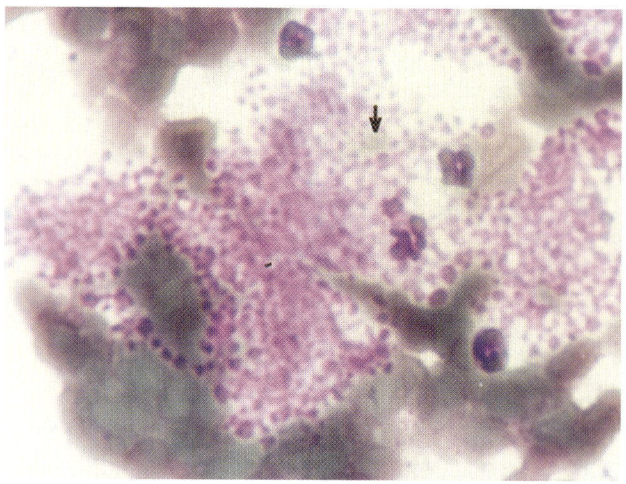

Figure 7.18, Wright x 800

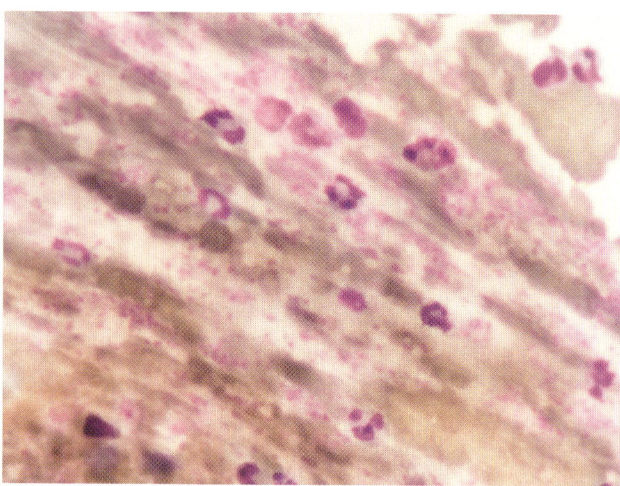

Figure 7.19, Wright x 520

Figure 7.18. This freshly prepared blood smear shows a single sheet of well-spread, similarly sized platelets replacing red cells and plasma. The arrow points to faintly remaining red cells. In the lower left, a single row of platelets is arising across the clumped red cells. Wright x 800

Platelets developing from red blood cells and marrow plasma in bone marrow (fig. 7.19)

Figure 7.19. Columns of vanishing red cells and the surrounding bone marrow plasma are giving rise to a large number of platelets. Bone marrow smear, Wright x 520

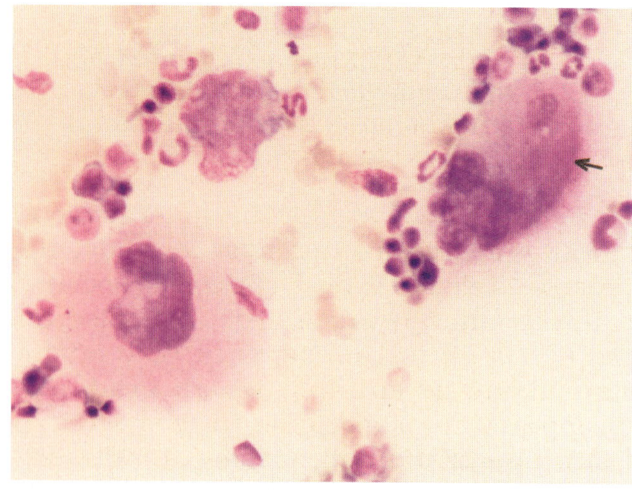

Figure 7.20, Wright x 400

Megakaryocytes showing no tendency towards platelet development (fig. 7.20)

Figure 7.20. In this bone marrow field, there are no platelets in the entire field despite the presence of three megakaryocytes. Even the fuzzy border of the megakaryocyte on the right exhibits no tendency towards platelet formation (arrow). Wright x 400

Chapter 8

LYMPH NODES AND LYMPHATIC CHANNELS (figs. 8.1-8.9)

1. Development and Transformation of Lymphocytes (figs. 8.1-8.6)

The development of lymphocytes can take place, in my experience, in any tissue and in any organ, as well as in individual cells, even though lymphocyte development occurs predominantly in the spleen and lymph nodes. Every tissue has the capacity to form lymphocytes whenever appropriate stimulation is present. Lymphocyte development can be observed in tissues with chronic inflammation and in tissues stimulated by nearby malignancy.

The origins of lymphocytes from local tissues are briefly demonstrated in Volume I (McDonald, 1989) from human mammary carcinoma, basal and squamous cell carcinoma of skin, malignant melanoma, prostate, colon and renal carcinomas, breast glands, adrenal cortex, adipose tissue, skeletal muscle, gastric mucosa, liver parenchyma, renal tubules, and spleen. Lymphocyte development from cardiac muscle is demonstrated in Volume II (McDonald, 1995). In Volume III (McDonald, 2001) the origins of lymphocytes are demonstrated in blood plasma and bone marrow acellular fluid.

In this volume, the origin of lymphocytes are further demonstrated from epidermis, papillary dermis, adipose tissue, cardiac muscle, skeletal muscle, smooth muscle, thyroid, adrenal cortex, spleen, gastric mucosa, kidney and bone marrow (in association with blackish reaction), as listed below. The transformation of lymphocytes into red cells is demonstrated in Volume I in human mammary carcinoma, basal cell carcinoma, malignant melanoma and prostate carcinoma. Examples of red cell development from lymphocytes demonstrated in this volume are listed below. Prevailing theory holds that red cells cannot develop within other cells and form only by passing through the nucleated phases of erythropoiesis. Any time erythrocytes are seen within any other cell the phenomenon has been described as erythrophagocytosis, and the cells containing erythrocytes have been called erythrophagocytes instead of erythrogenic cells (Listinsky, 1989).

I agree with the accepted theory of the transformation (transdifferentiation) of lymphocytes into plasma cells (Jordan, 1954, and Maximow and Bloom, 1957). In addition, I have demonstrated plasma cells transforming back into lymphocytes in Volume I (figs. 68, 73, 117 and 121). Moreover, in my studies I have observed that plasma cells may arise from the hyalinized tissue derived from the previously existing lymphocytes or vice versa. In fact, the lymphocytes themselves may arise from various hyalinized tissues. The nature of the life cycle of lymphocytes is demonstrated in this book and earlier volumes under various organs and tissues such as:

Formation of lymphocytes from
1. Bone marrow acellular fluid (Vol. III, figs. 13.41 and 13.42)
2. Coagulation products of blood plasma (Vol. III, fig. 13.38)
3. Benign and malignant tissues, as demonstrated in this volume (figs. 2.11-2.15, 2.21-2.23, 4.11, 4.27, 4.29, 4.31, 4.36, 5.22-5.24, 5.31, 5.33, 5.35, 5.36, 7.13, 8.7-8.9, 10.8, 14.1, 16.1, 16.3-16.5, 24.8, 24.11, 24.14, 26.5, 30.20, 34.4 and 34.7)
4. Benign and malignant tissues, as demonstrated in previous volumes (Volume I, figs. 34-37, 43, 46, 47, 54-

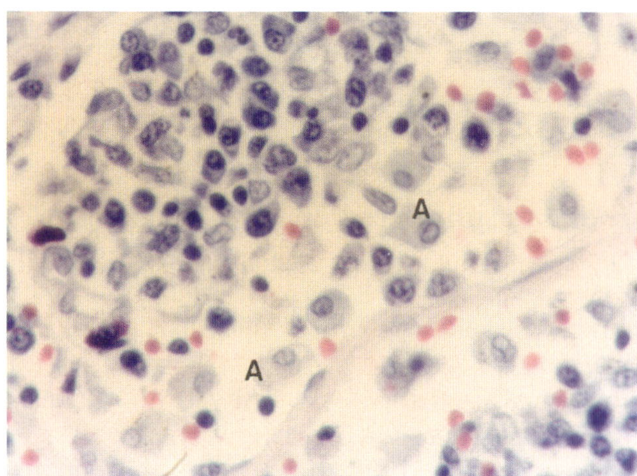

Figure 8.1, Giemsa x 520

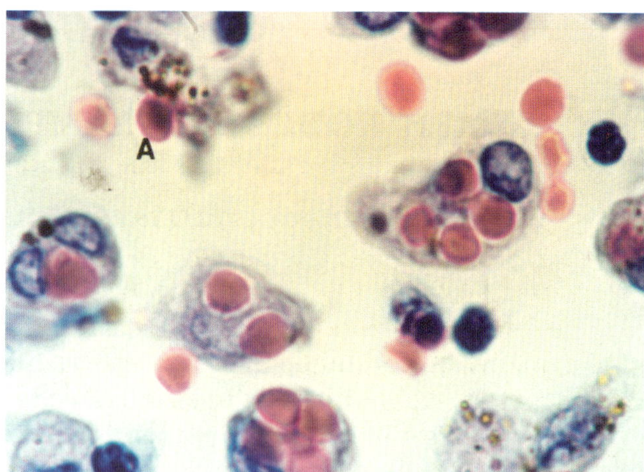

Figure 8.2, Giemsa x 1300

56, 62, 65, 67, 68, 72, 73, 76, 82, 86, 93, 106, 107, 109, 111, 114, 115, 125 and 127-1; Volume II, figs. 10, 29, 46 and 73)

Transformation of lymphocytes into
1. Plasma cells (figs. 4.36, 8.6 and 26.3; and Vol. I, figs. 68 and 73)
2. Connective tissue stroma (24.11)
3. Red cells (figs. 8.4, 8.5, 9.1-9.4, 10.1, 10.3, 13.4 and 22.17)

Figures 8.1-8.4 and 8.6 are taken from a mesenteric lymph node without metastasis from an autopsy of a patient with adenocarcinoma of the colon. Figure 8.5 is taken from a small peri-esophageal lymph node of a rabbit which was sacrificed by inhalation of ether. Figure 8.2 is reprinted from Volume III (McDonald, 2001).

Red cell development from erythrogenic mononuclear cells of lymph node (figs. 8.1-8.3)

Figure 8.1. In this section of a lymph node, several erythrogenic (plasma-cell-like) mononuclear cells (A) with large pale-blue cytoplasm are lying in the lumen of a wide developing subcapsular vascular channel. The erythrogenic capacity of these cells is demonstrated in figures 8.2 and 8.3. Giemsa x 520

Figure 8.2. Several bright-red developing red cells are shown within large

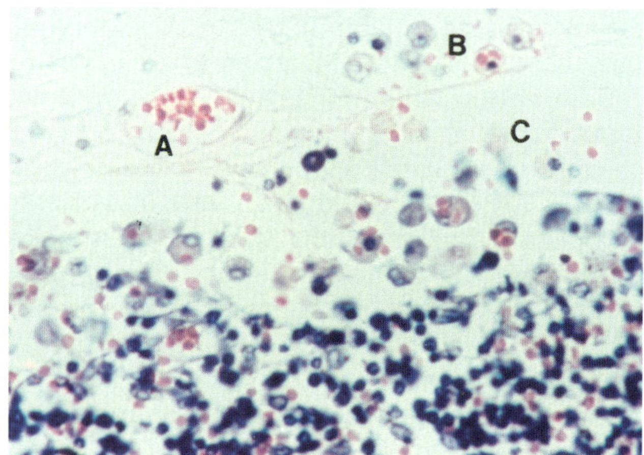

Figure 8.3, Giemsa x 200

(plasma-cell-like) mononuclear cells. Small remnants of bluish cytoplasm from the mother cells can still be seen attached to some of the red cells. Note the one cell with brownish hemosiderin pigment, commonly called a phagocyte, and the emerging red cell (A) also containing brownish pigmented areas. Giemsa x 1300

Figure 8.3. In this figure there are two intracapsular veins, one (A) with red cells and the other (B) with erythrogenic cells, and a few scattered red cells in the lumen. Developing red cells are seen in one of the erythrogenic cells in vein (B). In the subcapsular sinuses there are also erythrogenic mononuclear cells which are developing red cells. A few red cells and the dissolving cytoplasmic remnants of mother cells (C) are seen here. Giemsa x 200

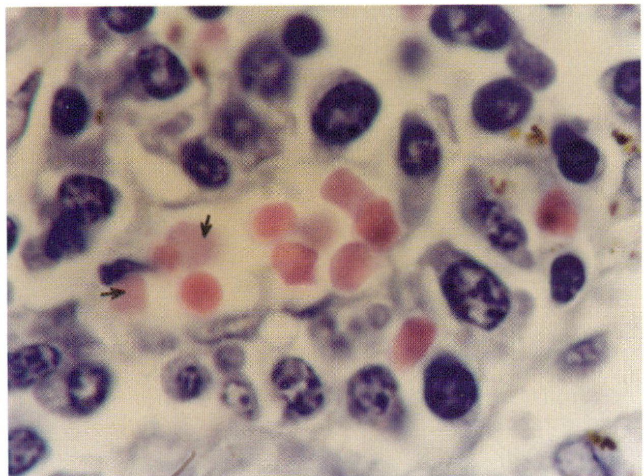

Figure 8.4, Giemsa x 1300

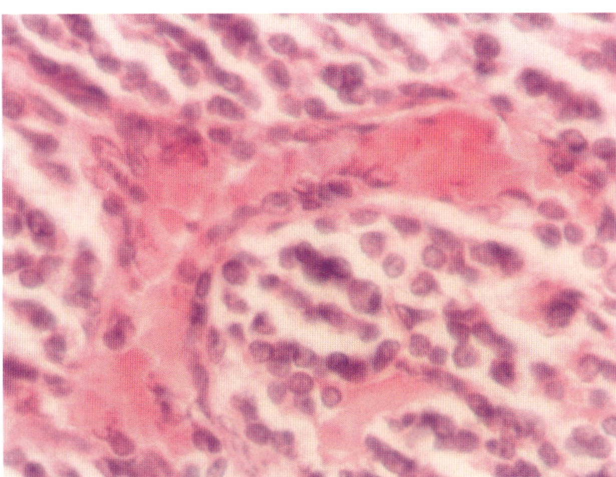

Figure 8.5, H&E x 800

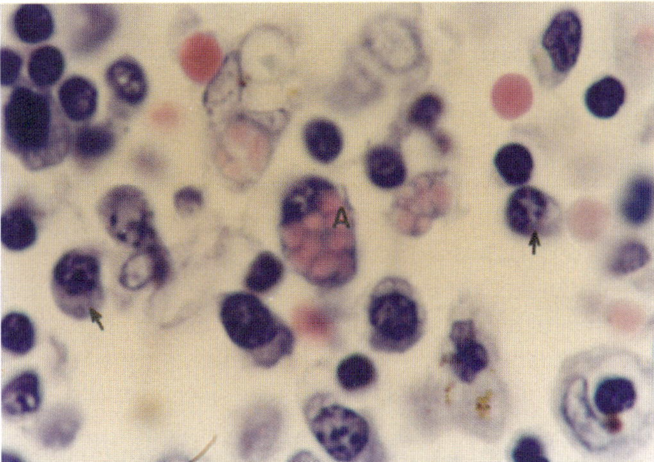

Figure 8.6, Giemsa x 1300

*Red blood cell and blood capillary
development from lymphoid tissue
(figs. 8.4 and 8.5)*

Figure 8.4. From local lymphoid tissue there is the beginning of a developing blood capillary, which is lying obliquely in this field of view. A few developing red cells still contain the blue-gray color of previous nuclear structures (arrows). Giemsa x 1300

*Red blood cell development from
lymphocytes (fig. 8.5)*

Figure 8.5. A blood capillary is developing directly from lymphoid tissue in a small peri-esophageal lymph node from a rabbit. Note that the solid sheet of lymphocytes is transforming into a column of packed red cells. The remnants of lymphocytes can still be recognized in this red cell column, and the endothelium is not yet well-formed. H&E x 800

*Grape cells (Russell body cells) developing
red cells in lymph node (fig. 8.6)*

Figure 8.6. This figure shows the development of many red cells from a grape cell or Russell body cell (A) formed from a plasma cell. This formation is shown in more detail in figures 14.1 and 14.9-14.11 in gastric mucosa in Chapter 14, Stomach. Several plasma cells with eccentric cytoplasm can also be seen in this figure (arrows). These plasma cells are developing from local lymphocytes. Giemsa x 1300

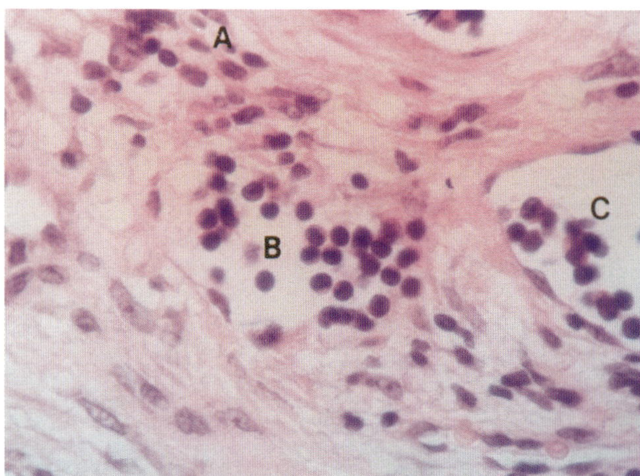

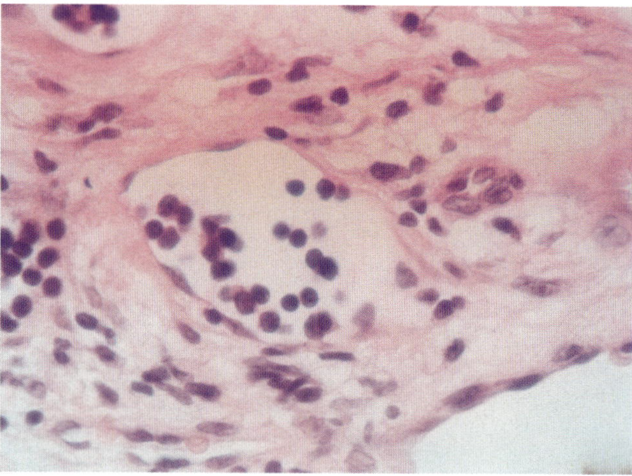

Figure 8.7, H&E x 520   **Figure 8.8, H&E x 520**

2. Development of Lymphatic Vessels (figs. 8.7-8.9)

Lymphatic vessels may arise from any tissue where lymphocytes are developed.[1] The endothelial lining of the vessels may arise from the local tissue. Lymphatic vessel formation is demonstrated below (figs. 8.7-8.9) in papillary dermis beneath a benign fibroepithelial polyp surgically removed from the ear of an elderly male.

Development of lymphocytes and lymphatic channels from papillary dermal tissue (figs. 8.7-8.9)

Figure 8.7. Upon careful examination, one can see the development of lymphocytes passing through various transitional stages (A) with regard to chromatism, size, and shape as they acquire the usual characteristic features of mature lymphocytes. Similar local development of lymphocytes in an appropriate site in the body could become the beginning of a lymph node. With the smooth rounding of the margins of the lymphocytes in the fluid environment (B), full development of lymphocytes can be seen. Area (B) is becoming a lymphatic channel through a process of liq-

uefaction of local original substance. An endothelial lining is just beginning to form on one side, and the other side is still continuous with the surrounding tissue. In the partially visible lymphatic channel (C), endothelial formation has progressed further. However, in the lumen, the differentiation of lymphocytes has not progressed to completion. H&E x 520

Figure 8.8. Similar to the previous figure, this figure shows lymphatic channels and lymphocytes arising from dermal tissue. The endothelial lining of these lymphatic channels is only partially formed. H&E x 520

Figure 8.9. Shown here is the development of a lymphatic channel containing plasma and lymphocytes from the upper developing papillary dermis. This figure beautifully demonstrates the undissolved cellular structures of dermal tissue (arrow) within the lumen of a formative lymphatic channel. The presence of local cells resisting lysis and lying within the lymphatic channel clearly demonstrates the local ori-

---

1. Lymphatic vessel formation from adipose tissue near a breast carcinoma as a determent to cancer spread is demonstrated in figures 53 and 54 of Volume I.

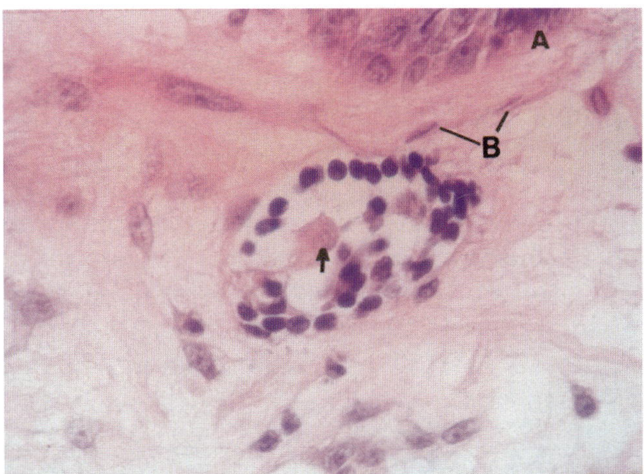

Figure 8.9, H&E x 520

gin of the lymphatic channel and its lymphocytes from dermal tissue. In the lumen, some of the lymphocytes are well-formed while others are still adhering to the remains of dermal tissue. (A) denotes the bottom layer of epidermal cells. The loose collagenous fibrils are forming from the lowest layer of these epidermal cells. Two narrow elongated fibrocyte nuclei (B) are probably arising from the newly developed connective tissue fibrils. H&E x 520

Chapter 9

THYMUS (figs. 9.1-9.4)

In humans the histological structure of the thymus varies with age and overall health of the individual. This organ consists mostly of lymphatic tissue and is closely related to the lymphatic and immune systems. Red blood cell and capillary development from lymphocytes of the thymus is demonstrated below. Figures 9.1-9.4 are from a surgical specimen of thymus tissue.

Red cell development, in a single column, from lymphocytes forming a developing capillary (figs. 9.1 and 9.2)

Figures 9.1 and 9.2. In figure 9.1, individual lymphocytes in a column (bracket) are transforming into red cells. This area is shown in higher magnification in figure 9.2. In order to form a single line, two individual lymphocytes (A) are eccentrically

transforming into red cells in the pathway of the forming blood capillary. Remnants of the mother lymphocytes can still be seen inside some of the red cells in this column. Figure 9.1 H&E x 520; and figure 9.2 H&E x 1300

Lymphocytes transforming into red cells (figs. 9.3 and 9.4)

Figure 9.3. Individual lymphocytes are transforming into red cells in a random pattern. Also, there are a few widely scattered reticulum cells (A) in this figure. H&E x 520

Figure 9.4. Areas such as this are commonly classified as so-called hemorrhage. However, this is actually a mass transformation of lymphocytes into red cells. H&E x 520

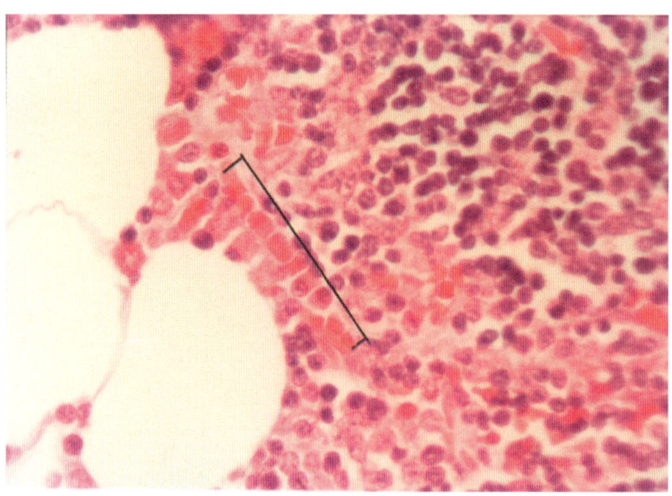

Figure 9.1, H&E x 520

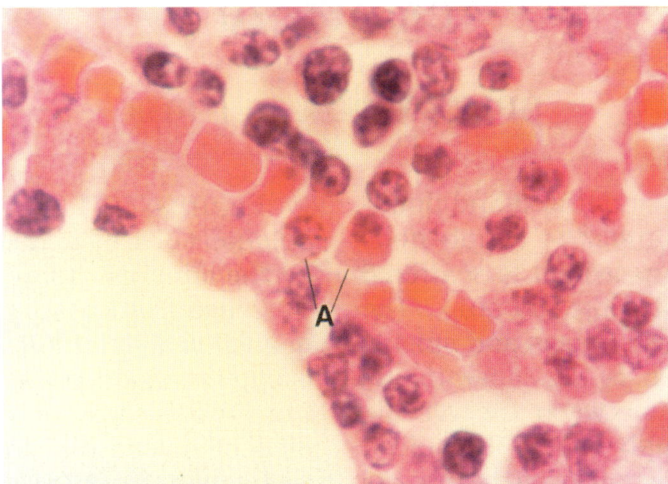

Figure 9.2, H&E x 1300

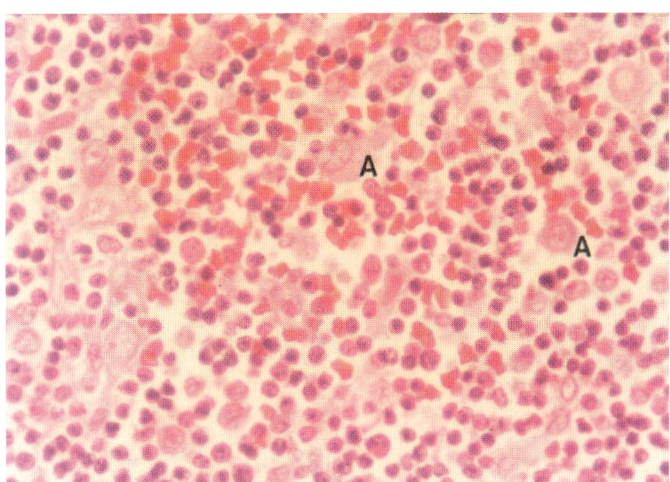

Figure 9.3, H&E x 520

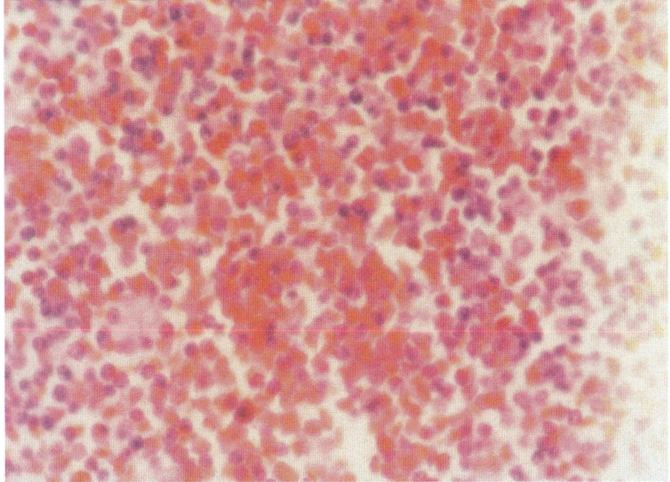

Figure 9.4, H&E x 520

Chapter 10

SPLEEN (figs. 10.1-10.9)

The spleen consists of diffused and nodular lymphoid tissue (similar to lymph node tissue) called white pulp, which is scattered throughout what is called the red pulp. The latter consists of venous sinuses and splenic cords of Bilroth (which is the loose tissue occupying the spaces between the venous sinuses). Like the lymph nodes, the spleen has a collagenous capsule with inward extensions called trabeculae (Bloom and Fawcett, 1975).

This chapter demonstrates the development of red cells from the lymphoid tissue of the white pulp, as well as red cell development from the splenic cords. In both areas, the red cells may develop directly from reticular cells and lymphocytes, as well as from locally derived SN cells, plasma cells, and eosinophils–all these cells originate locally in mature form with replacement of mother cells. The origin of a large number of plasma cells and their transformation into erythrogenic SN cells, as well as their direct development of red cells, is demonstrated in a congested spleen. The lymphocytes that are directly producing red cells are not prominent; however, lymphocytes transforming into erythrogenic plasma cells may be prominently seen at the periphery of the white pulp.

All the pathways of development of red cells described above prove the spleen to be a great hematopoietic organ. Even the trabecular fibrous tissue can be transformed into blood forming cells. Again, prevailing theory holds that red cells cannot develop within other cells and form only by passing through the nucleated phases of erythropoiesis. Any time the erythrocytes are seen within any other cells the phenomenon has been described as erythrophagocytosis, and the cells containing erythrocytes have been called erythrophagocytes instead of erythrogenic cells (Hernandez, et. al, 1984).

Figures 10.1-10.3 are taken from the splenic cord of a healthy rabbit sacrificed by the administration of ether. Figures 10.4-10.7 are from a surgically removed spleen of a 30-year-old male diagnosed with thrombocytopenia purpura. Figure 10.8 is taken from the autopsy of a 57-year-old male who died of myocardial infarction. Figure 10.9 is from a fresh autopsy of a sickle cell anemic child who died of sickle cell crisis. Figure 10.1 is reprinted from Volume I (McDonald, 1989).

Origin of red cells from reticular cells of the spleen (figs. 10.1-10.3 and 10.6); from eosinophils (figs. 10.1 and 10.5); by coalescence of eosinophilic granules (figs. 10.1 and 10.2); and by lymphocytes (figs. 10.1-10.3, 10.5, 10.8 and 10.9)

Figure 10.1. The origin of multiple red cells from locally developed eosinophils is shown in the splenic cord. (A) and (B) show red cells developing by the coalescence of eosinophilic granules. (C) shows red cells of various sizes arising from reticular cells, and (D) shows a red cell developing within a lymphocyte. H&E x 1300

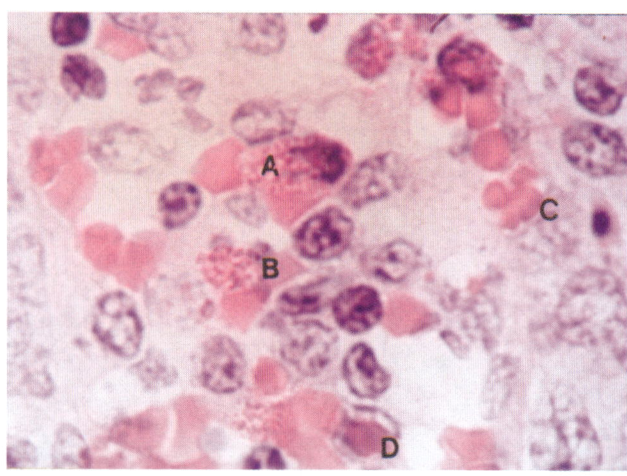

Figure 10.1, H&E x 1300

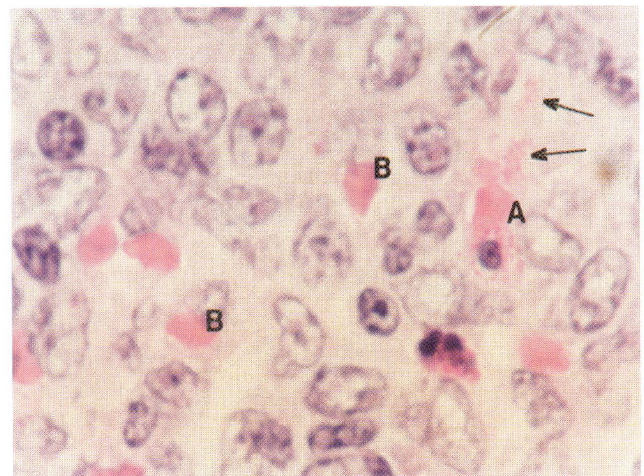

Figure 10.2, H&E x 1300

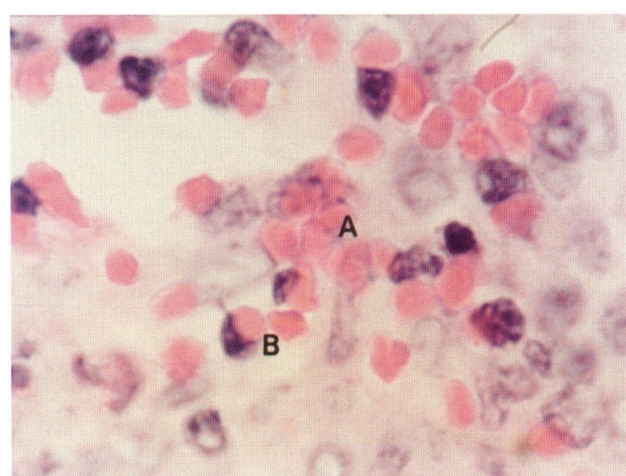

Figure 10.3, H&E x 1300

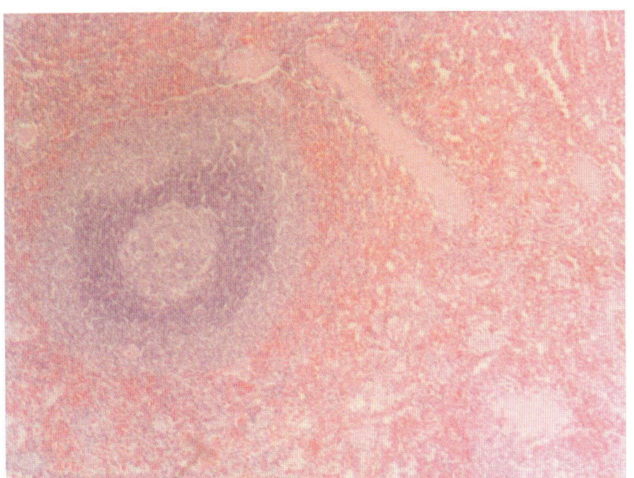

Figure 10.4, H&E x 52

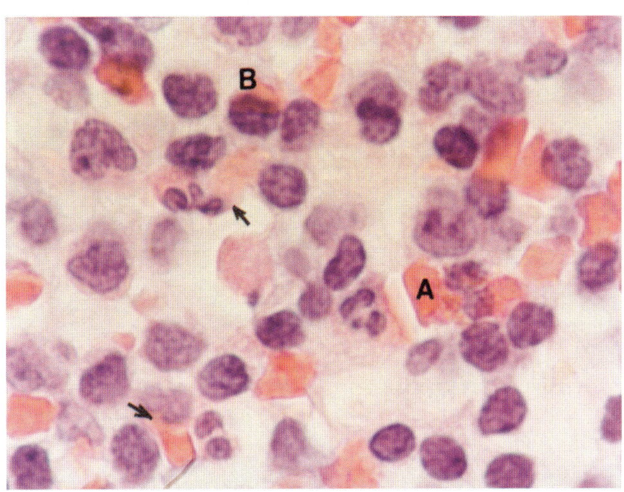

Figure 10.5, H&E x 1300

Figure 10.2. The development of hemoglobin globules (A) by coalescence of eosinophilic granules is seen here. Arrows point to the sprinkling of faintly visible eosinophilic granules formed from the vanishing cell substance. Two newly developed red cells (B) are directly derived from local cells. H&E x 1300

Figure 10.3. A cluster of newly formed red cells (A), lying in the center of this figure, are still covered by the visible remains of the disappearing mother cells. Note the lymphocyte (B) with an emerging red cell. H&E x 1300

Figure 10.4. This figure is a low-power view of the congested spleen of a patient who died of acute myocardial infarction. Figures 10.5, 10.6 and 10.7 are taken from this same patient. H&E x 52

Origin of red cells from segmented nuclear cells (figs. 10.5, 10.7 and 10.8)

Figure 10.5. Part of the cytoplasm of the SN cells (arrows) is transforming into hemoglobin. Also, an eosinophil (A) and a lymphocyte (B) are forming red cells. H&E x 1300

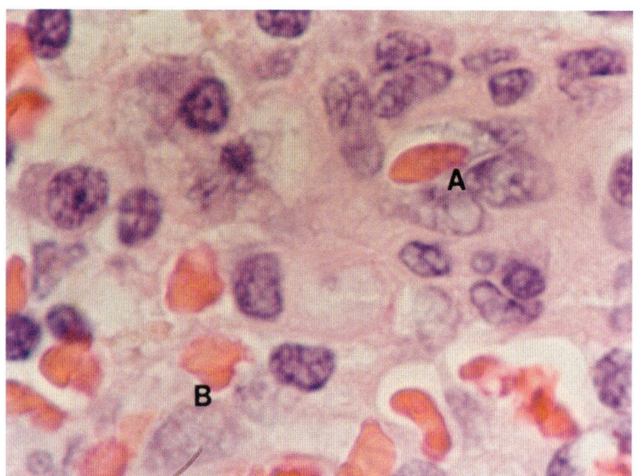

Figure 10.6, H&E x 1300

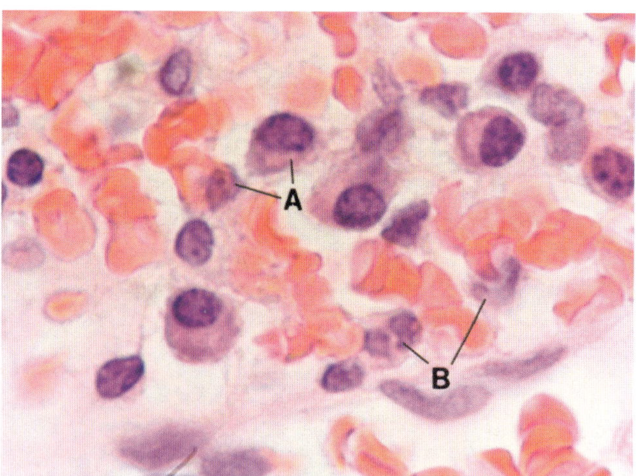

Figure 10.7, H&E x 1300

Figure 10.6. Reticulum cells (A) are forming red cells. Red cell (B) is arising from the cytoplasm of a vanishing reticulum cell. H&E x 1300

Red cell development from plasma cells (figs. 10.7 and 10.8)

Figure 10.7. Red cells, some in irregular rouleau formation, arise mostly from partially vanishing plasma cells (A), as well as from a few SN cells (B). H&E x 1300

Origin of reactive cells from fibrous connective tissue of spleen (fig. 10.8)

Figure 10.8. This figure demonstrates dissolution of fibrous tissue with development of plasma cells (A). The connective tissue of splenic trabeculae is being replaced by the origin of various reactive cells (predominantly plasma cells and lymphocytes, but also a few SN cells and eosinophils). Red cells develop from these reactive cells, particularly plasma cells (B). H&E x 520

Direct development of sickle cells from splenic tissue (fig. 10.9)

Figure 10.9. In approximately half of this figure, splenic tissue is replaced by the development of clumped sickle cells. Lymphocytes are separating and vanishing,

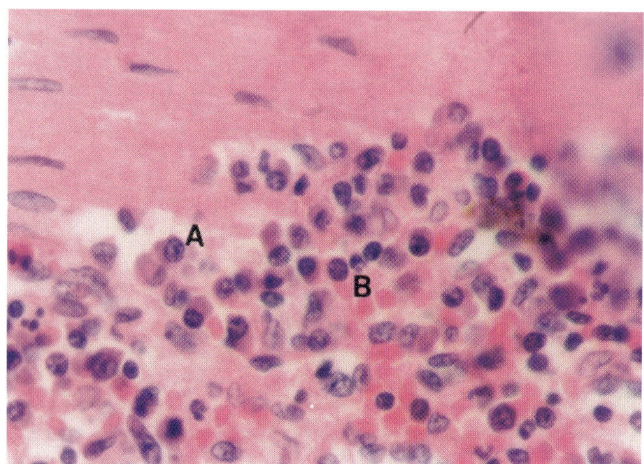

Figure 10.8, H&E x 520

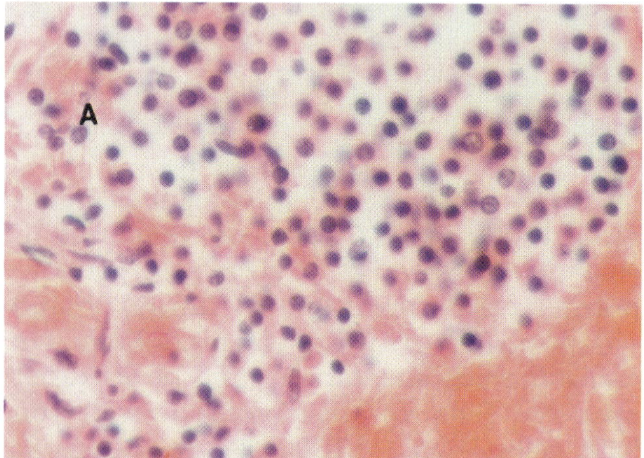

Figure 10.9, H&E x 520

and such areas are filling up with developing sickle cells (A).[1] H&E x 520

---

1. See Volume III (McDonald, 2001), Chapter 12, figs. 12.19, 12.20, 12.25 and 12.26, for additional figures showing replacement of splenic tissue by sickle cells.

Chapter 11

LUNG (figs. 11.1-11.8)

The lungs themselves form little more than a dense network of tubes (air passages). The functional unit of the lung parenchyma is the alveoli. The alveoli consist of alveolar septal cells and capillaries which surround and form the air passages.

Red blood cell and erythrogenic SN cell development from edema fluid within the alveolar lumens is shown below. Red blood cells can also develop directly from the alveolar septal cells. The development of hemosiderin pigment as an abortive attempt at red cell development by histiocytes (or phagocytes) is also presented below.[1] The origin of multiple red blood cells from histiocytes or phagocytes is also shown below. Cellular changes in malignancies of the lung are demonstrated in Chapter 27.

Figures 11.1-11.3 are from the autopsy of a 51-year-old male who died of essential hypertension with malignant nephrosclerosis and uremia. Figures 11.4 and 11.5 are taken from the autopsy of a 67-year-old male who died of advanced arteriosclerosis with thrombotic occlusion of left common and external iliac arteries. He also had moderately severe bronchopneumonia and apical fibrosis of both lungs. Figure 11.6 is from the autopsy of a 65-year-old male who died of complications from a recent radical gastrectomy for adenocarcinoma of the stomach and left lower lobe pneumonia. Figures 11.7 and 11.8 are from the autopsy of a 58-year-old male who died of a recent cardiovascular accident with heart failure and chronic congestion of the lungs.

Emphysematous lung (fig. 11.1)

Figure 11.1. In this emphysematous lung, the partitions between the distended alveoli are broken down in many areas (as is usual in chronic emphysema). H&E x 80

Origin of red cells within alveolar lumens from edema fluid (figs. 11.2 and 11.3)

Figures 11.2 and 11.3. These figures reveal alveolar lumens filled with mixtures of light-pink edema fluid and red cells. In figure 11.3, the higher magnification reveals

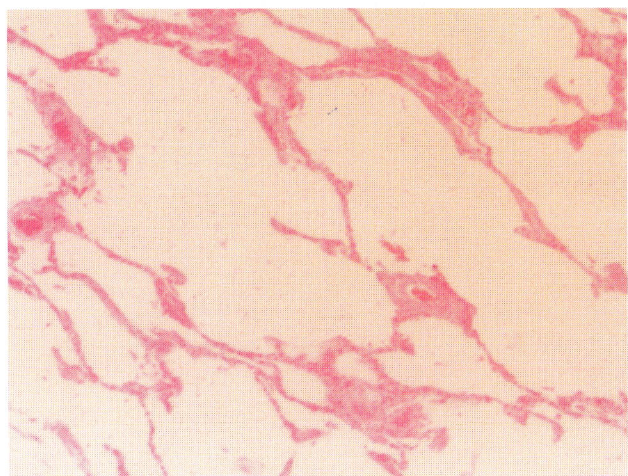

Figure 11.1, H&E x 80

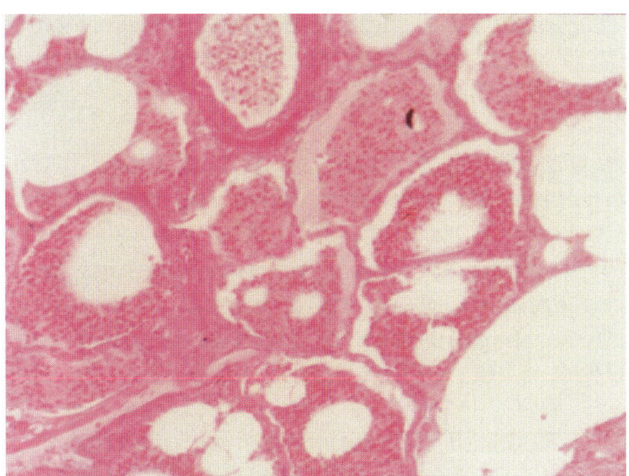

Figure 11.2, H&E x 130

---

1. This process is clearly described in renal carcinoma in Volume I (McDonald, 1989, figs. 101-105).

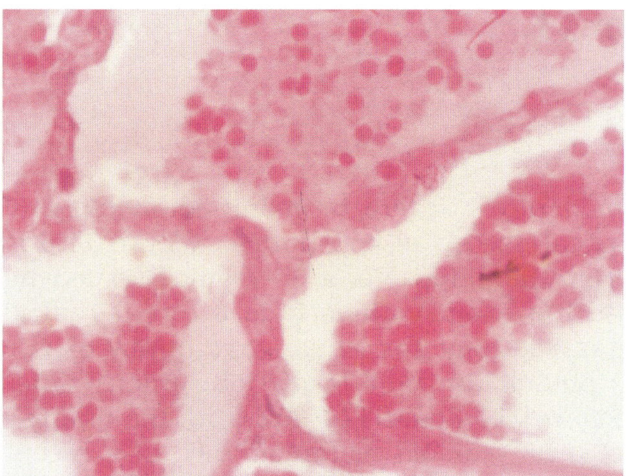

Figure 11.3, H&E x 520

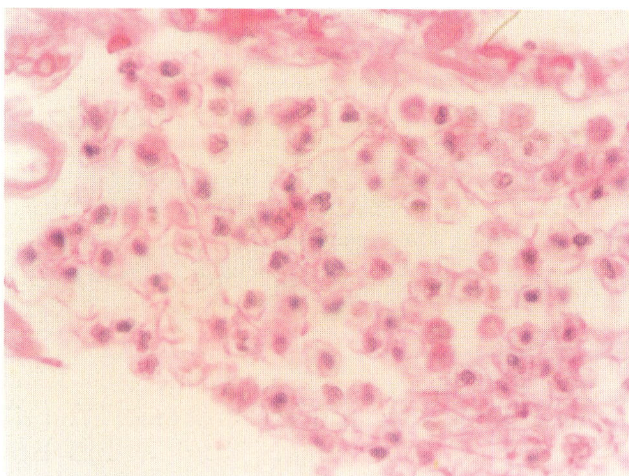

Figure 11.4, H&E x 520

red cells from faintly visible stage to well-formed and bright red stage developing from this light-pink colored edema fluid. There are hardly any capillaries in the alveolar septum showing red cells. Figure 11.2 H&E x 130; and figure 11.3 H&E x 520

*Origin of erythrogenic SN cells
(in sheet form) from edema fluid in alveoli
(figs. 11.4 and 11.5)*

Figure 11.4. Many erythrogenic SN cells (in a sheet form) in the alveolar lumen are derived from the edema fluid. H&E x 520

Figure 11.5. On the left of this figure, covering a large area from top to bottom, there are many SN cells in a sheet form, some of which are transforming into red cells. On the right half of this figure, there is evidence of fibrinous pleurisy. H&E x 260

*Red cell development from alveolar
septal cells (fig. 11.6)*

Figure 11.6. This figure demonstrates the erythrogenic capacity of the alveolar septal cells. H&E x 520

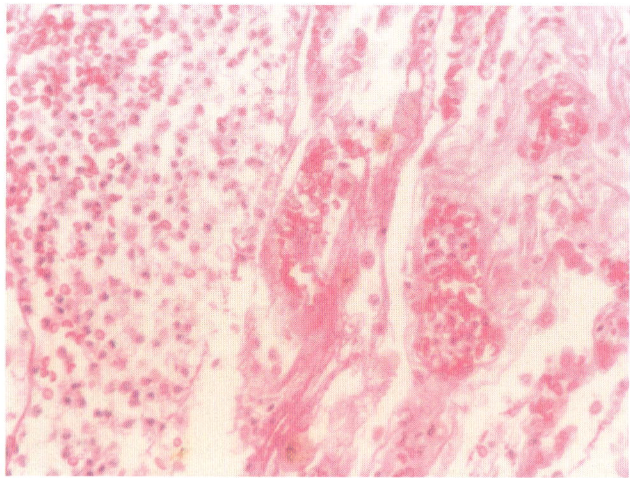

Figure 11.5, H&E x 260

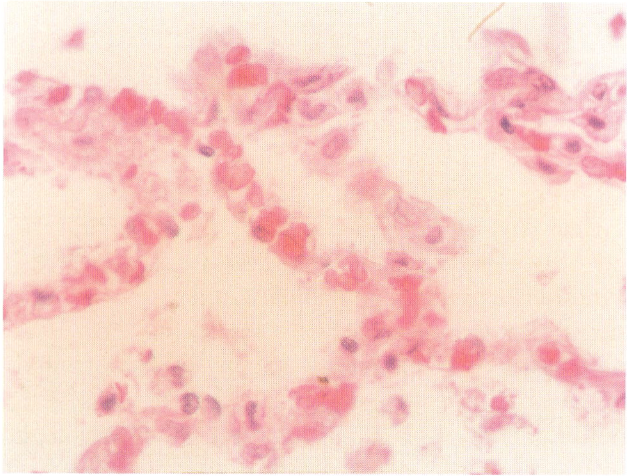

Figure 11.6, H&E x 520

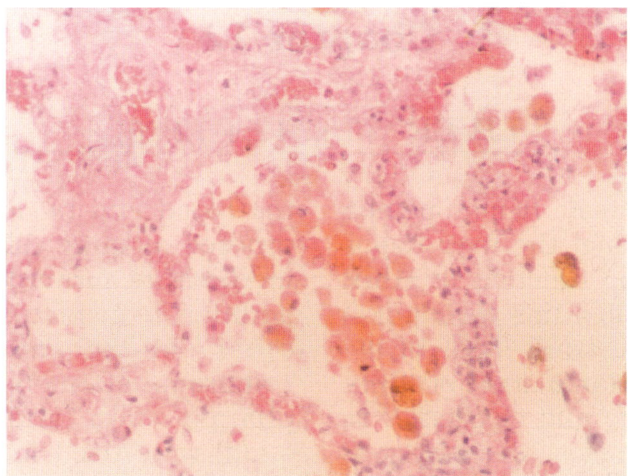

Figure 11.7, H&E x 260

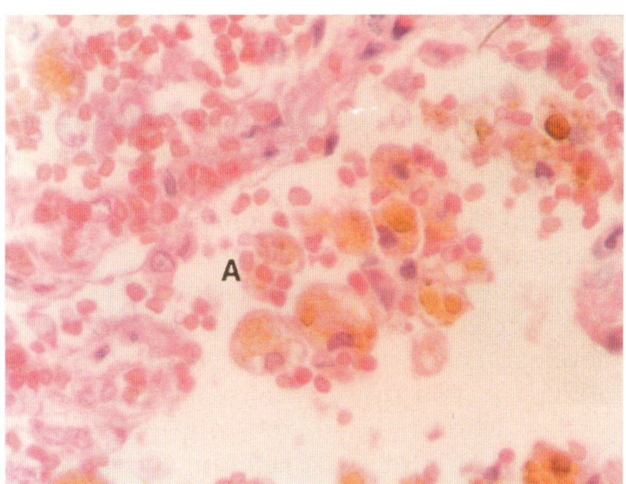

Figure 11.8, H&E x 520

Development of hemosiderin pigment as an abortive attempt at red cell development by so-called histiocytes/phagocytes (fig. 11.7)

Figure 11.7. Within the alveolar space there are many large golden-brown hemosiderin pigment-laden cells, commonly categorized as histiocytes or phagocytes. These large cells are often called *heart failure cells* and are commonly found in cases of chronic congestive heart failure. The hemosiderin pigment in these cells is evidence for their unsuccessful attempt at red cell development. This process is clearly described in renal carcinoma in Volume I (McDonald, 1989, figs. 101-105). The alveolar septae are thickened and there are many developing inflammatory infiltrates, mainly SN cells, present. H&E x 260

Origin of red cells from histiocytes/phagocytes (fig. 11.8)

Figure 11.8. In higher magnification, one can see the origin of multiple red cells within the histiocyte/phagocyte (A). Some of the red cells are coming out of the so-called histiocytes or phagocytes and are lying in the alveolar lumen. H&E 520

Chapter 12

LIVER (figs. 12.1-12.28)

In addition to its classic hematopoietic function in embryonic life, the liver has a prominent blood-forming function in adult life as well. I have verified this hematopoietic function during the adult stage of life through the thorough examination of livers in humans, mice, rabbits and chickens. In my study of all these cases, the predominant finding is the direct development of red cells as hemoglobin globules from local cells. Such occurrences may involve the nucleus and the cytoplasm, with or without dissolved mother substance.

The red cells may arise in large numbers from liver parenchyma undergoing liquefaction and other changes including granulation and vacuolation, with the vacuoles being about the size of red cells. The liquefied product of the remaining cellular elements takes part in the formation of blood plasma, while the adjacent liver cells form the endothelial lining in development of new vascular channels enclosing the columns of newly developed red cells. Endothelial cells may arise as tiny curved nuclei in lysing liver cells, or they may occasionally arise in nearly fully developed form.

Red cells arising from the liver parenchyma may be of different sizes and occasionally of different shapes. In patients with sickle cell disease, the liver produces a large number of sickle red cells out of liver cell substance. During sickle cell crisis, sheets of liver cells may transform into small or large blood vessels packed with masses of sickle cells.[1]

With application of various histochemical stains, the changing biochemical nature of the developing red cells is reflected by their color variations. In the same section, the blood and blood vessel formation may proceed at quite different speeds, and the methods of formation may also vary. In addition to hematoxylin and eosin stains, other histochemical stains (Masson Trichrome stain, fat stain, alkaline phosphatase stain, and non-specific esterase stain) were used for the detection of color variations reflective of the background biochemical variations from one place to another. On some occasions, serial sections were studied to confirm the particular structures in three dimensions.

In chick embryo livers, my earlier study of vessel formation from liver parenchyma (McDonald, 1968) demonstrated a glimpse of the biochemical changes that occur with application of a few special stains such as Giemsa, Chromotrope Aniline Blue, Gomori's aldehyde and Verhoeff's van Gieson. In this study, histochemical stains were also applied to frozen sections for detection of non-specific esterase using alpha naphthos technique, alkaline phosphatase by the calcium cobalt method, and detection of lipid by Oil red O reagent in propylene glycol. The capacity of liver tissue for producing a large number of red cells and fat cells was observed in tissue cultures of chick embryo liver cells.

The photomicrographs presented here are from sections of three mouse livers, four adult human autopsy livers, one human embryonic liver, one adult chicken liver, one chick embryo liver, one chick embryo liver *in vitro*, and one rabbit liver. The mouse livers are from C3H mice (with either spontaneous or transplanted breast carcinomas) used in a short-term experimental chemotherapy program. Figures 12.1, 12.3,

---

1. A markedly congested liver showing replacement of necrosed liver cells with developing red cells causing what is known as hemorrhagic necrosis is demonstrated in figure 76 of Volume II.

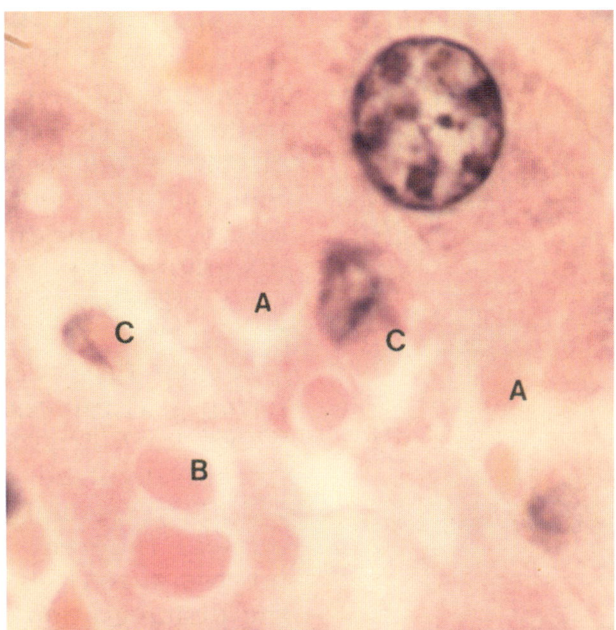

Figure 12.1, H&E x 2000

12.4, and 12.28 are taken from mouse M41, with a spontaneous breast carcinoma. Figures 12.2 and 12.5 are from mouse M2373, also with a spontaneous breast carcinoma. Figures 12.6 and 12.7 are taken from mouse M2291, with a transplanted breast carcinoma. Figures 12.8 and 12.9 are from a normal healthy adult rabbit sacrificed by the inhalation of ether for the purpose of detailed microscopic examination of various fresh tissues and organs. Figure 12.10 is from a normal healthy adult chicken. Figure 12.11 is from a six-day-old chick embryo, and figure 12.12 is from a three-day liver tissue culture of a six-day-old chick embryo.

Figure 12.13 is from a formalin-fixed human embryo weighing 7.8 grams that was a product of spontaneous abortion (miscarriage). Figures 12.14 and 12.20 are from the autopsy of a 67-year-old male with primary diagnosis of adenocarcinoma of left kidney, and congestion of the liver. Figures 12.15-12.17 are from the autopsy of a 67-year-old male who died of congestive heart failure. Figures 12.18 and 12.19 are from the autopsy of a 57-year-old male who died of a recent myocardial infarction with congestion of liver. Figures 12.21-12.25 are from a fresh autopsy of an adult who died of complications of sickle cell anemia. Figures 12.26 and 12.27 are taken from a biopsy section of the anterior edge of the right lobe of a liver obtained during exploratory laparotomy of a 58-year-old female with carcinoid of the ileum metastatic to the liver. Cellular changes in adenocarcinoma of the liver are presented in Chapter 29. Figure 12.10 is a magnified reprint from one of my previous articles (McDonald, 1968). Figures 12.1 and 12.28 are reprinted from Volume I (McDonald, 1989); and figures 12.11-12.13 and 12.24 are reprinted from Volume III (McDonald, 2001).

Liver cell nuclei and cytoplasm producing red cells (figs. 12.1-12.3)

Figure 12.1. Red cells are developing as hemoglobin globules from the slightly basophilic cytoplasm of liver cells (A). Some of the liver cell cytoplasm (B) has been utilized resulting in a red cell surrounded by a clearing of the cytoplasm. Red cells are also arising from dissolving liver nuclei (C). H&E x 2000

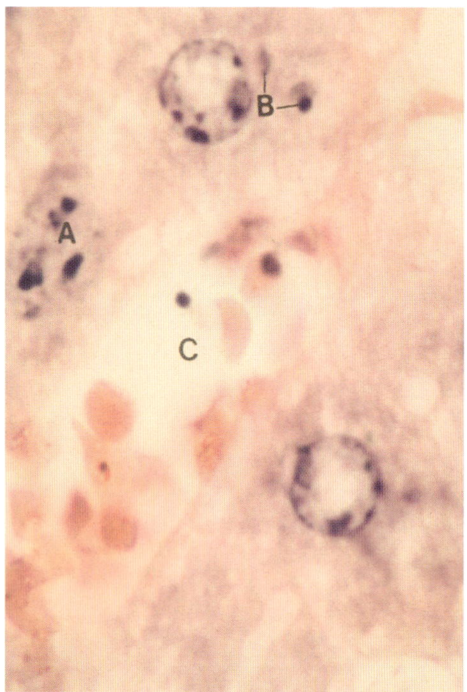

Figure 12.2, Giemsa x 2000

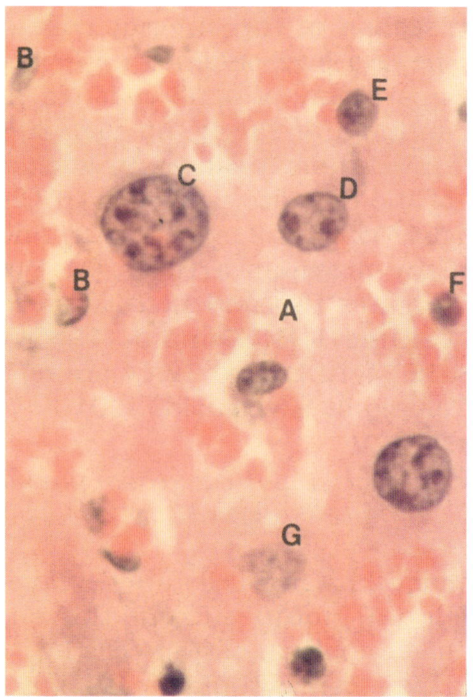

Figure 12.3, H&E x 900

Figure 12.2. The nucleus (A) has disintegrated into fragments. The nuclear fragments (B) were derived from a previous nucleus. The nearby nucleus is also dissolving, but its membrane is still intact. Sinusoid (C) is formed from liver cell dissolution and contains the contents of the dissolving nuclei and cytoplasm, and several developing red cells. Giemsa x 2000

Figure 12.3. Several red blood cells are forming within the developing sinusoids which have originated from liver parenchyma involved in vacuolar changes. Fibrillary remnants of cytoplasm (A) and nuclear remnants (B), which are in the process of red cell formation, remain after the disintegration of the cell. There is a successive reduction in the size of the liver cell nuclei (C-F) accompanied by a change in chromatin concentration. Note the barely visible outline of the vanishing nucleus (G). H&E x 900

Red blood cell and blood capillary development from liver parenchyma (figs. 12.4-12.6 and 12.9)

Figure 12.4. The developing capillary (A) originated from the liver cell and now contains the remains of the liver cell's cytoplasm and a developing red cell (arrow). The less-developed endothelial nucleus on the right side of this capillary is developing from the membrane of the liver cell's nucleus. The small hyperchromatic liver cell nucleus (B) is probably in the process of transforming into a lymphocyte. The nucleus of the liver cell (C) is fading away. An endothelial cell nucleus and red cells are forming from the liver cell substance (D). The endothelial cells are partially enclosing the red cells. H&E x 900

Figure 12.5. The right side of this developing blood capillary has a well formed membrane consisting of endothelial cells with typical nuclei, but the left side is less developed and consists of the shaggy remnants of lysing liver cells. Some of the same

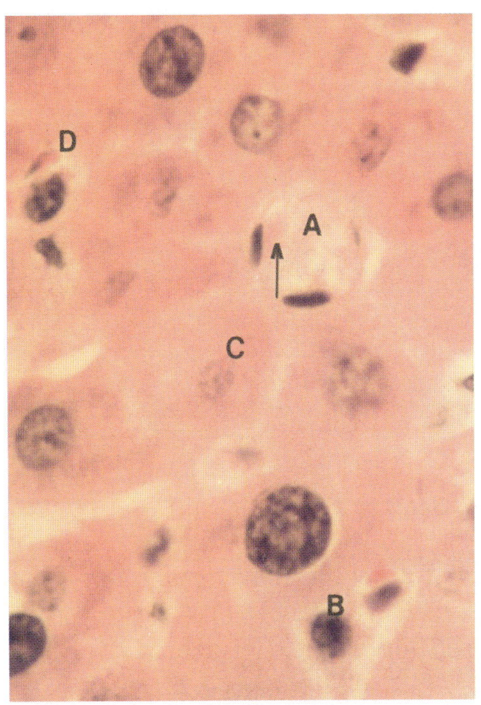

Figure 12.4, H&E x 900

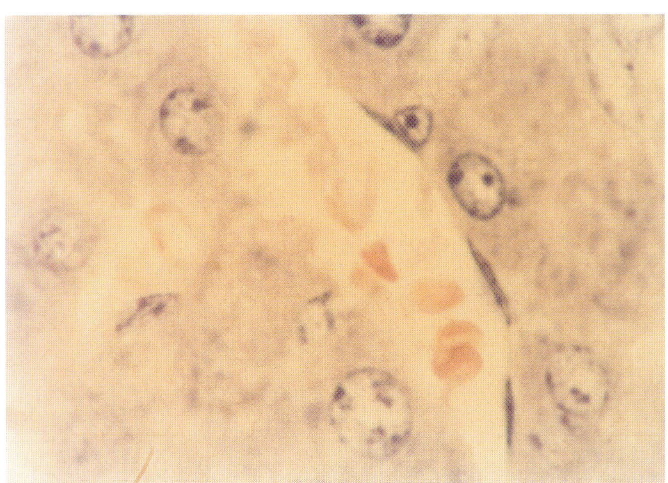

Figure 12.5, Giemsa x 1300

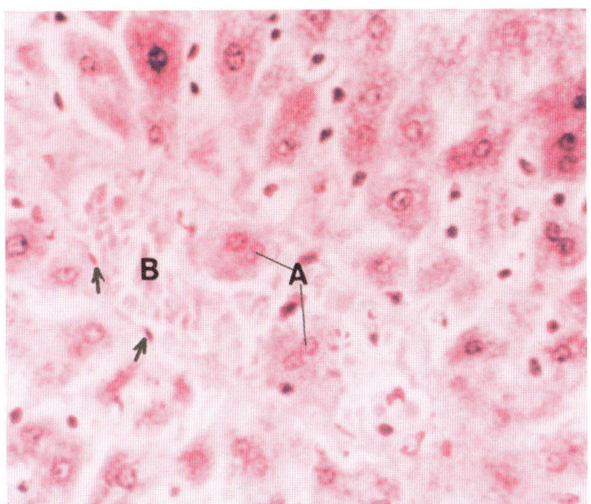

Figure 12.6, H&E x 300

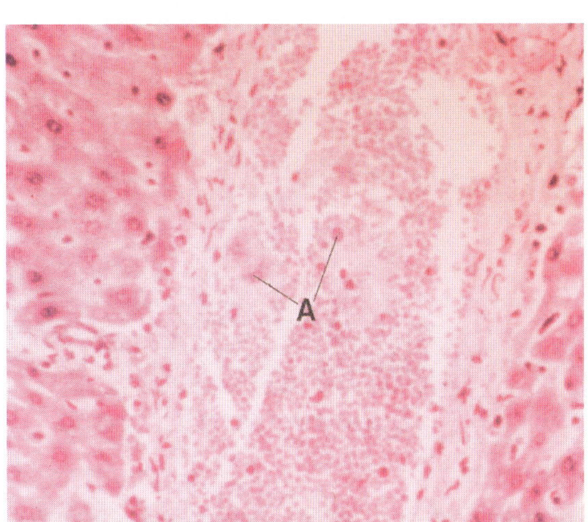

Figure 12.7, H&E x 215

materials that make up the wall can be found within the lumen of the capillary. Note the association between the red cells and the dissolving cell substance. Giemsa x 1300

Figure 12.6. Occasionally, a few liver cells (A) that resist lysis can be seen in the lumen of the formative capillary with the red cells (B) that were derived from the dissolving liver cells. A few tiny endothelial nuclei (arrows) are visible in the lining of the capillary wall. H&E x 300

Red cell and blood vessel formation from liver parenchyma (figs. 12.7 and 12.10)

Figure 12.7. A large blood vessel is forming from the liver parenchyma and is similar to the vessel in figure 12.6; however, this vessel has a more developed wall. Liver cells (A) can be seen within the lumen of the vessel along with the red cells. H&E x 215

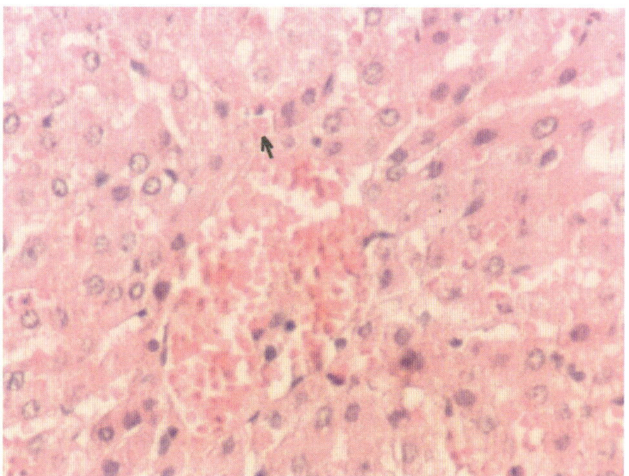

Figure 12.8, H&E x 400

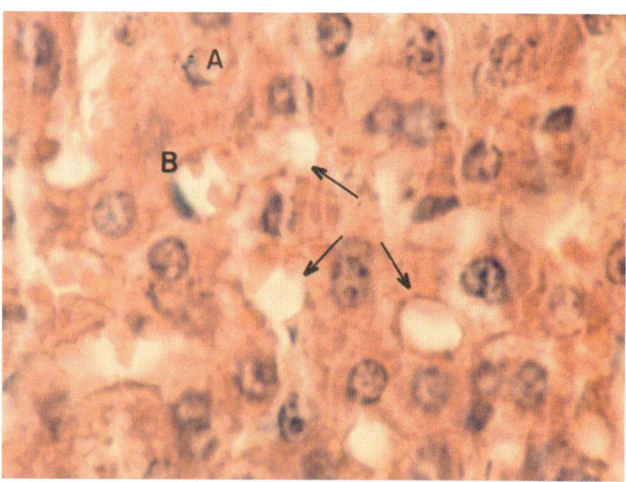

Figure 12.9, H&E x 1040

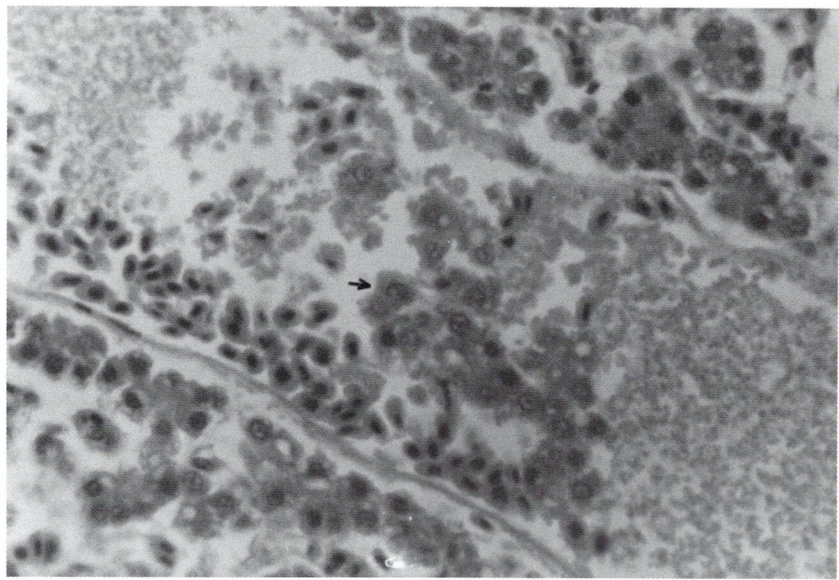

Figure 12.10, H&E x 600

Development of red cells and a central vein from dissolving liver cells (fig. 12.8)

Figure 12.8. In the developing central vein, liver cells, while dissolving, have turned into pink granules before becoming red cells. Hyperchromatic tiny lymphocyte-like nuclei have also formed from the dissolving liver cells within the vessel. This vessel is also partially lined, in the lower right, with endothelial cells that are also of liver cell origin. The liver parenchyma in other areas (mid-lower right) shows lysis and granular changes in the development of sinusoids, some of which contain red cells. Some of the surrounding sinusoids (arrow) are joining with the central vessel. H&E x 400

Figure 12.9. The lumen of a capillary may be started by the formation of large vacuolar spaces (arrows) resulting from the disappearance of liver cells. Two developing capillaries (A and B) and their contents are forming from disappearing liver cells and have the beginnings of an endothelial lining. Some of the vacuoles already contain formative red cells. The liver tissue between capillary (B) and adjacent capillaries (arrows) is breaking down to

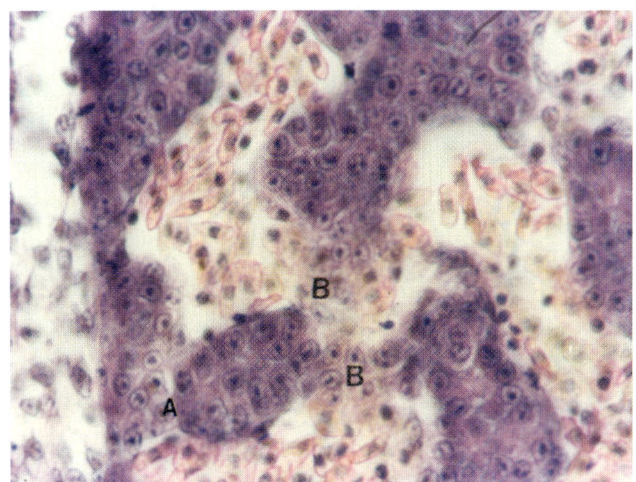

Figure 12.11, H&E x 520

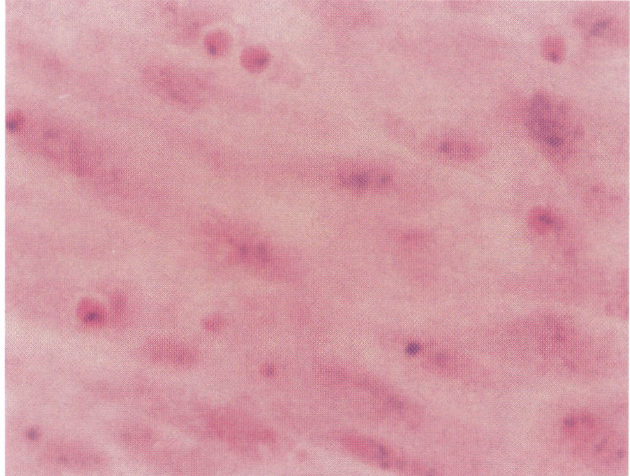

Figure 12.12, H&E x 800

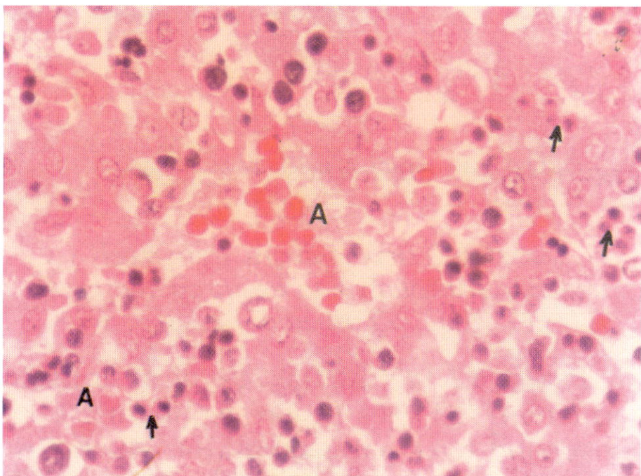

Figure 12.13, H&E x 520

allow the coalescence of the two lumens to form a larger vessel. H&E x 1040

Figure 12.10. A large blood vessel is developing from the dissolving liver parenchyma of an adult chicken. Within this lumen, there are liver cells (arrow) that have not yet dissolved. These red cells are nucleated and spindle-shaped, as is typical for an adult chicken red cell.[2] H&E x 600

Red cell development from chick embryonic liver (fig. 12.11)

Figure 12.11. This six-day-old chick embryo liver shows the transformation of liver cells into spindle-shaped red cells with tiny hyperchromatic nuclei (A). The liver tissue (B) between the sinusoids is dissolving to allow coalescence. Within the sinusoids there are many well formed red cells. (See also figure 1.4 in Chapter 1, Embryology.) H&E x 520

Red cells developing from chick embryonic liver in-vitro (fig. 12.12)

Figure 12.12. This three-day *in vitro* culture of a six-day-old chick embryo's liver shows the development of red cells. These red cells are developing from the nuclei as well as the cytoplasm of the liver cells. (See also figure 1.8 in Chapter 1, Embryology.) H&E x 800

Red cells and erythrogenic nucleated cells directly developing from lysing liver cells of human embryonic tissue (fig. 12.13)

Figure 12.13. In this human embryonic liver, red cells are mainly forming directly from dissolving liver cells (A). There are many erythrogenic nucleated cells (arrows) of various sizes which fall under the category of normoblastic erythropoietic cells; however, the majority of red cells are formed from the process described in (A). (See also figure 1.7 in Chapter 1, Embryology.) H&E x 520

---

2.   Figure 12.10 is a magnified reproduction of a photomicrograph which appeared in one of my earlier scientific articles (McDonald, 1968). I greatly appreciate the permission for reproduction given by the editor and publisher of the *Journal of the American Medical Women's Association.*

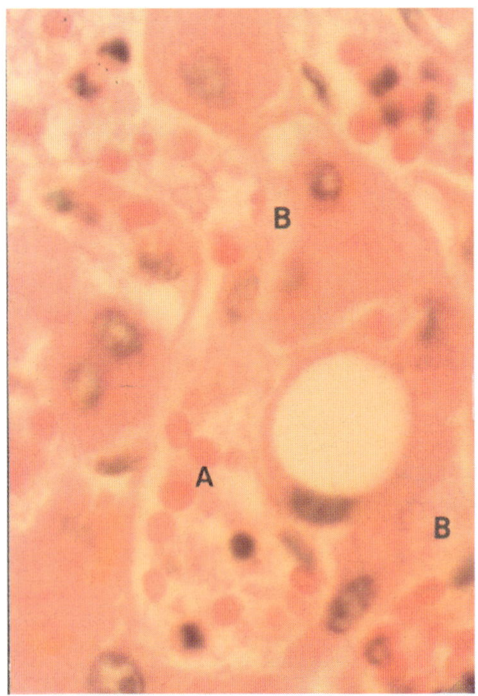

Figure 12.14, H&E x 800

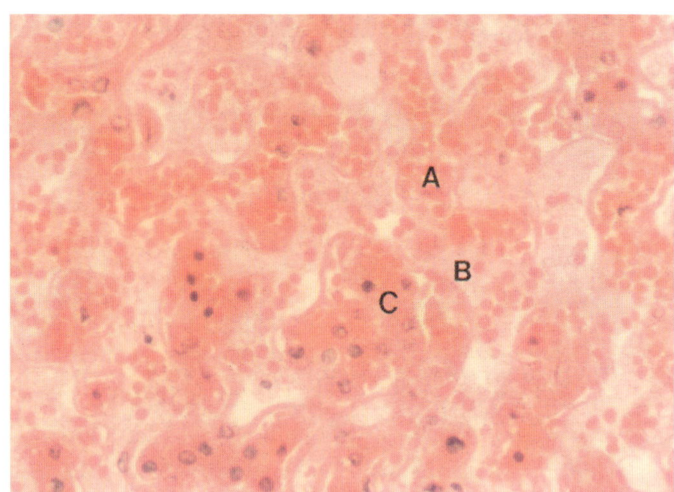

Figure 12.15, H&E x 400

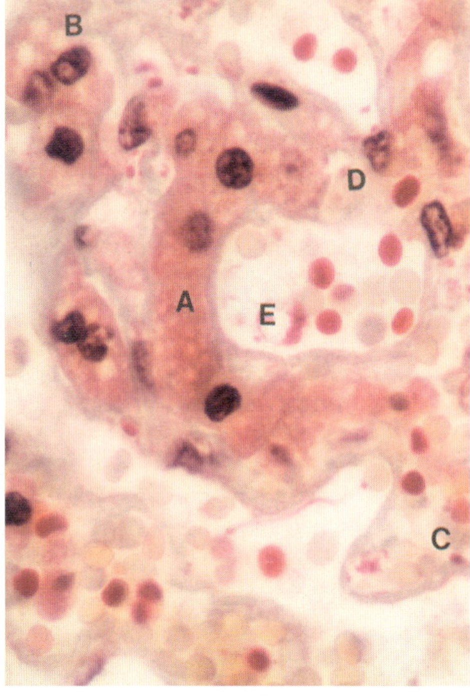

Figure 12.16, Chromotrope x 800

*Red cell development from liver
sinusoids through fine erythrogenic mesh
(figs. 12.14 and 12.15)*

Figure 12.14. At the site of a sinusoid formation, red cells (A) are originating from dissolving liver parenchyma of which fine mesh-like remains are visible. The erosion lines (B) of liver cells expand as the sinusoid enlarges. H&E x 800

*Liver cell cords developing red cells
(figs. 12.15 and 12.16)*

Figure 12.15. Many of the liver cell cords are mostly filled with red cells (A) that are replacing the liver parenchyma. In the sinusoidal spaces, fully hemoglobinized red cells are sparse and the spaces are filled with an erythrogenic mesh of fine uniform red cell-sized vacuoles, from which red cells (B) develop (use a magnifying glass). Some of the liver cells (C) are developing red cells as well. H&E x 400

Figure 12.16. These liver cell cords are in the processes of filling themselves with red cells. In the early stage, fine vacuoles (A) appear in the cytoplasm and the nuclei may become small and pyknotic or disappear. Next, the vacuoles coalesce and grow to the size of a red cell (B) and subsequently develop into red cells (C). The liver cells around the sinusoid are breaking down to allow coalescence with the surrounding sinusoids and begin red cell development (D). With the use of this Chromotrope Aniline Blue Stain, red cells in various stages of development are shown by their staining differences, from fine colorless vacuoles, to

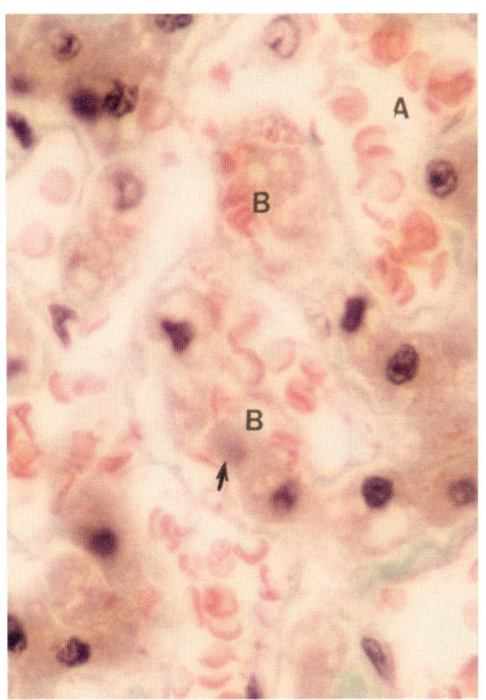

Figure 12.17, Masson Trichrome x 800

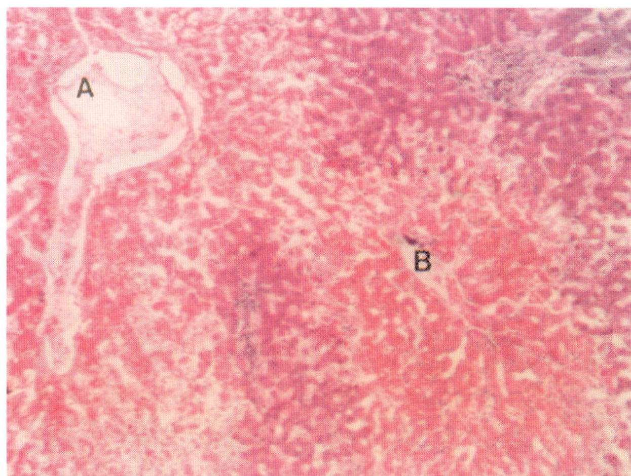

Figure 12.18, H&E x 80

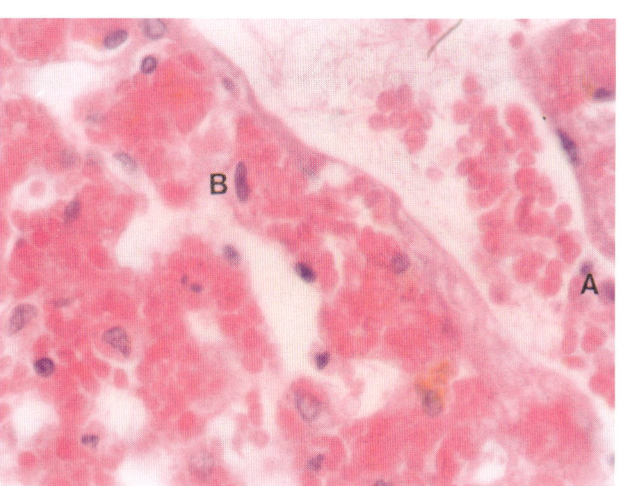

Figure 12.19, H&E x 520

grayish-green, to bright red.   Different colored red cells and ghost red cells within the sinusoid (E) show variations in hemoglobin development.   Chromotrope Aniline Blue x 800

Development of irregularly shaped red cells from liver tissue (fig. 12.17)

Figure 12.17. The liver sinusoid (A) is filled with stringy or twisted red cells. The liver cell (B) shows the origin of red cells in this twisted form. Note the pyknosis and small size of liver cell nuclei, as well as nuclei in stages of dissolution. One nucleus (arrow) is almost completely dissolved. Masson Trichrome x 800

Large central veins with developing red cells arising from lysing liver parenchyma (figs. 12.18-12.20)

Figure 12.18. There are many liver cell cords undergoing lysis resulting in a downward extension of this large central vein (A). The liver parenchyma surrounding the small central vein (B) is brick red in coloration.

This is usually designated as hemorrhagic necrosis, but it actually indicates that this anoxic parenchyma is filled with developing red cells. H&E x 80

Figure 12.19. In a higher magnification of figure 12.18, the liver cell cords are found to be filled with developing red cells. The lumen of the central vein contains the remains of lysed liver cells (A).   The connecting tributary (B) of the central vein also contains a dissolving liver cell within its lumen. H&E x 520

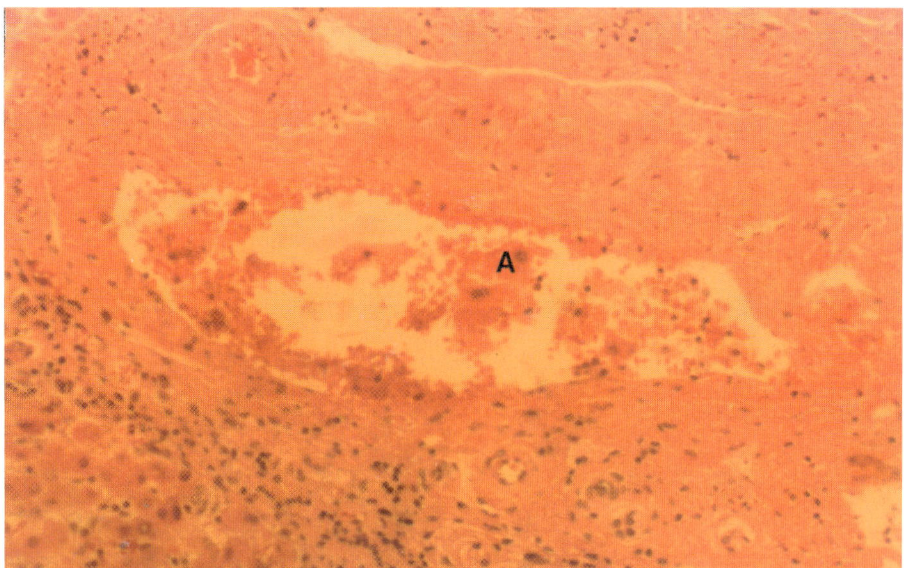

Figure 12.20, H&E x 200

Figure 12.20. On casual observation, this large blood vessel may appear to be a well-established artery or vein with a thick wall, but it is actually a developing vascular channel of liver cell origin that has not yet connected with the rest of the circulatory system. The facts that there are many undissolved liver cells (A) within the lumen and that liver cells are in various stages of transformation within the wall, support this assertion. At this point, the red cells are nothing more than deformed hemoglobinized particles of liver cell origin. H&E x 200

Development of sickle cells from liver tissue (figs. 12.21-12.25)

Figure 12.21. The portal system and surrounding area of this liver tissue are massively congested with sickle cells. H&E x 260

Figure 12.22. This figure is a magnification of the portal system in figure 12.21. The red cells in the portal system have all converted to sickle cells, and the portal system is becoming larger with widening of the vascular area. H&E x 520

Figure 12.23. This figure is a magnification of the surrounding sinusoids of figure 12.21. These liver cells are disappearing

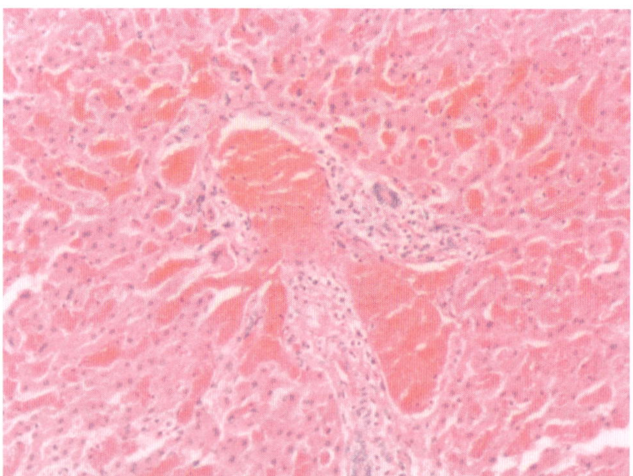

Figure 12.21, H&E x 260

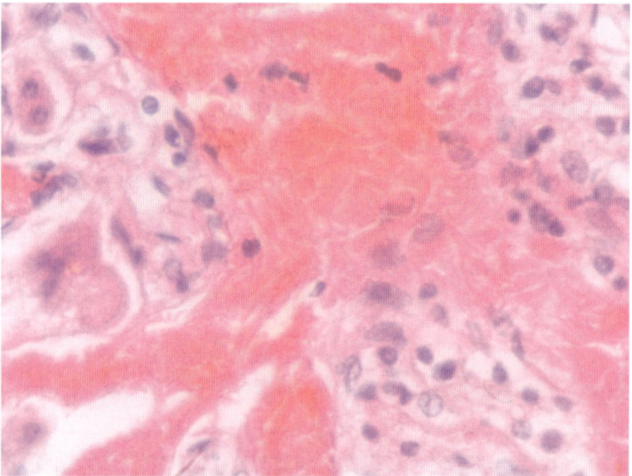

Figure 12.22, H&E x 520

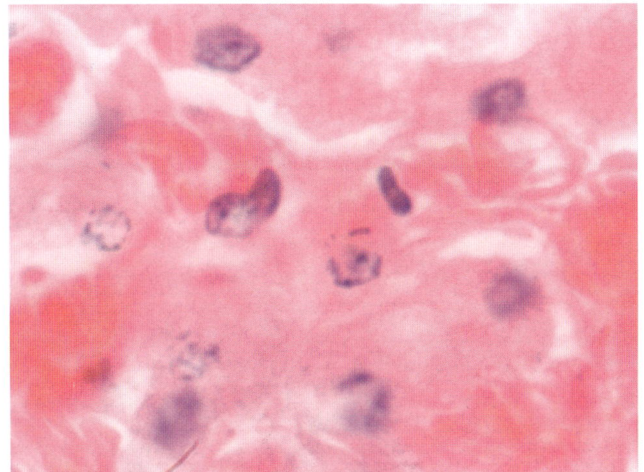

Figure 12.23, H&E x 1300

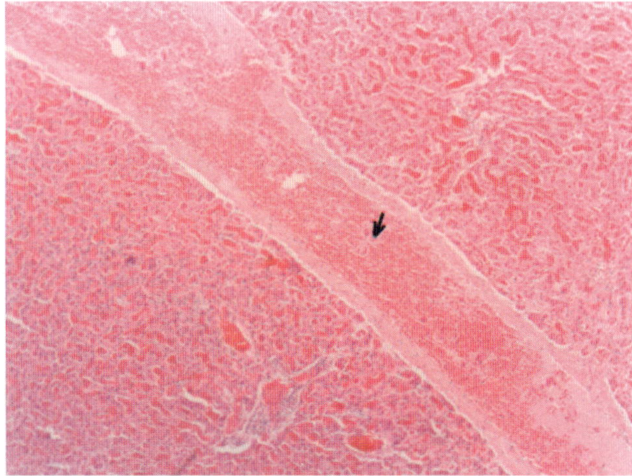

Figure 12.24, H&E x 52

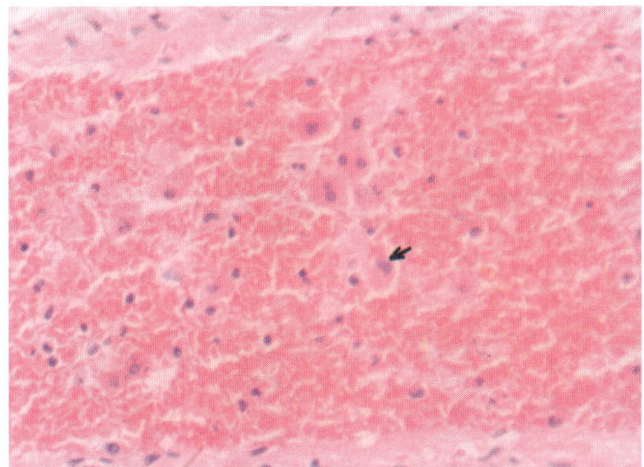

Figure 12.25, H&E x 260

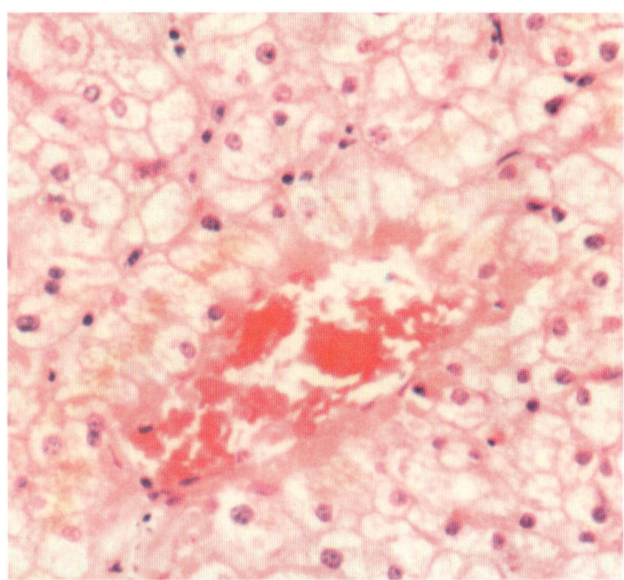

Figure 12.26, H&E x 400

as they are replaced by sickle cells developing out of them. H&E x 1300

Figure 12.24. A large blood vessel, with
what appears to be a well-developed wall,
is formed following the conversion of a
column of liver cells into sickle cells.
Lysing liver cells (arrow) are barely identified at this magnification. This development
is shown in a higher magnification in
figure 12.25. H&E x 52

Figure 12.25. Lysing liver cells (arrow)
lying in the lumen of the vessel are clearly
visible at this magnification. This large
formative blood vessel appears to have a
well-developed wall in figure 12.24, but
under higher magnification it is apparent
that the wall is made of nothing more than

a parenchyma derived fibrinoid material in
which scattered small nuclei can be seen.
The endothelium and the smooth muscle are
not yet present. H&E x 260

*Blood and blood vessel formation
from lysing liver parenchyma
(figs. 12.26 and 12.27)*

Figures 12.26 and 12.27. In figure 12.26,
normal liver tissue undergoing liquefaction
is transforming into large clumps of
hemoglobinized substance and clear plasma
in development of a blood vessel. A few
endothelial cells are developing from the
peripheral remaining liver tissue. Figure

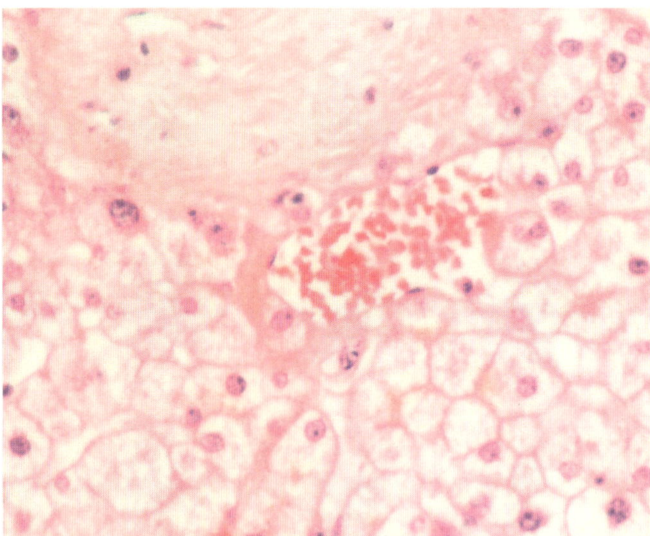

Figure 12.27, H&E x 400

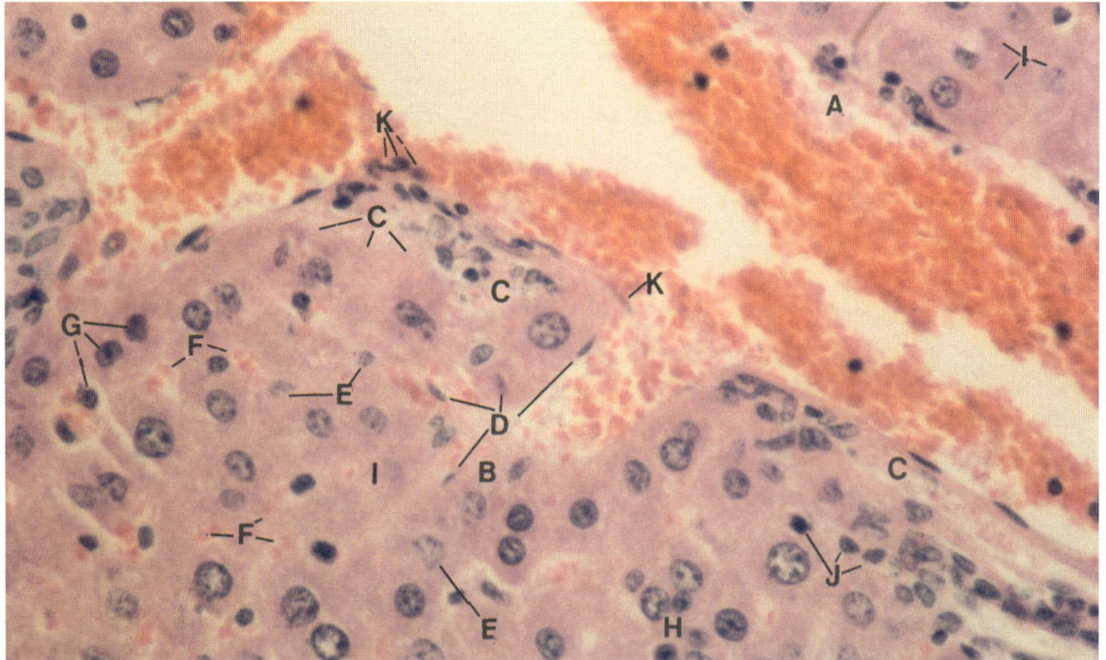

Figure 12.28, H&E x 520

12.27 shows a further development of an area similar to the one described above. In this developing blood vessel the clumps of hemoglobinized substance have broken down into red cell sized fragments. These red cells are not yet well rounded or well separated. Several endothelial cells are seen surrounding the lumen of this developing blood vessel. H&E x 400

Rapid and extensive mobilization of liver tissue in the formation of a large blood vessel packed with developing red cells (fig. 12.28)

Figure 12.28. A large, branching blood vessel is rapidly developing from liver tissue undergoing fragmentation. Within this formative vascular channel not yet connected with the circulatory system, red cells are arising as angular hemoglobinized particles transformed from dissolving minute liver cell fragments (A). In (B) liver cells lying in the path of vascular progression show similar early stages of red cell formation. In (C) liver tissue at the periphery of the main vessel shows very early changes of red cell formation, and (C), on the left, points out the anticipated margin of further expansion of the vessel. (D) denotes endothelial nuclei of liver cell origin lining the vascular wall. (E) shows early stages of endothelial nuclei development from liver cell nuclei. In (F) red cells are originating in a single column or more from liver cells undergoing lysis in development of sinusoids. (Note: sinusoids are not usually lined by endothelium.) The liver cell nuclei (G) are changing toward formation of lymphocytes. Multiple small nuclei (H), including those of lymphocyte size, are developing from liver cells. In (I) at the upper right corner and near the center, the remains of liver cell nuclei are fading away in the cytoplasm. (J) shows lymphocytes developing from liver cells. (Note a few lymphocytes among the developing red cells in the lumen of the main vessel.) In (K) endothelial nuclei and lymphocytes, at the margin of the vanishing vascular wall, are in the process of being transformed into red cells and added to the developing red cell mass in the lumen.

The observations demonstrated in this figure on the origin of red cells agree with Heitzman (1872), who believed that the protoplasm of hematoblastic substance can break into fragments which themselves become red cells.[3] H&E x 520

---

3. Rapid mobilization of tumor tissue in formation of blood in spontaneous breast carcinomas of two of this same group of mice is presented in figures 24.1-24.3.

Chapter 13

GALL BLADDER (figs. 13.1-13.9)

The gall bladder is a small organ that stores and excretes bile. The wall of this organ consists of four layers. From the inside out these layers are: 1. a mucus layer consisting of surface epithelium and a lamina propria; 2. a layer of smooth muscle; 3. a perimuscular connective tissue layer; and 4. a serous layer. In chronic inflammatory conditions, the epithelium and smooth muscle layers can transform into erythrogenic inflammatory cells, which then develop into red cells. Red blood cells can also develop directly from the epithelium, smooth muscle and connective tissue of the gall bladder wall.

Figures 13.1-13.9 are taken from a gall bladder wall removed by cholecystectomy of a 59-year-old male with polyarteritis nodosa, cholecystitis and cholelithiasis.

Reactive/inflammatory cells arising from gall bladder wall and forming red blood cells and blood capillaries (figs. 13.1-13.9); red cell development from locally derived eosinophils (figs. 13.2, 13.3, 13.8 and 13.9), SN cells (fig. 13.3), and lymphocytes (fig. 13.4)

Figure 13.1. In this section of the gall bladder wall, inflammatory cells, many of which are eosinophils, are present in both the mucosa and the edematous muscular coat. A magnification of area (A) is seen in the next figure. H&E x 52

Figure 13.2. In this group of inflammatory cells there are more eosinophils than SN cells (segmented nuclear cells). The two ends of this developing blood vessel (A) are merging into the surrounding inflammatory cells present in the muscle coat. Incompletely hemoglobinized purplish red cells (arrow) are arising from the inflammatory cells of the smooth muscle. In the center of the lumen, there are a few

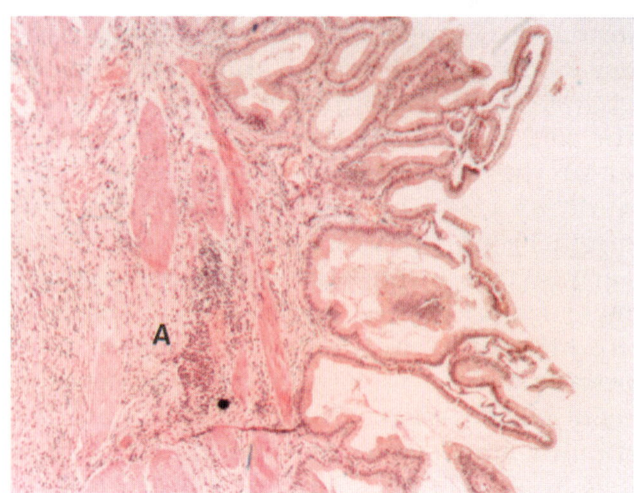

Figure 13.1, H&E x 52

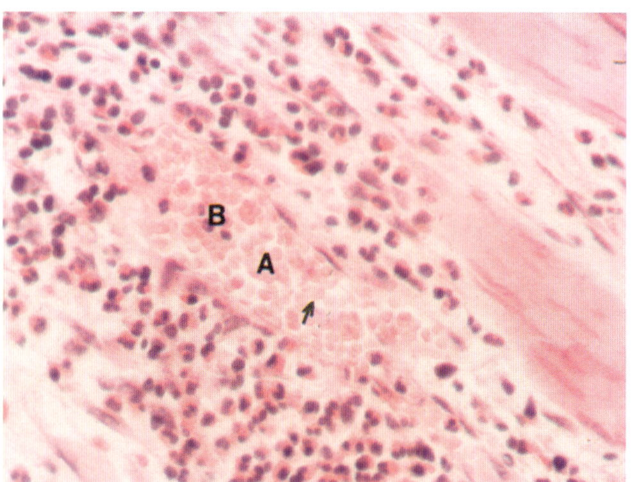

Figure 13.2, H&E x 400

erythrogenic reactive cells (B) that have not been completely transformed into red cells. There are also a few endothelial cells that have developed along the upper wall of this vessel. H&E x 400

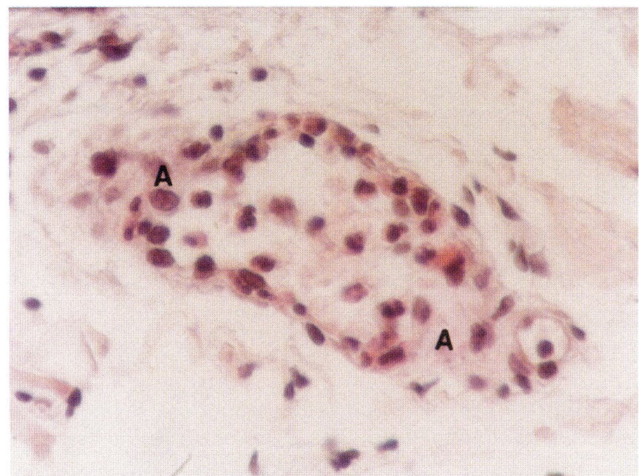

Figure 13.3, H&E x 520

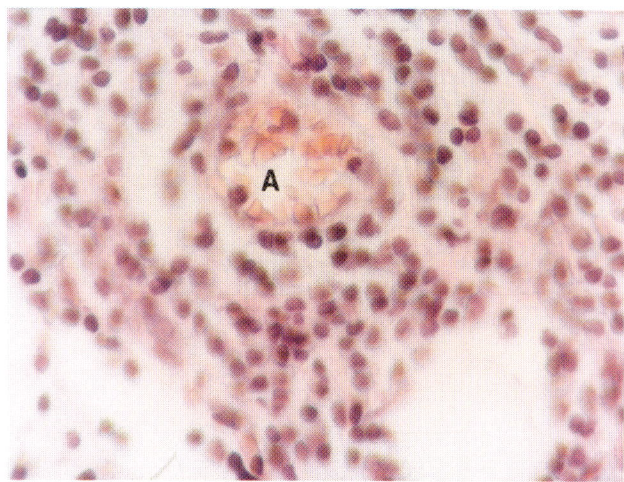

Figure 13.4, H&E x 520

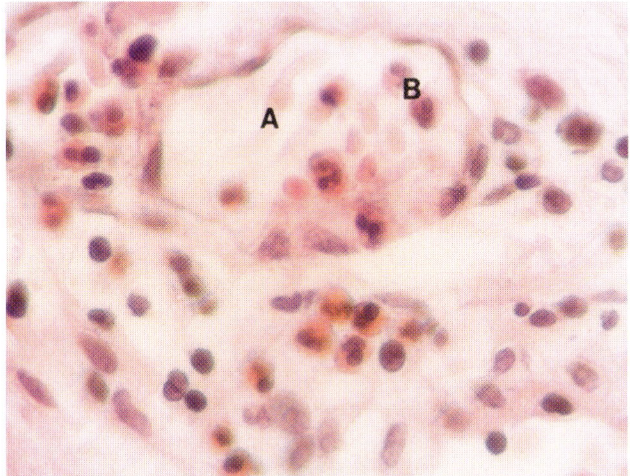

Figure 13.5, H&E x 800

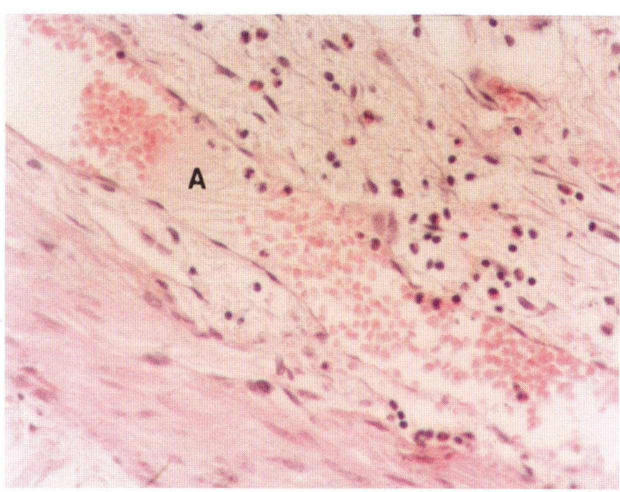

Figure 13.6, H&E x 260

Figure 13.3. This blood capillary is mainly developing from locally derived eosinophils and SN cells. The lumen lengthens by dissolving and utilizing the edematous inflammatory gall bladder wall (A) surrounding both ends of the vessel. H&E x 520

Figure 13.4. The red cells and endothelial membrane of this blood capillary (A) are developing from chronic inflammatory cells, mainly dissolving erythrogenic lymphocytes. The bluish-gray coloration present in some of the formative red cells within the capillary shows that they have not completely hemoglobinized. H&E x 520

Red blood cells and blood capillaries developing from dissolving smooth muscle (figs. 13.5-13.7); and from connective tissue (figs. 13.7 and 13.8)

Figure 13.5. A blood capillary is forming from the dissolving muscle fibers (A) and inflammatory cells (B) of the gall bladder wall. H&E x 800

Figure 13.6. Within this locally-formed large blood capillary (formed by liquefaction of local tissue) there are faint remnants of the original muscle fibers (A). The developing red cells contain different concentrations of hemoglobin. H&E x 260

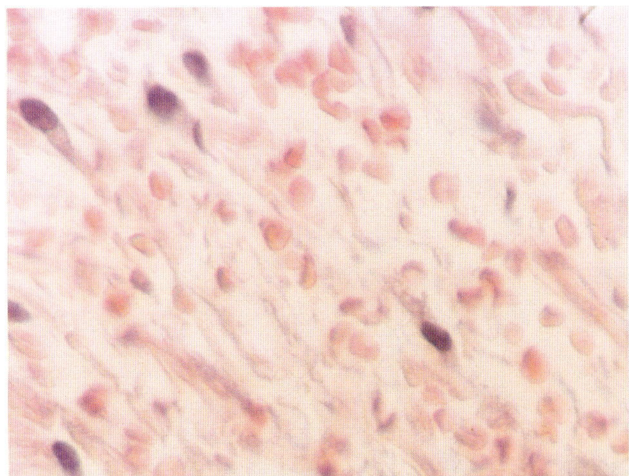

Figure 13.7, H&E x 800

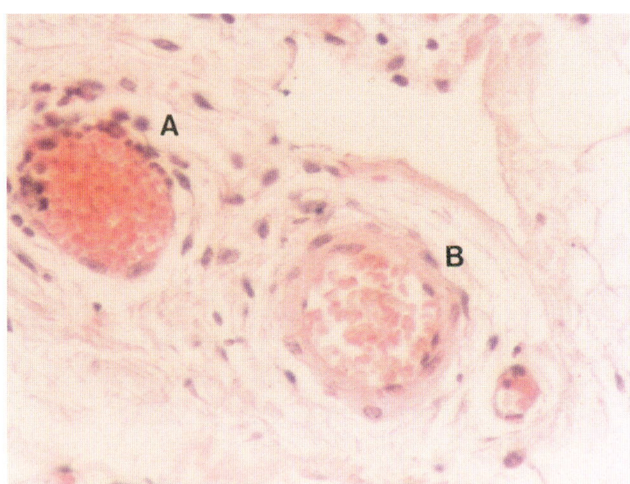

Figure 13.8, H&E x 400

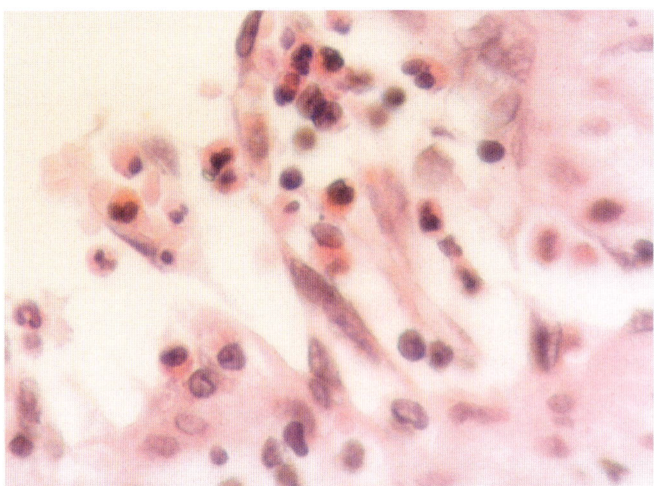

Figure 13.9, H&E x 520

Figure 13.7. Several locally-derived red cells lie among the remnants of the dissolving smooth muscle and connective tissue fibers.   H&E x 800

Figure 13.8. The blood capillary (A) is developing mainly from eosinophils and is continuing to grow towards the upper left. The blood vessel (B), on the other hand, is derived mainly from connective tissue.   As usual, the eosinophils in (A) appear to have formed red cells with a much deeper red coloration than the red cells derived from the connective tissue (B).   H&E x 400

Red cell development from desquamating epithelial cells (fig. 13.9)

Figure 13.9. Both red cells and erythrogenic eosinophils are arising from desquamating epithelial cells of this chronically inflamed gallbladder wall.   H&E x 520

Chapter 14

STOMACH (figs. 14.1-14.14)

(Inflammatory Lesions of the Gastric Wall Including Mucosa, Submucosa and Smooth Muscle Coat)

The wall of the stomach consists of the usual layers of the digestive tube: mucosa, submucosa, muscularis, and adventitia. I have observed and studied blood and blood vessel formation from all four layers of the gastric wall.[1]

Blood and blood vessel formation and the origin of inflammatory or reactive cells from local tissues of the gastric wall are presented here. Red blood cells arising as hemoglobin globules from the tissues of the mucosa, submucosa, and muscular coat are demonstrated below. Red cells also develop from locally-formed inflammatory (reactive) cells such as plasma cells, Russell body cells, segmented nuclear cells (SN cells), lymphocytes and eosinophils. Plasma cells, besides giving rise to red cells directly or through the stages of grape cells (Russell body cells), may act as originators of SN cells, eosinophils, lymphocytes and endothelial cells.

The development of blood and blood capillaries by proliferating Redefined Vasoformative cells of Ranvier in the subserosa of the stomach is described in figures 21.18-21.22 in Chapter 21, Inflammatory Tissue. Also, the development of erythrogenic SN cells and endothelium from the smooth muscle coat of the gastric wall is described in figure 4.38 in Chapter 4, subsection 3, Smooth Muscle, and in figures 30.1-30.4 in Chapter 30, subsection 1, Adenocarcinoma of the Stomach. For cellular changes in malignant stomach tissue see Chapter 30, subsections 1 and 2.

Figures 14.1-14.5 and 14.9-14.12 are taken from the gastric mucosa, and figure 14.13 from the smooth muscle coat near a chronic ulcer surgically removed from one patient. Figure 14.6 is from the gastric mucosa near a similar lesion surgically removed from another patient. Figures 14.7 and 14.8 are taken from the autopsy of a 73-year-old male suffering from chronic gastritis. Figure 14.14 is from the gastric wall of subtotal gastrectomy specimen surgically removed from a 47-year-old female with chronic eosinophilic gastritis. Figure 14.1 is reprinted from Volume I (McDonald, 1989).

The development of red blood cells, blood vessels and chronic erythrogenic inflammatory cells in the gastric mucosa (figs. 14.1, 14.3 and 14.4); and red cell development from erythrogenic segmented nuclear cells (figs. 14.1 and 14.2)

Figure 14.1. This figure shows the development of plasma cells and lymphocytes from the gastric mucosa. The further transformation of plasma cells into SN cells is shown at (A), and from SN cells to red cells (B) within the formative vessel (1). Many erythrogenic SN cells are producing red cells within a larger formative vascular channel (2). The origin of eosinophils from plasma cells is pointed out by (C), where

---

1. These findings were briefly demonstrated in Volume I (McDonald, 1989), figures 117-121.

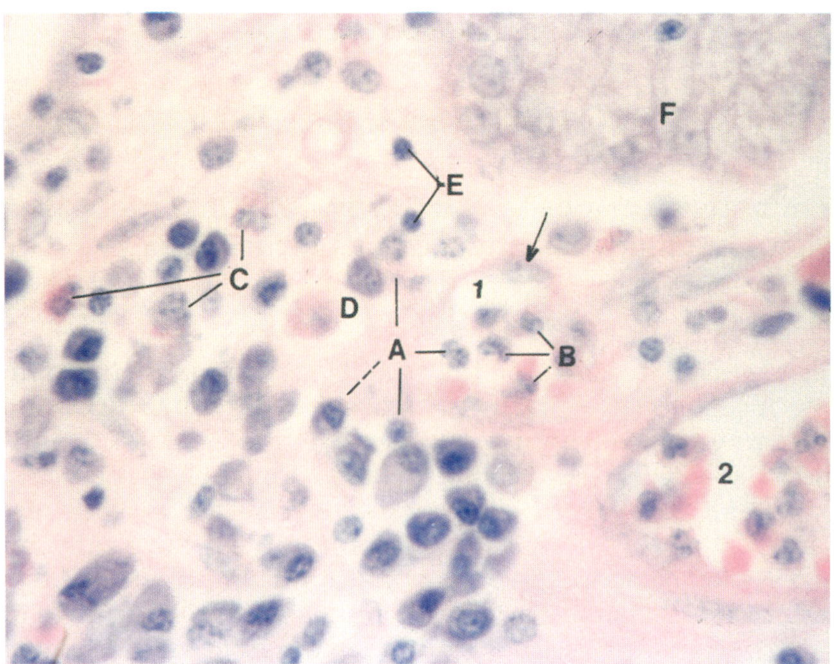

Figure 14.1, Giemsa x 800

red granules are developing adjacent to the nuclear structures. (D) points to a faintly visible grape cell (Russell body cell) developing from a plasma cell. (E) points to lymphocytes locally derived from plasma cells. The arrow points to a transitional stage of endothelial nuclear formation from a plasma cell. (F) denotes normal gastric epithelium. Giemsa x 800

Figure 14.2. In this figure, there is red cell development from erythrogenic SN cells within a large blood vessel. A few red cells (A) have not yet separated from their mother cells (SN cells). In area (B), red cells are still developing from this group of erythrogenic SN cells. Note the margins of all the red cells are fuzzy and have not yet become rounded, suggesting their recent origin. Giemsa x 800

Figure 14.3. Many plasma cells of different sizes are transforming into blood capillaries filled with developing red cells. A narrow developing blood capillary (A) is arising from vanishing plasma cells. A further developed stage is seen in capillary (B). There is a similarly formed, somewhat

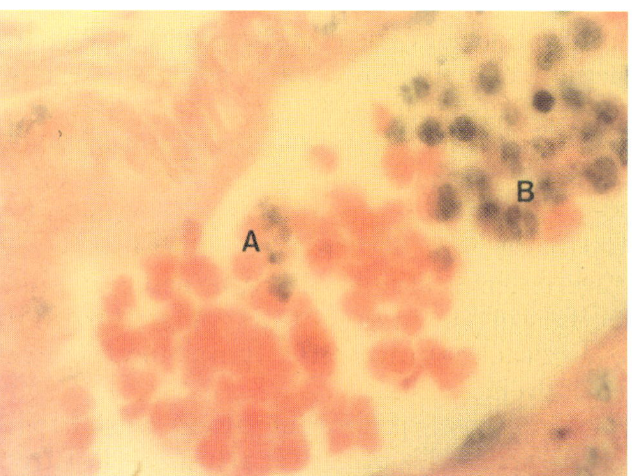

Figure 14.2, Giemsa x 800

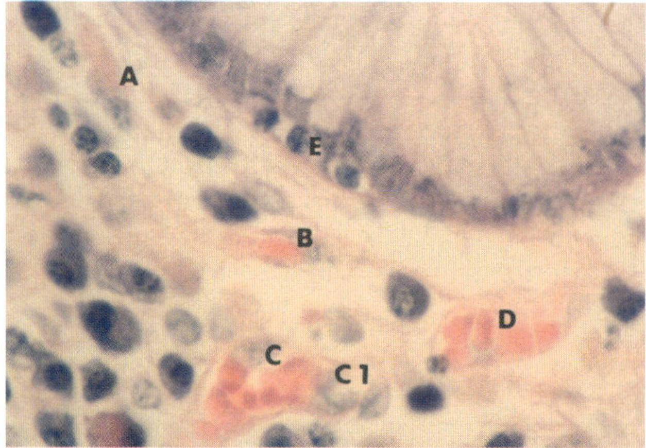

Figure 14.3, Giemsa x 800

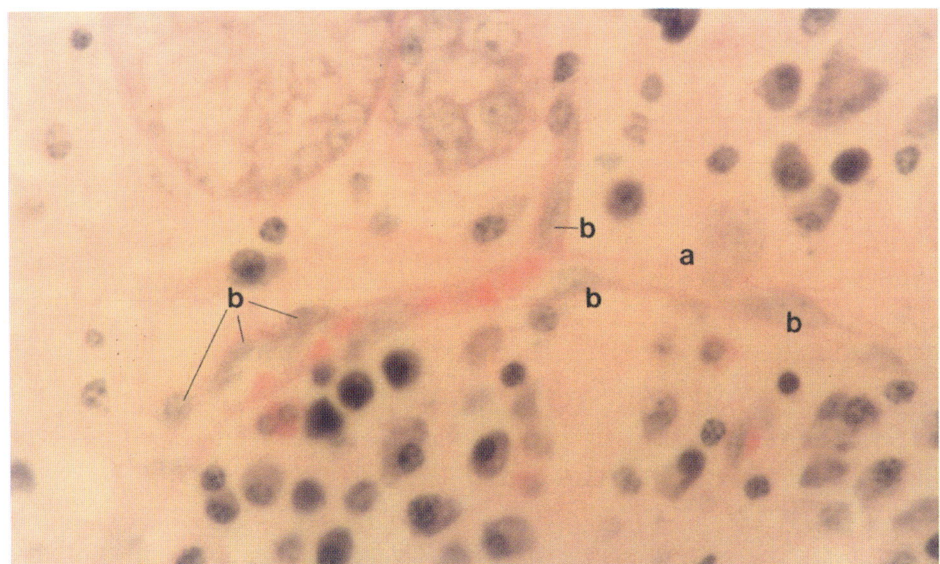

Figure 14.4, Giemsa x 800

Y-shaped, capillary in which remains of plasma cells can be seen at (C) and (C1). In blood capillary (D), evidence of plasma cells are no longer present. (E) is the edge of a gastric gland. Giemsa x 800

Figure 14.4. Below a gastric gland there is a long, narrow, Y-shaped capillary developing from fading and disappearing plasma cells. This capillary is barely canalized and contains only a few developing red cells. A major part of this developing capillary wall is formed by a very thin, delicate, reddish membrane (a). Red cells have not yet formed in the lumen of this branch of the capillary. Some parts of the capillary wall are formed by faintly visible developing endothelial cells (b) derived from plasma cells. Giemsa x 800

Origin of blood and blood vessels in gastric submucosa (fig. 14.5)

Figure 14.5. There are three, oblique lying, narrow, developing blood capillaries (A, B and C) each with developing red cells. The capillaries have not yet been fully canalized and the lining endothelium has hardly begun to form. In capillary (A) there are at least three deformed, isolated, hemoglobinized red cells. The capillary wall is

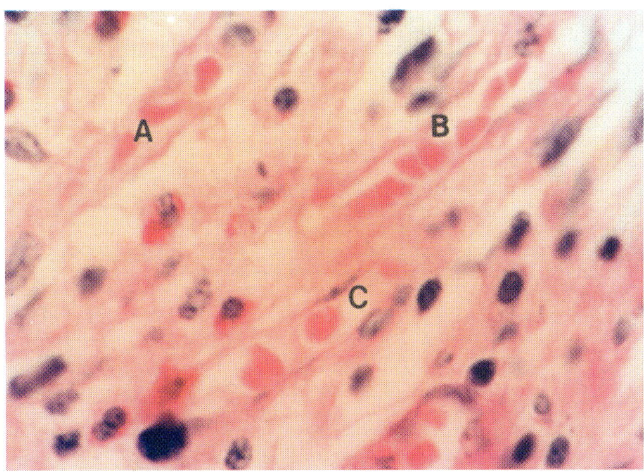

Figure 14.5, Giemsa x 780

mainly formed of a delicate, acellular membrane representing the beginning of endothelial lining formation. The red cells in the lower part of capillary (B) are arranged like a stack of pressed biscuits as a result of partitioning of a narrow column of hemoglobinized substance. This type of formation has been studied in other tissues. In capillary (C) there is some liquefaction surrounding the small, isolated red cell groups. This capillary has a more developed endothelial lining. Giemsa x 780

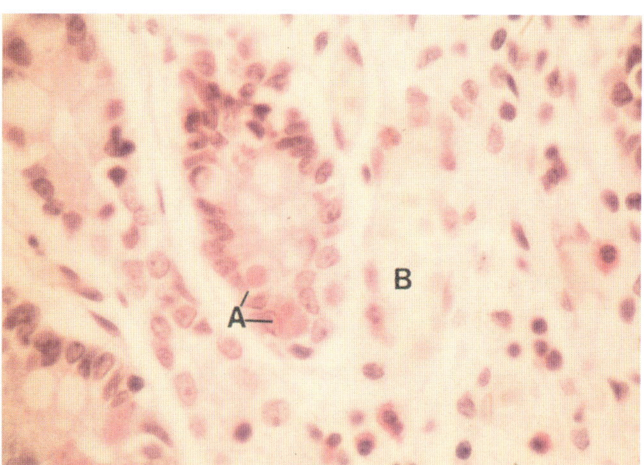

Figure 14.6, H&E x 520

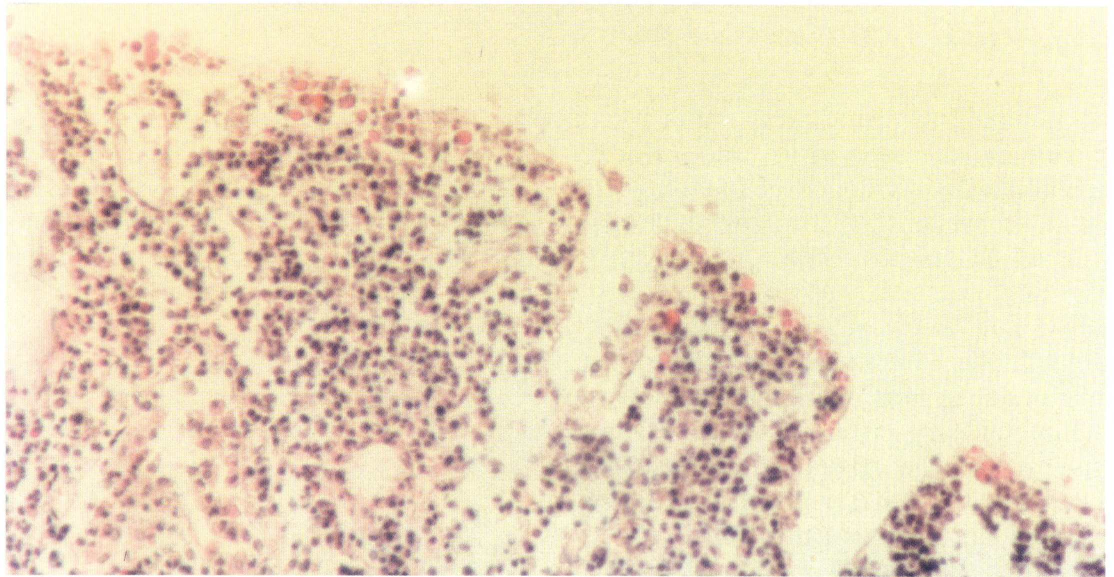

Figure 14.7, H&E x 260

Red blood cell development from epithelial cells of gastric gland (fig. 14.6)

Figure 14.6. Red cells (A) are originating from the lining epithelial cells of this gastric gland. (B) appears to be a partially hemoglobinized fluid inside a gastric gland which is almost denuded. H&E x 520

Red blood cell development directly from the gastric lining epithelium (figs. 14.7 and 14.8)

Figure 14.7. Widespread development of hemoglobin globules, some are huge, from the lining of the gastric mucous membrane (including the very edge) is shown here. Deliverance of such locally formed red cells into the gastric lumen (see also figure 14.8) may be one of the causes of positive occult blood in the stool. H&E x 260

Figure 14.8. Development of red blood cells from the surface of the denuded gastric mucosa is presented here. Note the variation in size and color of these red cells (A and B), reflecting different stages of hemoglobinization. H&E x 1040

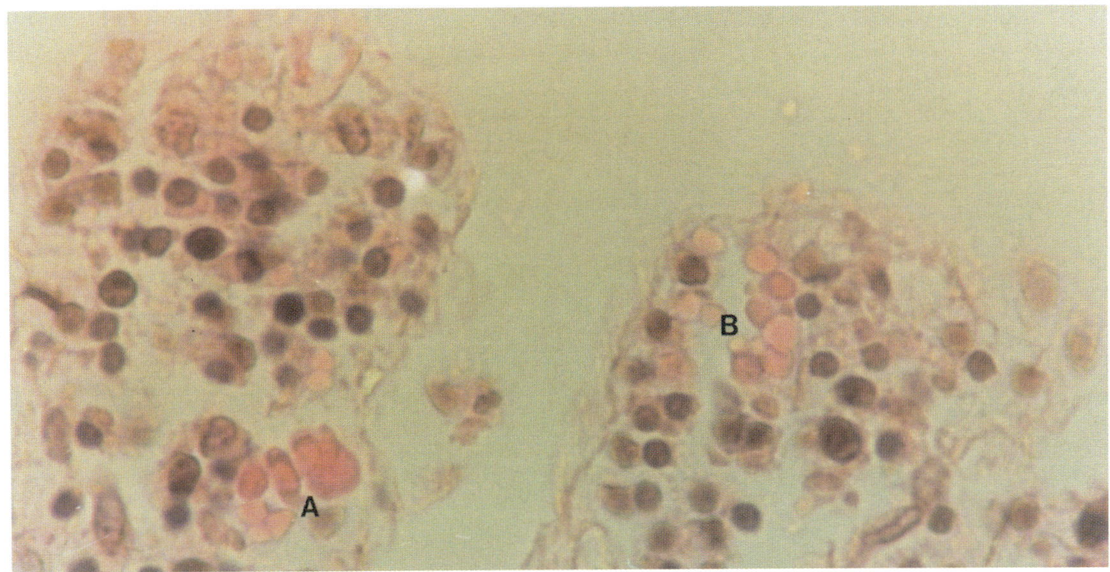

Figure 14.8, H&E x 1040

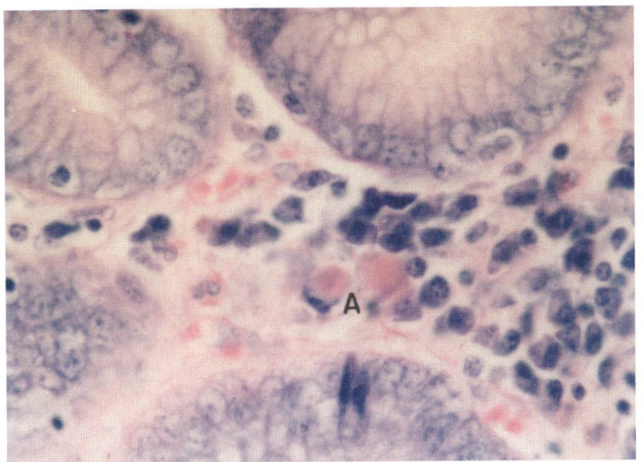

Figure 14.9, Giemsa x 520

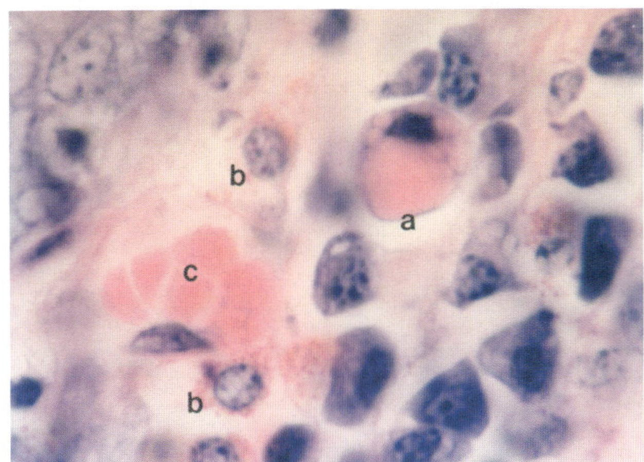

Figure 14.10, Giemsa x 1300

Red cell development through Russell body cells (grape cells) in the gastric mucosa (figs. 14.9-14.12)

Plasma cells may give rise to red cells either directly or through the development of a distinctive form called a Russell body cell or grape cell. Latta (1921) noted red cell formation in the connective tissue of the intestine. By applying various diagnostic tools, Keasby (1923) concluded that hemoglobin is present in the globules of the globular leukocytes (same as Russell body cells or grape cells) derived from plasma cells. Presented below (figs. 14.10 and 14.11) are some examples of red cell formation from

Russell body cells of plasma cell origin in the gastric mucosa. Other examples of red cell development from grape cells can be seen in figures 8.6, 21.16 and 21.17.

Figures 14.9. Between the gastric glands there is marked proliferation of plasma cells associated with development of large hemoglobin globules (A). Giemsa x 520

Figure 14.10. There is an irregular Russell body cell (a) formed from a plasma cell. Two eosinophils (b) with faintly visible small eosinophilic granules are present. Blood capillary (c) is partially filled with red cells. Giemsa x 1300

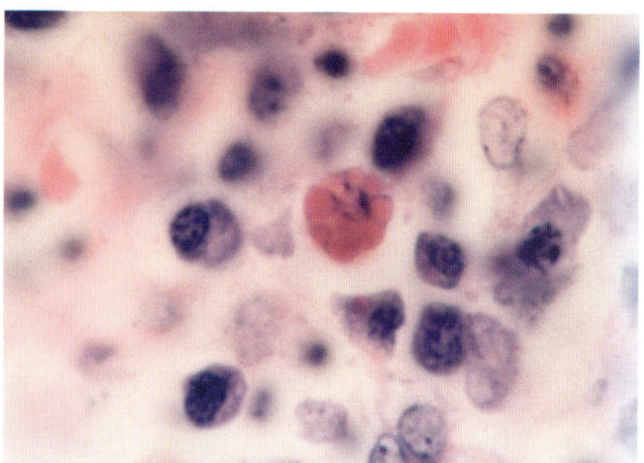

Figure 14.11, Giemsa x 1300

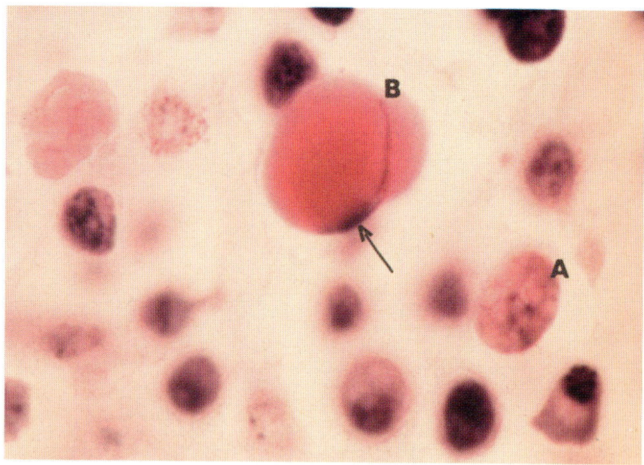

Figure 14.12, Giemsa x 1300

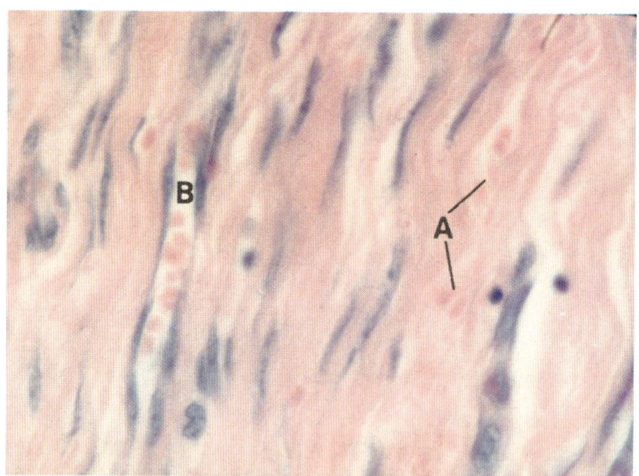

Figure 14.13, Giemsa x 520

Figure 14.11. In this figure, there are mainly fading plasma cells. In the center there is a Russell body cell with only a few remnants of nuclear material remaining. Giemsa x 1300

Figure 14.12. In this inflamed gastric mucosa, multiple tiny red cells (A) are developing from a Russell body cell. Also, abnormally large hemoglobin globules (B) are apparently arising from a nucleated cell. The compressed rim-like remnants of the nucleus is present at one edge (arrow). Red cells of a size comparable to these hemoglo-bin globules are only rarely found in the peripheral blood. A prize winning name *stupendocyte* was once given to such a large red cell found in peripheral blood. Giemsa x 1300

Red blood cell and endothelium development directly from smooth muscle coat of gastric wall (fig. 14.13)

Figure 14.13. Red blood cells and blood capillaries are developing directly from the smooth muscle coat of the stomach. On the right, isolated deformed red cells (A) are originating in muscle fibers along the

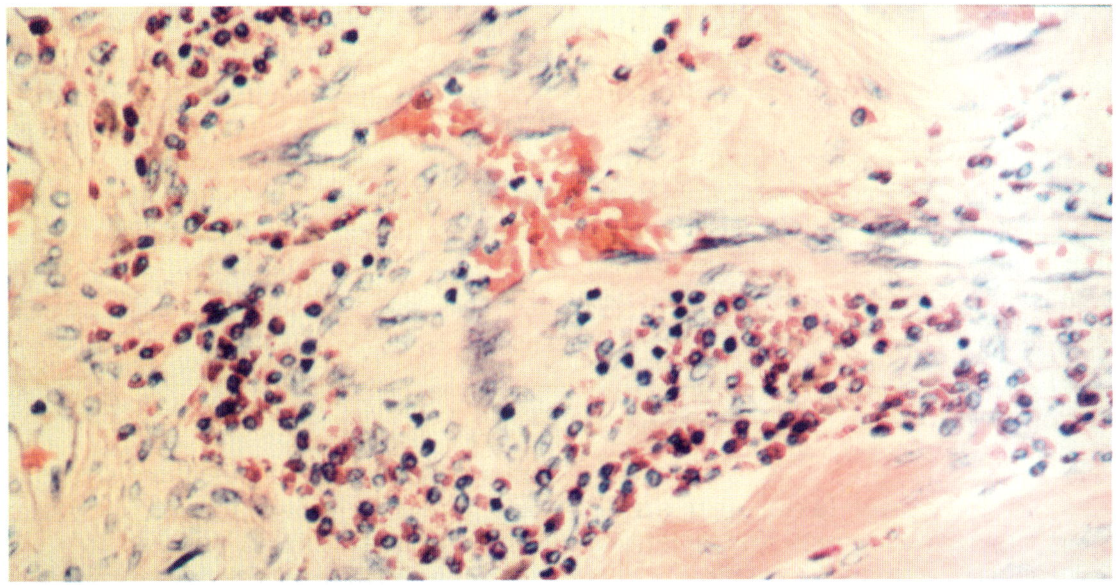

Figure 14.14, Giemsa x 520

longitudinal plane. On the left, there is a developed capillary (B) with well recognized endothelial nuclei apparently arising from smooth muscle and with several well-formed red cells in the lumen. The upper part of the capillary is still progressing in its development. Giemsa x 520

Eosinophil development from smooth muscle of gastric wall (fig. 14.14)

Figure 14.14. In this figure, there is prominent proliferation of eosinophils replacing a major portion of the disappearing smooth muscle. Giemsa x 520

Chapter 15

RECTAL TISSUE IN INTERNAL HEMORRHOIDS (figs. 15.1-15.4)

The rectal wall consists of mucosa, submucosa and smooth muscle layer, *muscularis mucosa.* Hemorrhoids are prominent varicosities of the hemorrhoidal venous plexuses. In the specimen shown below there is blood and blood vessel formation from the edematous mucosa, submucosa and smooth muscle of the rectal wall. Cellular changes in tumors of the gastrointestinal tract are demonstrated in Chapters 30 and 31. Figures 15.1-15.4 are taken from the rectal wall, including mucosa and underlying tissue, obtained via hemorrhoidectomy of a 62-year-old male.

Blood and blood vessel formation
from rectal mucosa and underlying tissue
(figs. 15.1-15.4)

Figure 15.1. Many large and small thin-walled vascular channels are distended with red cells in this edematous rectal mucosa. The endothelium of the wall of these vascular channels is poorly developed.

Lying transversely, in a corkscrew shape, individual capillary buds are beginning to join each other. H&E x 200

Figure 15.2. A large irregular blood capillary is not fully developed in the edematous submucosal tissue. The strands of smooth muscle of muscularis mucosa are seen on both sides of this vessel. Some of the strands are intimately connected with undissolved fibrillary material which can be seen within the lumen (arrow). H&E x 400

Figure 15.3. A large developing blood vessel, similar to the one in figure 15.2, is shown in transverse plane. Another narrow developing capillary, in longitudinal plane, is shown in the left side of this figure. There is one red cell in its lumen. The canalization of the lumen is interrupted by the remaining collagenous fibrils which form bridge-like structures. H&E x 800

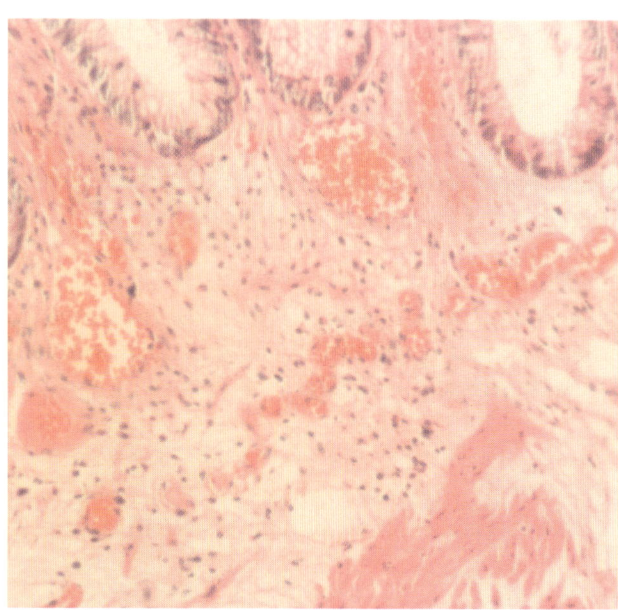

Figure 15.1, H&E x 200

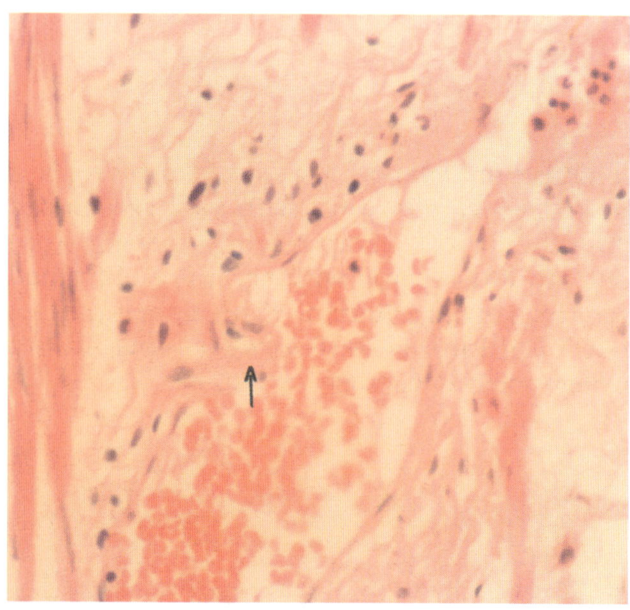

Figure 15.2, H&E x 400

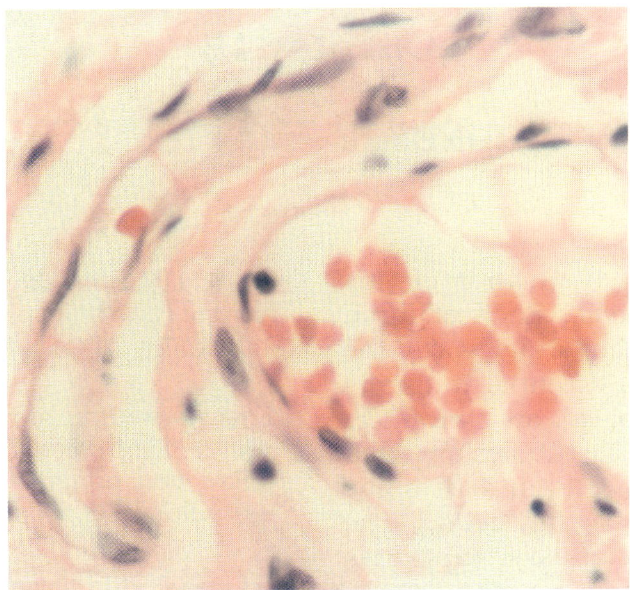

Figure 15.3, H&E x 800

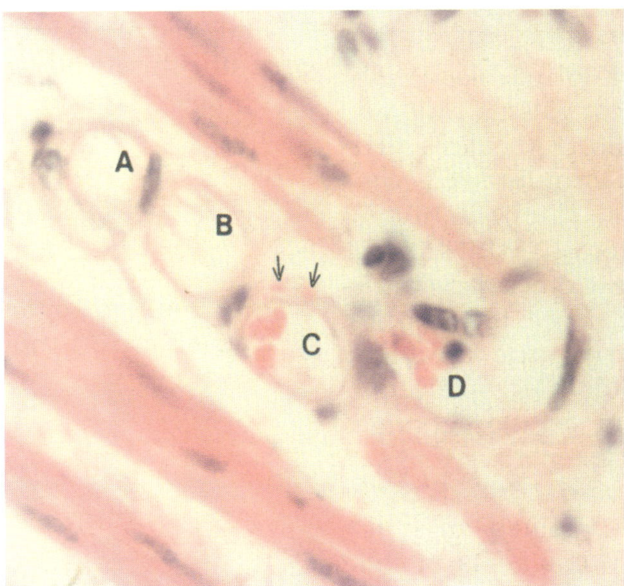

Figure 15.4, H&E x 800

Figure 15.4. A narrow capillary is developing by the unification of small vascular units (A, B, C and D). The individual units have not been completely canalized and the endothelium is only partially developed. From this view, red cells can be seen only in the two lower units (C and D). In unit (C), more red cells (arrows) are forming in its slightly thickened upper wall. These red cells will be included in the lumen after the dissolution of the inner margin of the present wall. The capillary (D) may itself be the result of the joining of small individual units because of its shape and a faint partition line, which can barely be seen in its center. This capillary contains a few red cells and also a lymphocyte-like cell. Similar cells are present in the surrounding tissues. H&E x 800

Chapter 16

KIDNEY (figs. 16.1-16.13)

The kidneys contain thousands of tiny filtering units, or *nephrons*. Each nephron can be divided into two important parts: the *tubule*, where water and essential nutrients are extracted from the blood, and the filtering part or *glomerulus*. The tubules are lined by cuboidal epithelial cells and in some places by flattened epithelial cells. The glomerulus is a network of blood capillaries and glomerular epithelium.

Under certain conditions, as demonstrated here, the renal tubules may transform into blood and blood vessels. During this process the epithelial cells of the tubules may transform into red blood cells, endothelium, lymphocytes, plasma cells and segmented nuclear cells (SN cells). The glomeruli may also transform into red blood cells and blood vessels. Liquefaction or breakdown of the glomeruli results in the hemoglobinization of the glomerular epithelium in the formation of red blood cells. These products are carried into the blood circulation and the glomerular space may be re-epithelialized. For cellular changes in renal cancers see Chapter 33.

Figures 16.1, 16.2 and 16.4-16.6 are taken from a five-hour postmortem specimen of kidney, with the diagnosis of mild interstitial nephritis, from a 68-year-old male who died of necrotizing pancreatitis. Figure 16.3 is from the autopsy of a 63-year-old male who died of hepatoma with metastasis. Figures 16.7-16.11 are from a specimen taken from the kidney of a healthy rabbit sacrificed by inhalation of ether. Because of extreme asphyxiation there is urgent need for rapid blood formation; this is evidenced in every organ. Figures 16.12 and 16.13 are from an autopsy specimen of an adult who died of complications of sickle cell anemia. Figure 16.4 is reprinted from

Volume I (McDonald, 1989), and figures 16.12 and 16.13 are reprinted from Volume III (McDonald, 2001).

Transformation of renal tubules into blood and blood vessels (figs.16.1-16.6 and 16.11); and origin of lymphocytes from renal tubular cells (figs. 16.1 and 16.3-16.5)

Figure 16.1. A narrow renal tubule is transforming into a vascular channel (A) that is mostly filled with red cells of various sizes and degrees of hemoglobinization. Some red cells are vacuolated with thin membranes and a few are tiny deep-red globules. Most of the vascular wall of this capillary is formed by a narrow delicate membrane, with only one endothelial cell present (arrow). Note that the epithelial cells of tubule (B) are changing into lymphocyte-like cells. The transition of renal cells into endothelial-like cells is better shown in tubule (C). H&E x 800

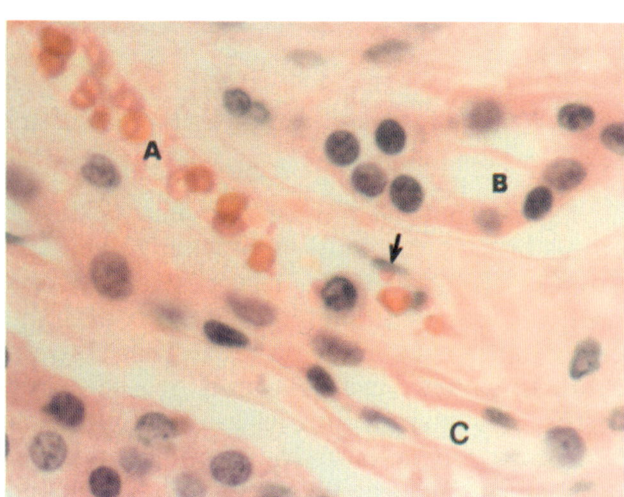

Figure 16.1, H&E x 800

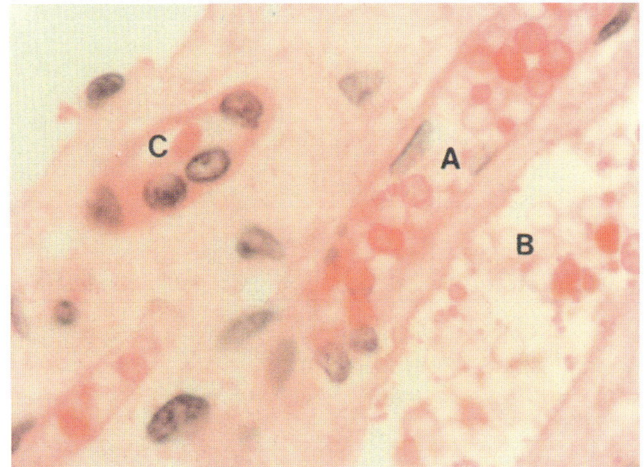

Figure 16.2, H&E x 800

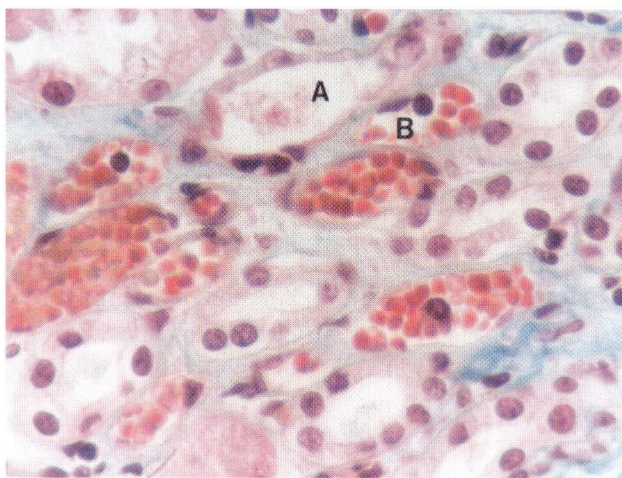

Figure 16.3, Chromotrope x 520

The development of red cells through the stages of vacuolar red cell (ghost red cell) formation (figs. 16.2, 16.4 and 16.5)

Figure 16.2.   Red cells and blood capillaries are developing from renal tubules in the medullary region.   The developing blood capillary (A) has several fully formed red cells and several vacuolated red cells, and there are indications of endothelial cell development on both sides of the lumen. The vacuolated mesh found in the tubular structure (B) is a preliminary step for red cell production.   There is a small denuded capillary (C) containing a single red cell. H&E x 800

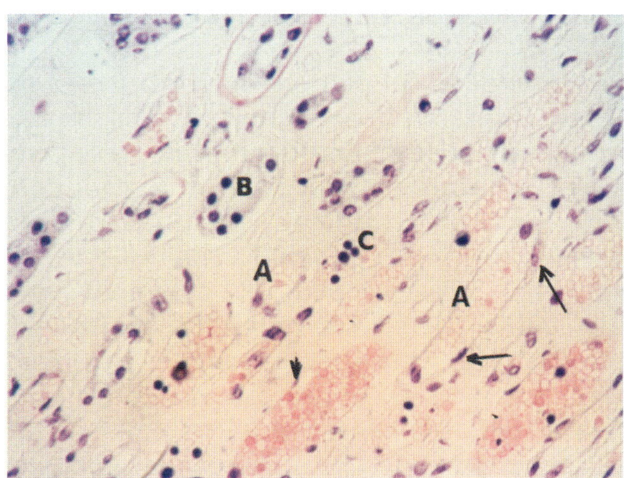

Figure 16.4, Giemsa x 260

Figure 16.3.   Several of these kidney tubules are transforming into blood channels.   In tubule (A), the lining epithelium is being denuded in preparation for the development of a blood capillary, and in the lumen there is a suggestion of red cell formation.   One small tubule (B) is transforming into a blood capillary in which one lymphocyte-like cell and one endothelial cell have transformed from the lining epithelium. Chromotrope Aniline Blue x 520

Figure 16.4.   These renal tubules are hyalinizing and vascularizing.   Small vascular channels (A) are developing from small denuded renal tubules filled with vacuolated red cells.   (B) is a cross section of a renal tubule in which the epithelial cells have transformed into lymphocyte-like cells. (C) is a cross section of a small denuded tubule filled with lymphocyte-like cells and rare developing red cells.   There are several developing vascular channels filled with vacuolated red cells and lined with rare, barely developed endothelial cells (arrows). Giemsa x 260

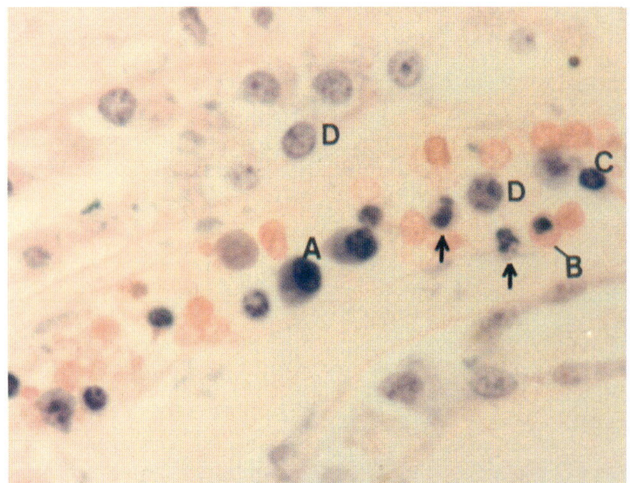

Figure 16.5, Giemsa x 800

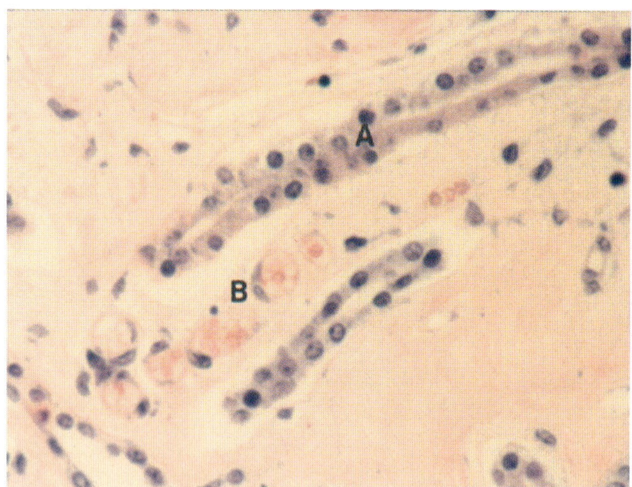

Figure 16.6, Giemsa x 400

*Plasma cell development from renal
tubular epithelium (fig. 16.5)*

Figure 16.5. This developing blood
vessel is arising from a renal tubule. In the
lumen lie a few small lymphocytes, two
plasmacytoid cells (A), and a red cell
with pyknotic nuclear remains (B), all
developing from epithelial cells. (C) is an
epithelial cell developing into a lymphocyte-
like cell. (D) shows partially hyalinized
tubular cells with a few chromatic particles.
The arrows point to small, partially
hyalinized, segmented nuclear cells. The
origin of red cells from SN cells (segmented
nuclear cells) has been demonstrated by
many figures in this volume and earlier
volumes. There are also several vacuolar
red cells present. Giemsa x 800

Figure 16.6. In tubule (A), the lumen is
collapsing and the lining epithelial cells are
vanishing and leaving behind cellular
remnants with faintly stained nuclear
particles. From this substance, endothelial cells may develop to line small round
vascular units (B) that enclose red cells that
were lying in the path of the vanishing
tubule. The unification of similar units in
the formation of a narrow blood vessel is
demonstrated in hemorrhoidal tissue in
Chapter 15, Rectal Tissue, figure 15.4.
Giemsa x 400

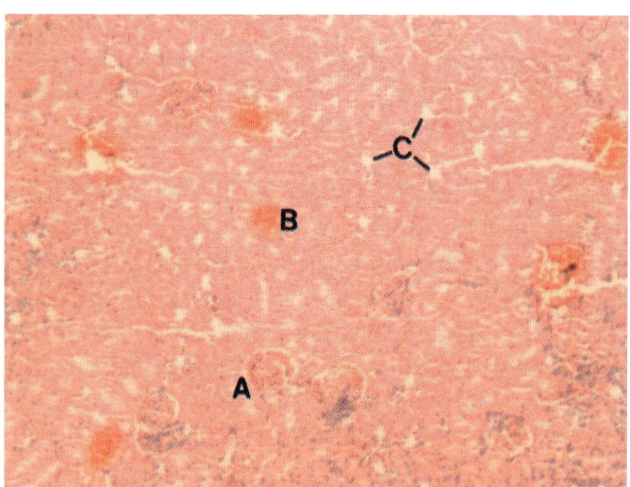

Figure 16.7, H&E x 80

*Transformation of glomeruli into blood and
blood vessels (figs. 16.7-16.10)*

Figure 16.7. In the cortical region there
is rapid red cell production by the glomeruli, and a few glomeruli are almost completely replaced by blood (A and B). When
this newly developed blood is carried away
by the circulation empty clear spaces (C)
appear, and these empty spaces may be re-
epithelialized. These empty clear spaces give
the appearance of a *starry sky,* and their
development is seen in higher magnification in figures 16.8-16.10. H&E x 80

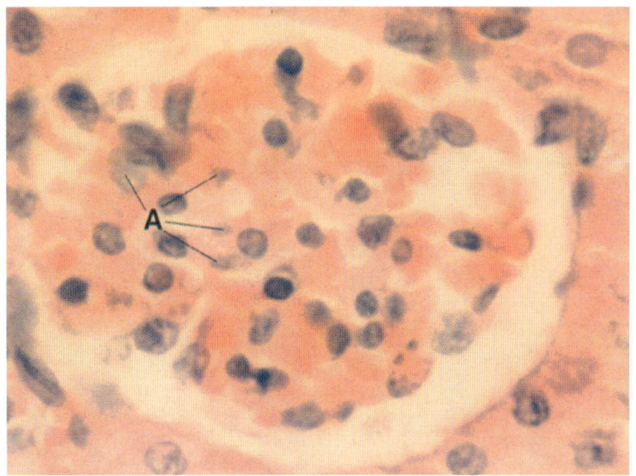

Figure 16.8, H&E x 800

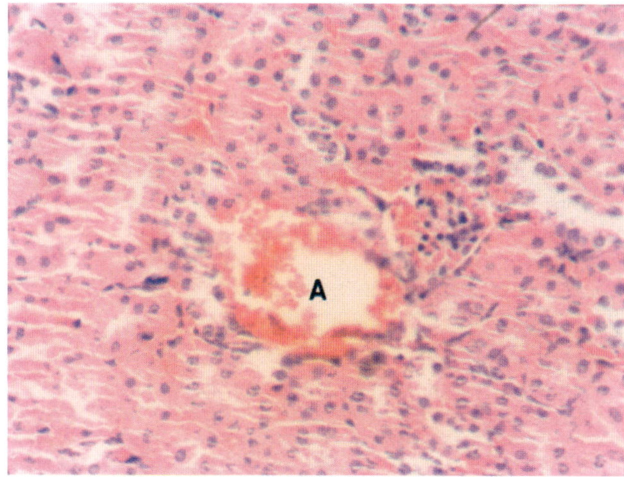

Figure 16.9, H&E x 260

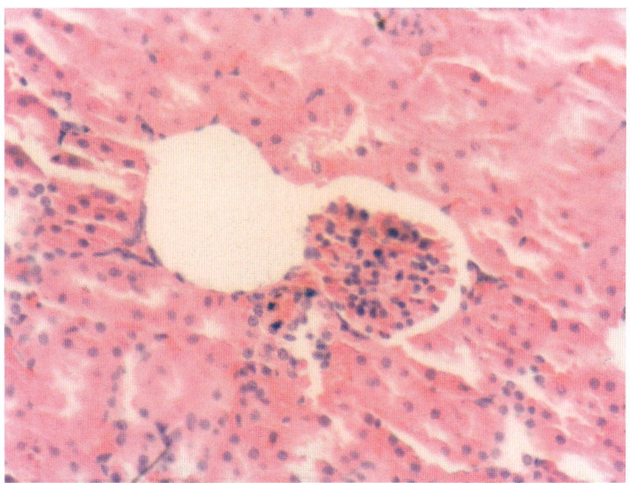

Figure 16.10, H&E x 260

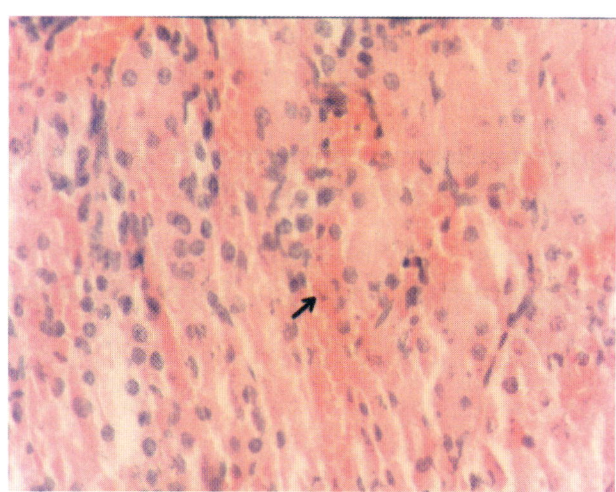

Figure 16.11, H&E x 400

Figure 16.8. This is a higher magnification of a less hemorrhagic glomerulus. The nucleated glomerular cells are hemoglobinizing in the production of red cells. The faintly visible remnants of the former nucleated cells (A) can be seen among the red cells. H&E x 800

Figures 16.9 and 16.10. In figure 16.9, the left side of the glomerulus (A) is almost completely transformed into hemoglobin particles which will be carried away by the circulation to form the empty glomerulus space seen in figure 16.10. The right side of the glomerulus has some cellular remnants of the original glomerulus and some hemoglobin particles. In figure 16.10, the right side of this glomerular structure is undergoing hemoglobinization similar to what is described in figure 16.8. H&E x 260

Figure 16.11. In the same kidney as shown above, some renal tubules are also transforming into columns of red cells in which remnants of tubular epithelial cell substance are noticed (arrow). H&E x 400

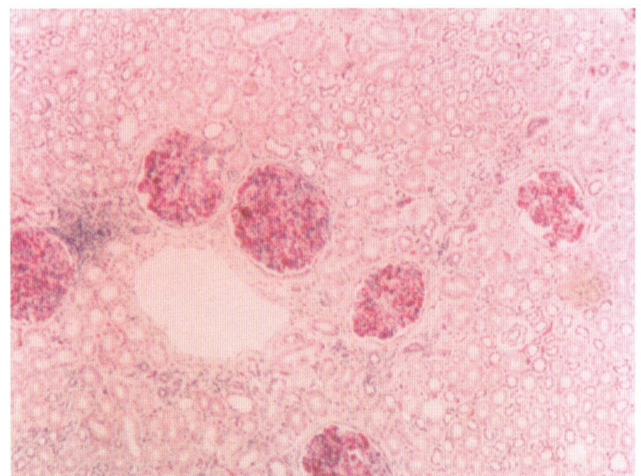

Figure 16.12, H&E x 52

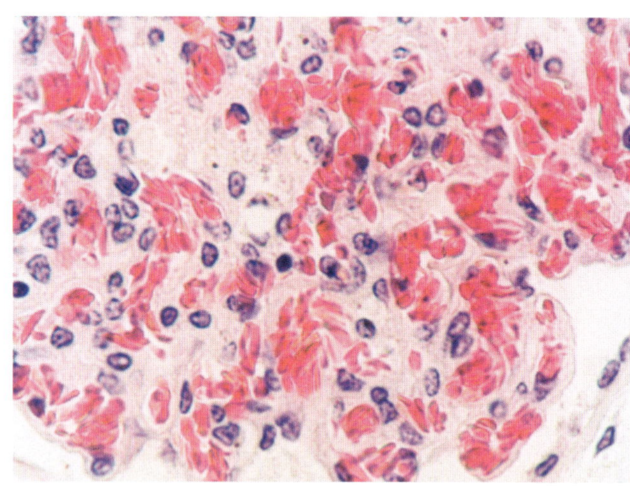

Figure 16.13, H&E x 520

Direct development of sickle cells from renal tissue (figs. 16.12 and 16.13)

Figures 16.12 and 16.13. Figure 16.12 demonstrates a section of kidney showing multiple glomeruli engorged with developing blood capillaries full of sickle cells, which are better seen in higher magnification in figure 16.13. Figure 16.12 H&E x 52; and figure 16.13 H&E x 520

Chapter 17

PROSTATE (figs. 17.1-17.3)

The prostate gland consists of epithelial glandular tissue together with smooth muscle and connective tissue stroma. Painless gross hematuria associated with prostatic hyperplasia or prostatic cancer is a common clinical finding. Shown here is the development of red cells from both glandular epithelial tissue as well as from stromal tissue. For development of red cells from smooth muscle of the prostate see figure 4.34 in Chapter 4, Smooth Muscle. Cellular changes in adenocarcinoma of the prostate are demonstrated in Chapter 34. Figures 17.1-17.3 are from the autopsy of a 72-year-old male with mild nodular hyperplasia of the prostate, the patient died of hypertensive cardiovascular disease.

Red blood cell and blood capillary formation from partly hyalinized stromal tissue of prostate (fig. 17.1); and from glandular epithelial cells (figs. 17.1-17.3)

Figure 17.1. From proliferating lining glandular epithelial cells there is prominent development of red cells, mainly starting as tiny hemoglobin globules and maturing to regular size red cells. This process is more prominent in the left half of this figure. Such occurrence is the cause of hematuria in prostatic hyperplasia or cancer of the prostate. In the middle of this figure, there are several single columns of red cells (of varying length) developing from the hyalinized stroma, probably in anticipation of narrow blood capillary formation (endothelial cells have not formed yet). H&E x 400

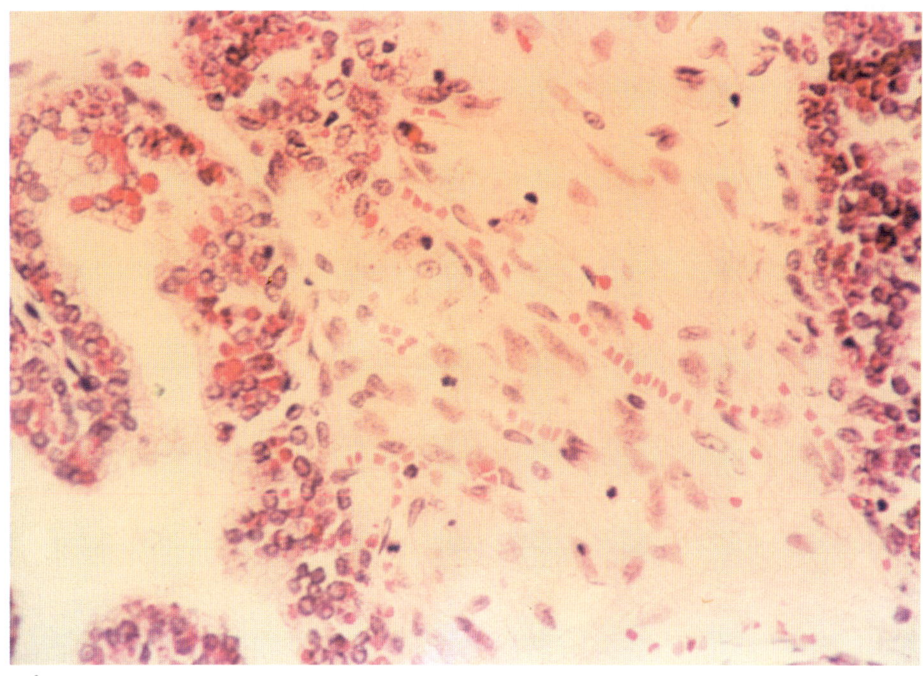

Figure 17.1, H&E x 400

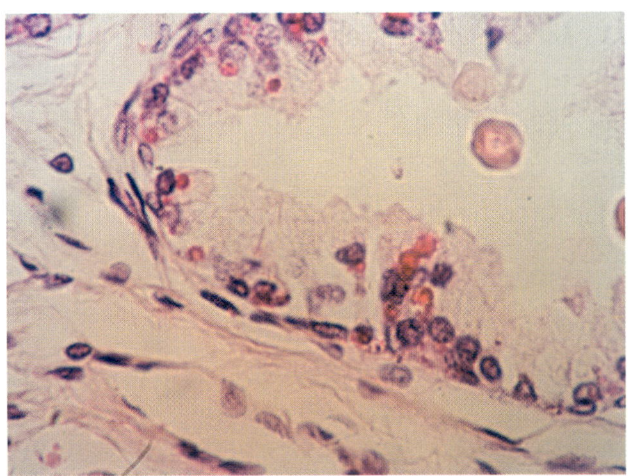

Figure 17.2, H&E x 520

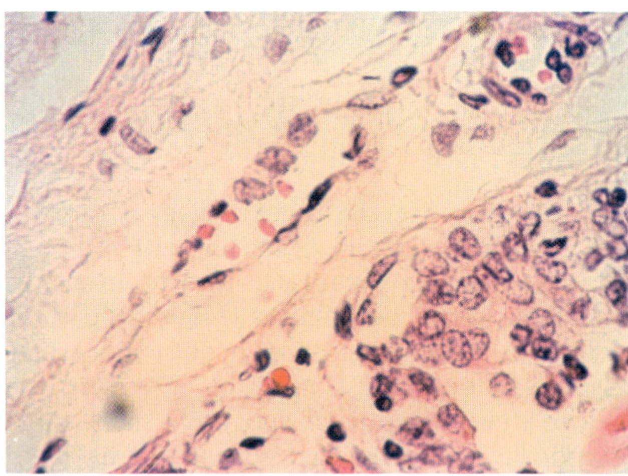

Figure 17.3, H&E x 520

Figure 17.2. Red cell development directly from epithelial cells is demonstrated in this hyperplastic glandular epithelium. H&E x 520

Figure 17.3. In the center of this figure, a blood capillary is developing from prostatic epithelial tissue. In the upper left side of the capillary wall there are three clumped prostatic epithelial cells which have not yet been transformed into endothelium. Whereas, in the lower right wall the endothelium is well formed. In the lumen there are a few red cells and plasma derived from glandular tissue. H&E x 520

Chapter 18

BREAST (figs. 18.1-18.9)

1. Mammary Gland (figs. 18.1-18.3)

The mammary gland undergoes extensive structural changes depending on age, sex and functional condition of the individual. In adult females each mammary gland has multiple lobes consisting of ducts, connective tissue and secretory structures of acinar epithelium. In the hyperplastic mammary gland of pregnancy presented below, red blood cells are developing directly from the glandular epithelium and from erythrogenic gel derived from the glandular epithelium. Cellular changes in mammary carcinoma in mice and humans are presented in Chapter 24. Figures 18.1-18.3 are taken from a surgical specimen of hyperplastic mammary gland of pregnancy.

Red cells arising from glandular epithelium (figs. 18.1 and 18.2)

Figure 18.1. This figure shows a hyperplastic mammary gland of pregnancy packed with acinar structures containing secretions. There are also scattered areas of groups of red cells. The area in the center of this figure is shown in higher magnification in figure 18.2. H&E x 130

Figure 18.2. Here, the small glandular epithelial cells are being transformed into hemoglobin globules in the development of red cells. H&E x 520

Red cells developing from erythrogenic gel derived from glandular epithelium (fig. 18.3)

Figure 18.3. In the lake-like area, reddish glandular secretion (erythrogenic gel) is produced by the liquefying epithelial cells. Many red cells are appearing from this erythrogenic gel. Various ways of red cell development from mammary carcinoma in mice and humans are demonstrated in detail in Volume I (McDonald, 1989) figures 1-56, and in Chapter 24 of this volume. H&E x 260

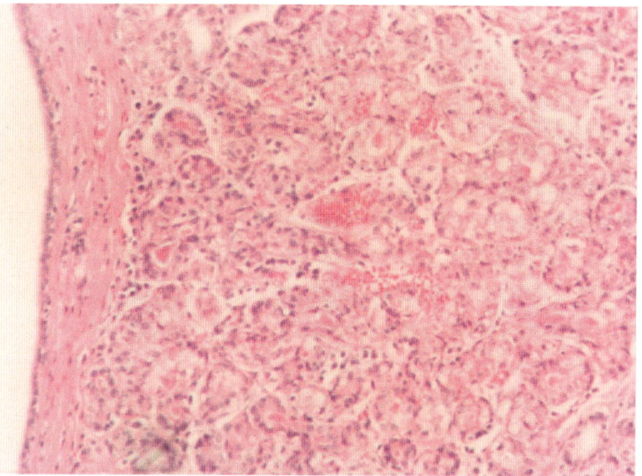

Figure 18.1, H&E x 130

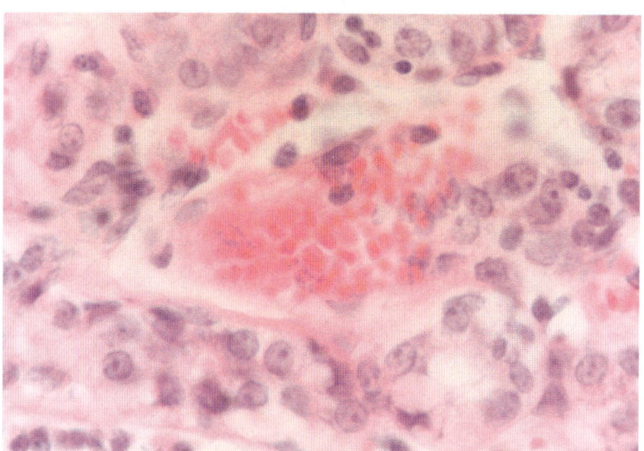

Figure 18.2, H&E x 520

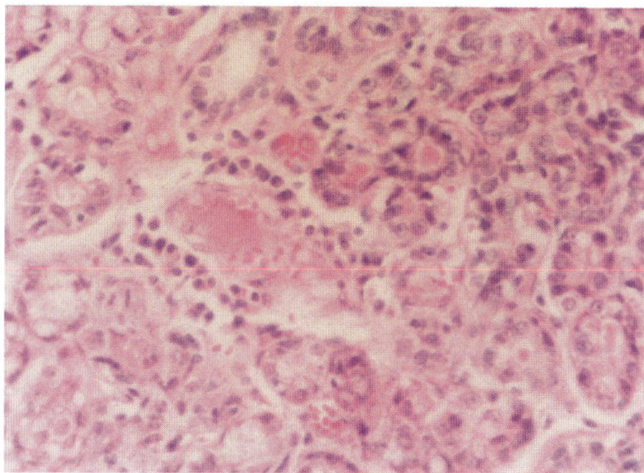

Figure 18.3, H&E x 260

2. Intraductal Papilloma of the Breast (figs. 18.4-18.9)

Figures 18.4-18.9 are taken from breast tissue of an 85-year-old female with a large intraductal papilloma obtained by simple mastectomy of the left breast.

Typical intraductal papilloma of the breast (figs. 18.4 and 18.5)

Figures 18.4 and 18.5. The beautiful papillary structures of this intraductal papilloma of the breast can be seen in Hematoxylin and Eosin stain (figure 18.4) and Giemsa stain (figure 18.5). Additional cellular structures cannot be seen at these magnifications. Figure 18.4 H&E x 260; and figure 18.5 Giemsa x 104

Red cell and blood vessel development from dissolving tumor tissue (figs. 18.6-18.9)

Figure 18.6. In the center of this figure, compact red cells are developing directly from the tumor cells. Remnants of nuclear material can still be seen attached to many of the developing red cells. Giemsa x 520

Figure 18.7. In the center of this figure, the tumor cells have transformed into small nucleated cells which further develop into red cells. Giemsa x 520

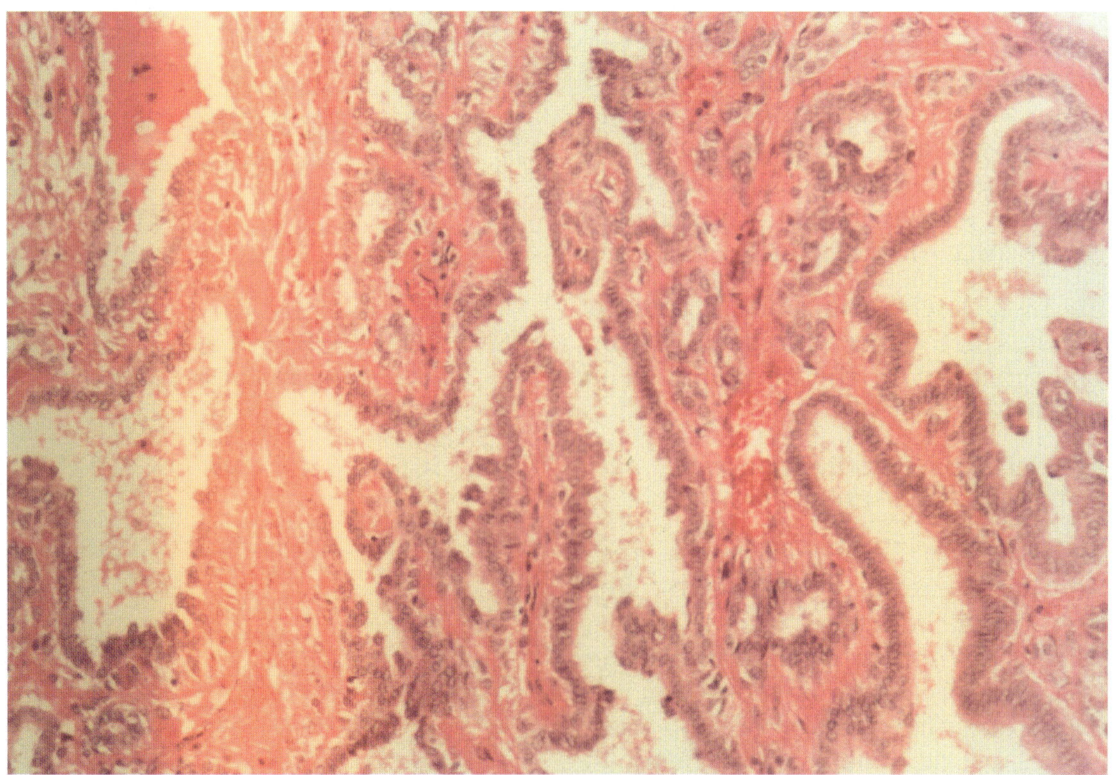

Figure 18.4, H&E x 260

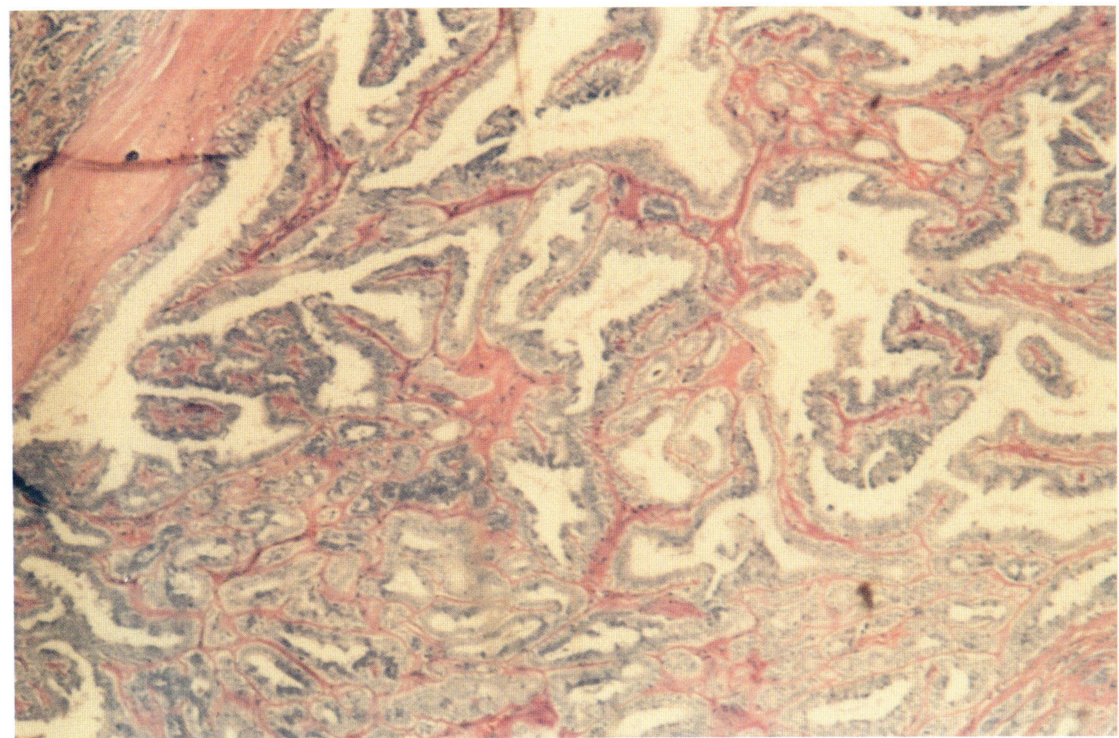

Figure 18.5, Giemsa x 104

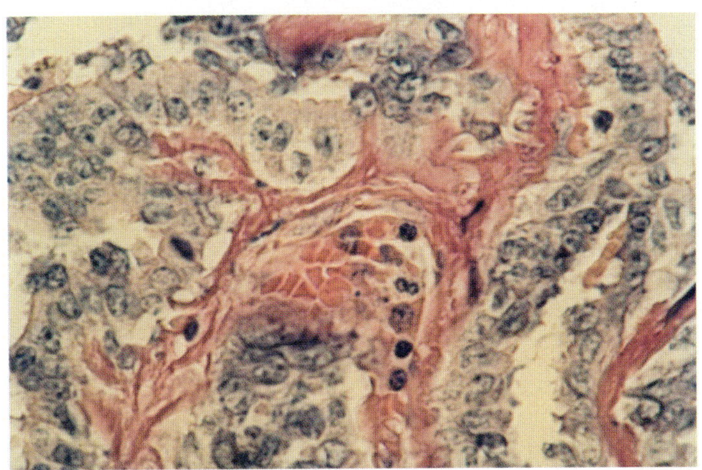

Figure 18.6, Giemsa x 520

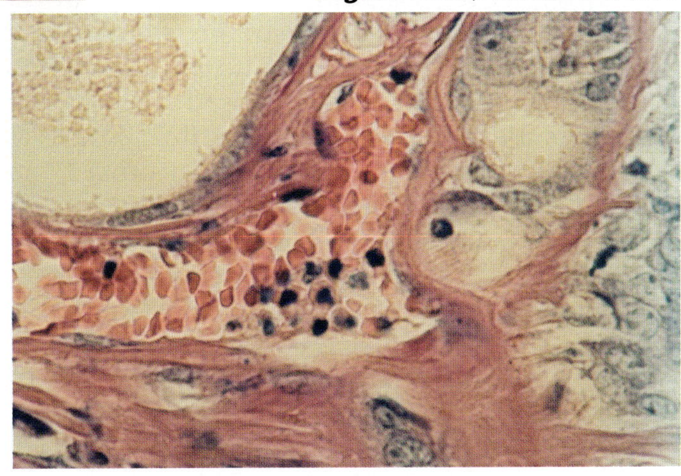

Figure 18.7, Giemsa x 520

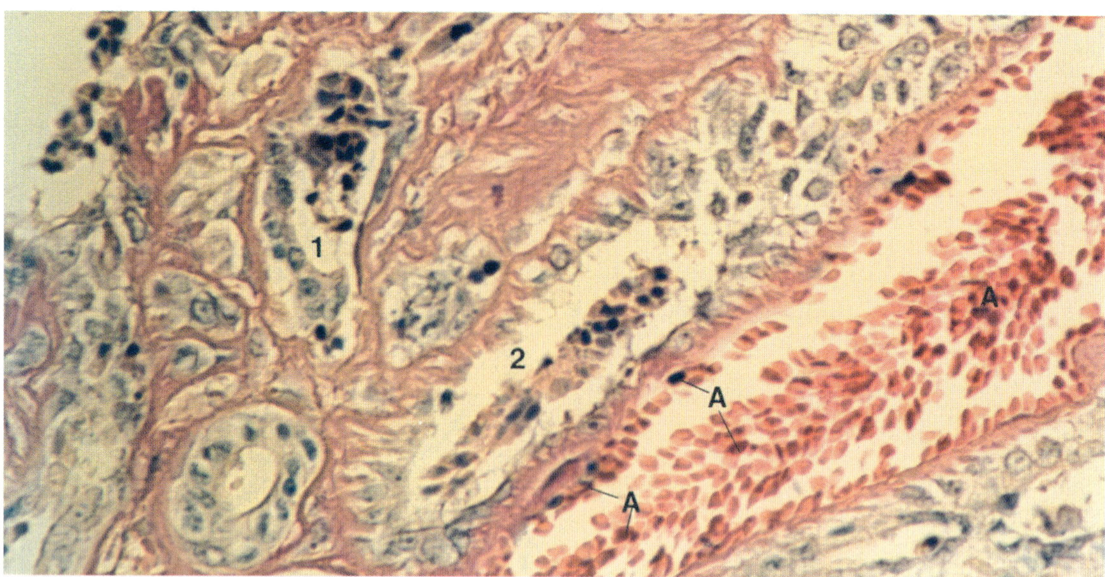

Figure 18.8, Giemsa x 520

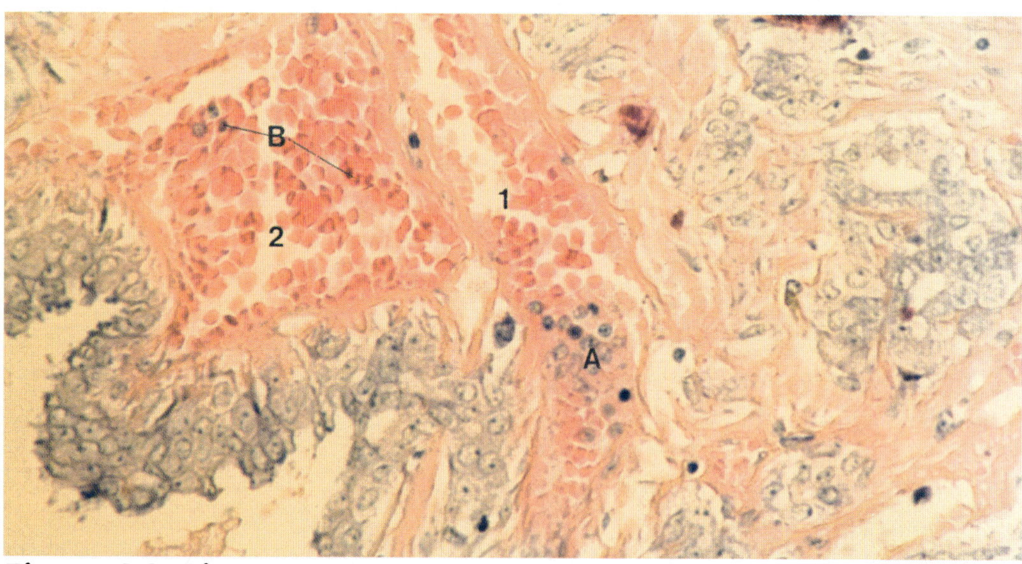

Figure 18.9, Giemsa x 520

Figure 18.8. Columns of necrosed dissolving tumor tissue can be seen in lumens (1) and (2) of the cystic pattern of this benign tumor. This necrosed dissolving tumor tissue is forming clear fluid. No definite blood formation from this necrosed tumor tissue can be identified. In the lower right of this figure, a section of a large developing blood vessel is seen filled with already developed elongated red cells. Remnants of tumor nuclear substances (A) can be barely identified still attached to some of the red cells. The endothelium of this blood vessel has not yet formed. Giemsa x 520

Figure 18.9. Lying vertically in the center, there is a large blood vessel (1) developing from dissolving tumor tissue. In the lower part of this vessel, transformation of tumor cells (A) into red cells can be seen. A large rhomboid-shaped developing blood vessel (2) can be seen in cross section. Red cells in this blood vessel are more developed and only a few remnants of nuclear structures (B) can still be identified. The endothelium of both these blood vessels has not yet developed. Giemsa x 520

Chapter 19

BRAIN [Cerebrum] (figs. 19.1-19.8)

Brain tissue is not commonly considered erythrogenic. However, in autopsies of patients who have died of cerebral vascular accidents, it is difficult to find any ruptured blood vessels which could be the source of the hemorrhage found in brain tissue. As described below, I have found that liquefying brain tissue directly transforms into red blood cells, which is the source of this so-called hemorrhage. Blood development from brain tissues has been observed and reported by Bjornson (1999). Also, directly from brain tissue there is formation of erythrogenic SN cells (segmented nuclear cells) which can transform into red blood cells. The last two figures show the direct transformation of brain tissue into sickle cells in a case of sickle cell anemia.[1] For cellular changes in malignant tissues of the brain see Chapter 36.

Figures 19.1-19.5 are all from the autopsy of a 58-year-old male who died of cerebral vascular thrombosis with massive hemorrhage of the left cerebrum involving all the lobes and basal nuclei. Figure 19.6 is from the autopsy of a 67-year-old male who died of gastric adenocarcinoma with massive hemorrhage, necrosis, and metastasis to lymph nodes, liver and kidneys with portal vein obstruction by metastatic nodes. Figures 19.7 and 19.8 are both from the autopsy of an adult who died of complications of sickle cell anemia. Figures 19.7 and 19.8 are reprinted from Volume III.

Development of red cells directly from the cerebrum (figs. 19.1, 19.2 and 19.6-19.8); as well as through erythrogenic SN cells arising from cerebral tissue (figs. 19.3-19.5)

Figure 19.1. The large scattered areas commonly considered to be hemorrhage are actually red cells which replace the lique-

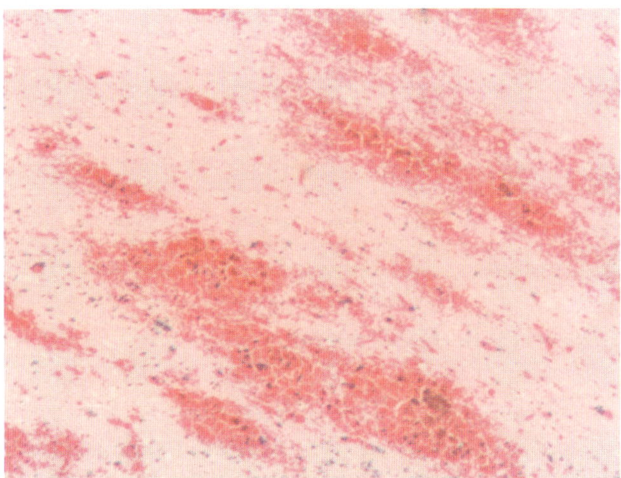

Figure 19.1, H&E x 80

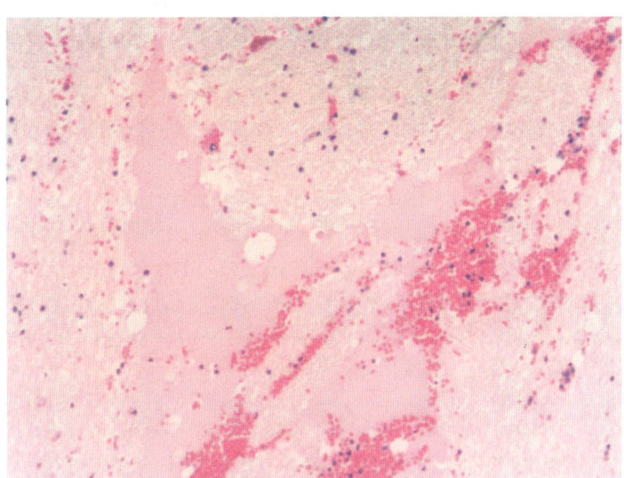

Figure 19.2, H&E x 130

fying brain tissue from which they directly develop. H&E x 80

Figure 19.2. In this figure, brain tissue is dissolving into a light pink fluid. Also, there are many red cells developing directly from the dissolving brain tissue. H&E x 130

---

1. Direct development of red cells from human embryonic cerebellum tissue is demonstrated in figures 1.24 and 1.25 of Volume III (McDonald, 2001).

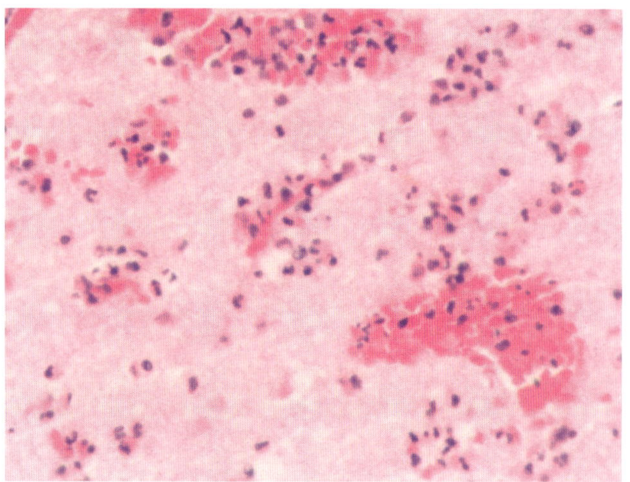

Figure 19.3, H&E x 400

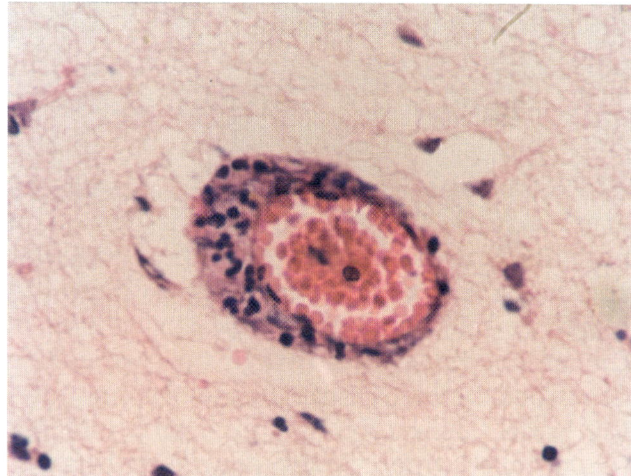

Figure 19.4, H&E x 520

Figure 19.3. This figure shows scattered clusters of SN cells (segmented nuclear cells) developing from brain tissue. Some of these erythrogenic SN cells are turning into red cells. H&E x 400

Eccentric enlargement of a vascular lumen by erythrogenic SN cells (fig. 19.4)

Figure 19.4. This is a cross section of columns of red cells derived from, and surrounded by, layers of erythrogenic SN cells. On the left side of this capillary, there are multiple layers of erythrogenic SN cells which are participating in the eccentric widening of the blood vessel. On the right side, there appears to be only a single layer of SN cells. H&E x 520

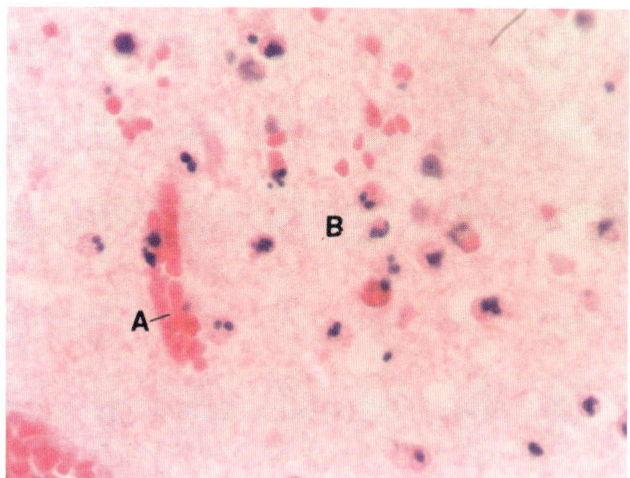

Figure 19.5, H&E x 520

Figure 19.5. A developing capillary (A), whose endothelium has not yet formed, is filled with columns of red cells. Widely scattered in this figure, erythrogenic SN cells (B)–which arise locally from brain tissue–are forming red cells. The remains of some of these erythrogenic cells can be seen at the edge of capillary (A). H&E x 520

Figure 19.6. In the lumen of artery (A), there is entrapped brain tissue mixed with developing red cells. In capillary (B), the wall is composed of only a single layer of endothelial cells which are not fully developed. In the lumen of this vein there are

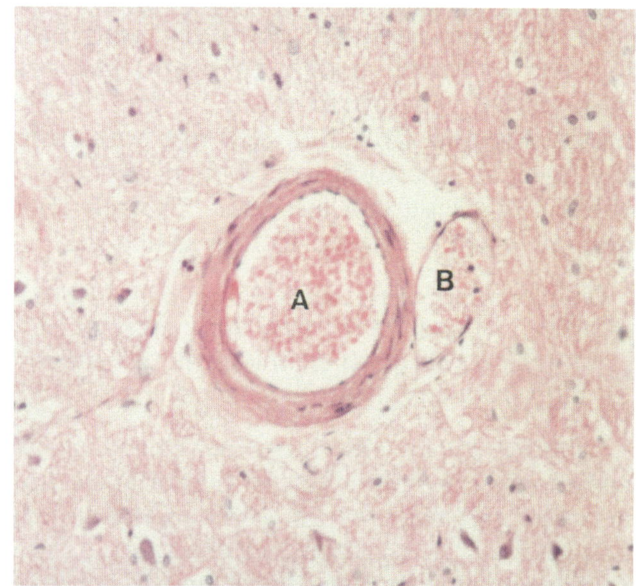

Figure 19.6, H&E x 260

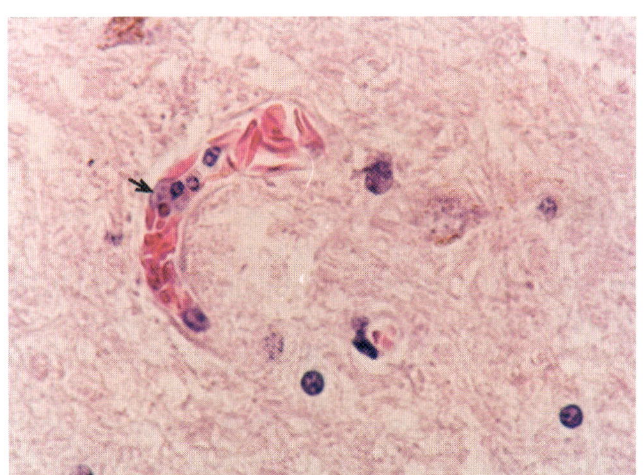

Figure 19.7, H&E x 520

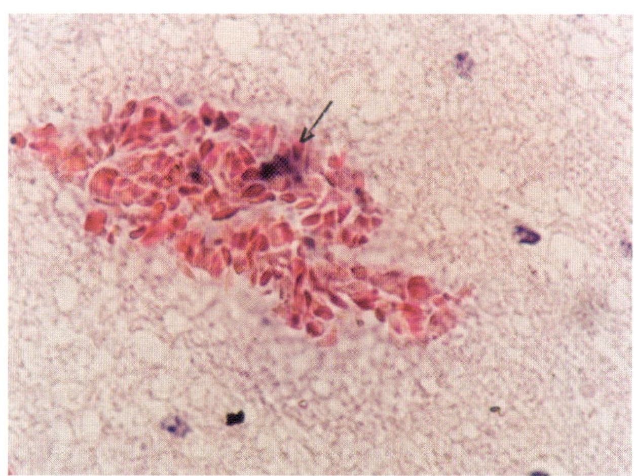

Figure 19.8, H&E x 520

diluted brain tissue and only a few developing red cells. Note that the developing red cells are less hemoglobinized and fewer in number than the red cells in figures 19.1-19.4. This is probably due to the patient's slow death and less urgent need for oxygen. H&E x 260

Direct development of sickle cells from brain cells (figs. 19.7 and 19.8)

Figure 19.7. A developing blood capillary is filled with sickle cells. The arrow points to a nucleated cell of brain tissue origin transforming into SN cells. The developmental process is continuing at the two ends of the capillary. H&E x 520

Figure 19.8. A large group of sickle cells is developing from disappearing brain tissue. The arrow points to remnants of nucleated brain cells. This irregular area of brain tissue replaced with blood is usually referred to as a hemorrhage. H&E x 520

Chapter 20

EYE [Retina and Choroid] (figs. 20.1-20.4)

Hemorrhagic lesions of the retina and choroid are one of the common causes of loss of vision in elderly patients. The eye is made up of several distinct layers; however, the hemorrhagic lesions occur mostly in the choroid and the plexiform layers of the retina. Described below are blood and blood capillary development from the choroid and the plexiform layers of the retina. Figures 20.1-20.4 are from a surgically removed eyeball from a clinically blind patient with painful phthisis bulbi.

Development of blood and blood capillaries from choroid layer (figs. 20.1 and 20.2); and from the plexiform layer of the retina (figs. 20.3 and 20.4)

Figure 20.1. Pictured here in low-magnification is a retina widely detached from a prominently vascularized choroid. A higher magnification of the middle part of the choroid is shown in the next figure. H&E x 130

Figure 20.2. The origin of red cells and vascular channels is preceded by hyalinization of columns of local substances. Here, different stages of blood and blood capillary development are shown beginning with light hyalinization in area (A), and following with the development of blood capillaries (B, C and finally D). Even in the most fully developed capillary (D), the red cells have not completely rounded and the background choroid substance has not completely dissolved. H&E x 520

Figure 20.3. A developing blood capillary is arising from dissolving retinal tissue of the plexiform layer. The red cells in the lumen are not separated from each other, and the endothelium is only beginning to develop. H&E x 520

Figure 20.4. A developing blood capillary is forming from dissolving local tissue of the plexiform layer of the retina. The majority of these red cells are still clumped together. In the upper-left-inside area of this blood vessel, locally formed eosinophils are also participating in the development of red cells. H&E x 520

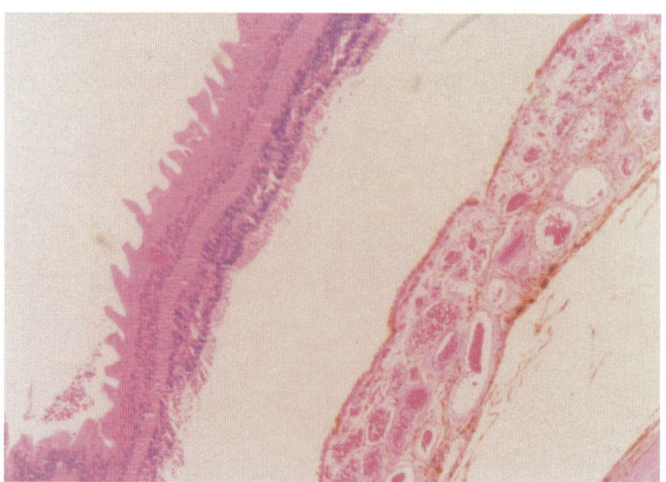

Figure 20.1, H&E x 130

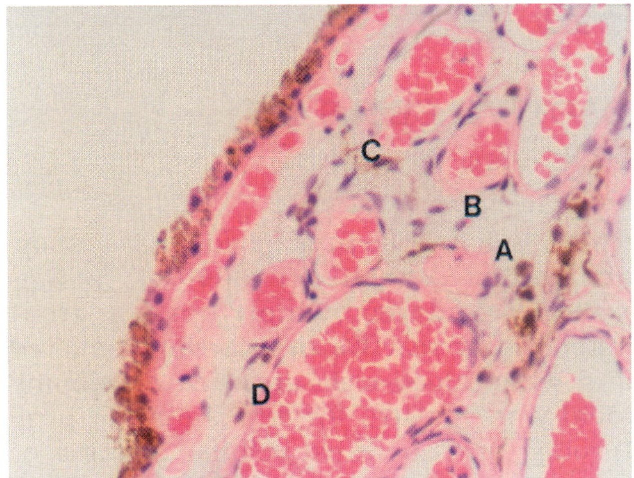

Figure 20.2, H&E x 520

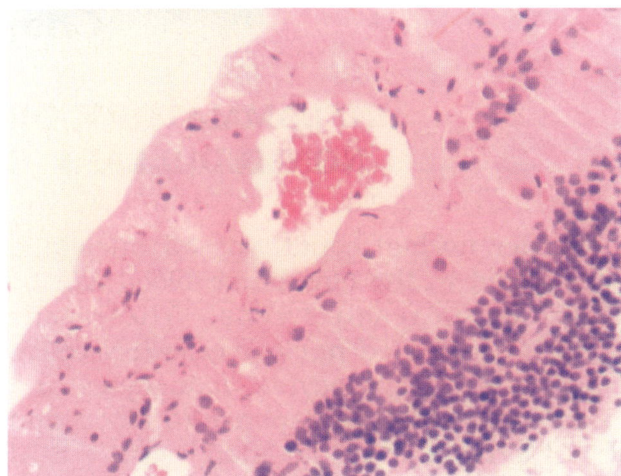

Figure 20.3, H&E x 520

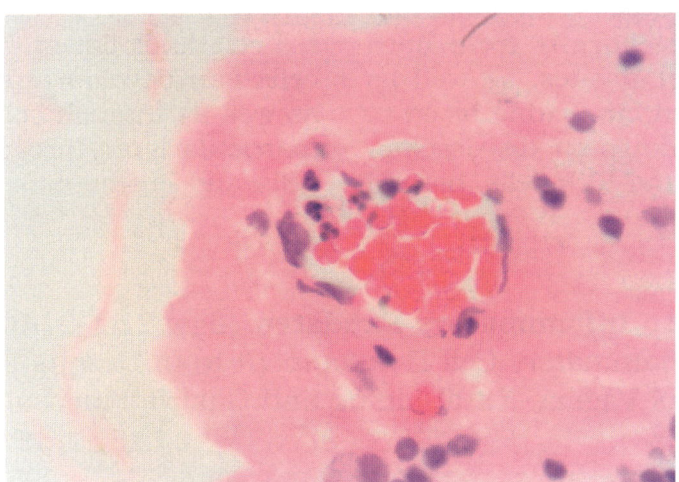

Figure 20.4, H&E x 520

Chapter 21

INFLAMMATORY OR REACTIVE CELLS OF NONMEDULLARY ORIGIN (figs. 21.1-21.34)

Under normal conditions all nucleated cells found in the circulating blood can be classified as reactive cells. These cells include segmented nuclear cells (SN cells), eosinophils, lymphocytes, monocytes and endothelial cells (in response to blood and vessel formation). In addition, reactive cells that are found primarily in tissues include plasma cells and Redefined Vasoformative Cells of Ranvier (RVR cells or *fish-head* cells, the latter name is given by this author because of its appearance under the microscope). Similar to red cells, inflammatory or reactive cells arise locally from both benign and malignant tissues, as well as from bone marrow and circulating blood.[1]

Important new findings on the development of plasma cells and RVR cells, both of which mainly arise in nonmedullary tissues, are presented in this chapter. In inflammatory exudates, which have a great potential for reactive cell development, red and white blood cells are seen developing locally in mature form without passing through any nucleated cellular changes or through any mitosis. This is because inflammatory or reactive cells can develop and transform into other types of reactive cells, as well as into red cells. In the latter case, reactive cells mainly function as intermediary cells between local tissues and red cell development. Whole columns of red cells can develop directly from local cells, as well as through columns of reactive cells (particularly erythrogenic SN cells, plasma cells and eosinophils).

1. Inflammatory Exudates (figs. 21.1-21.9)
(Red Cell and Inflammatory Cell Development from Desquamated Epithelium of Vagina, Nasal Lining and Gastric Mucosa)

The vaginal smear examination, commonly known as a *Pap smear*, has been a common practice for many years to rule out cervical cancer. The presence of inflammatory cells, mainly segmented nuclear cells (SN cells), and red blood cells is the most common finding. It has always been assumed that these cells are coming from cervical or vaginal lesions. Evidence is presented here that shedded vaginal epithelial cells, as well as other exudates, have the capacity to be transformed into such inflammatory cells, as well as red blood cells. Similar findings are also presented in nasal lining cells and gastric lining mucosal cells. Red cell development from desquamating epithelial cells of the gall bladder wall is demonstrated in figure 13.9 in Chapter 13.

Figures 21.1 to 21.6 are taken from a Pap smear, negative for cancer, from a 37-year-old female. Figure 21.7 is from a Pap smear from a 42-year-old female with a trichomonas infection. Figure 21.8 is from a nasal smear taken from a 31-year-

---

1. Numerous examples of red cells and inflammatory cells developing from nonmedullary (extramedullary) tissues are given in Volumes I, II and III (McDonald, 1989, 1995 and 2001), and throughout this volume.

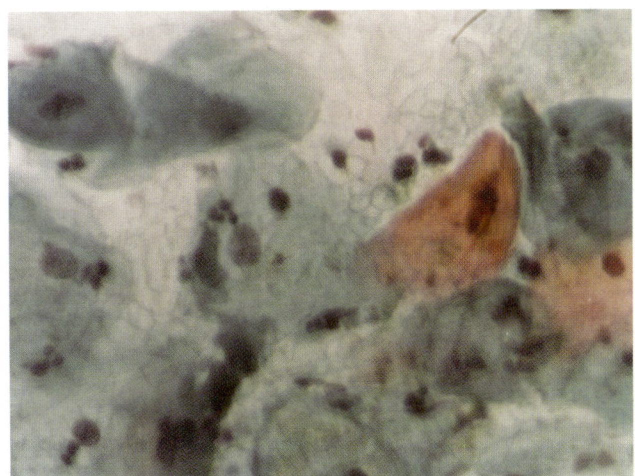

Figure 21.1, Pap Stain x 520

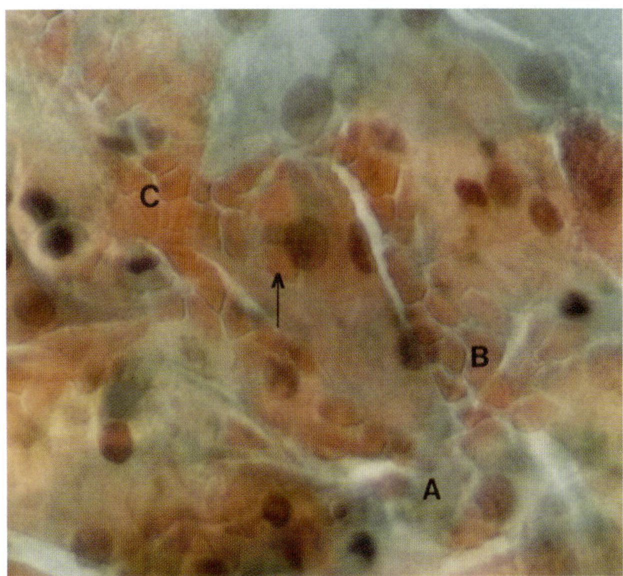

Figure 21.2, Pap Stain x 800

old female with allergy symptoms. And figure 21.9 is from a surgical excision of a distal segment of stomach, with a large deep ulcer, taken from a 72-year-old male.

In desquamated vaginal epithelium red cell formation through vacuolated mesh (figs. 21.1 and 21.6)

Figure 21.1. Vacuolated mesh (net-like structure consisting of red cell sized vacuoles) is developing from and replacing desquamated vaginal epithelial cells and the acellular material. This vacuolated mesh develops red cells similar to the area seen in figure 21.6. Pap Stain x 520

Granular changes towards development of red cells from vaginal epithelial cells and intercellular substance (fig. 21.2)

Figure 21.2. This vaginal smear shows granular changes towards the development of red cells in both the vaginal epithelial cells and the intercellular substance. This development starts as fine gray-green, clumpy granular changes (A), proceeding to greenish-orange granules (B), and then into what appears to be compressed orange-red granules (C). The next stage will possibly be compressed red cells. A few red cell sized orange-red globules (arrow) are forming within the epithelial cells. Pap Stain x 800

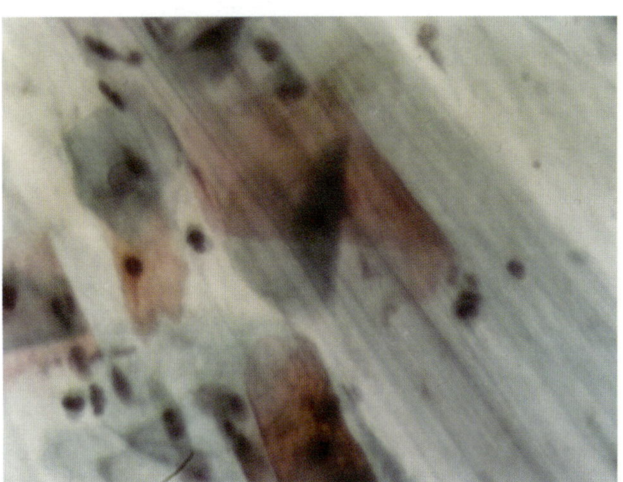

Figure 21.3, Pap Stain x 520

Stages of red cell development from linearly locally formed SN cells in vaginal exudate (figs. 21.3-21.6)

Figures 21.3-21.6. These figures show the stages of red cell development from linearly locally-formed SN cells and vaginal

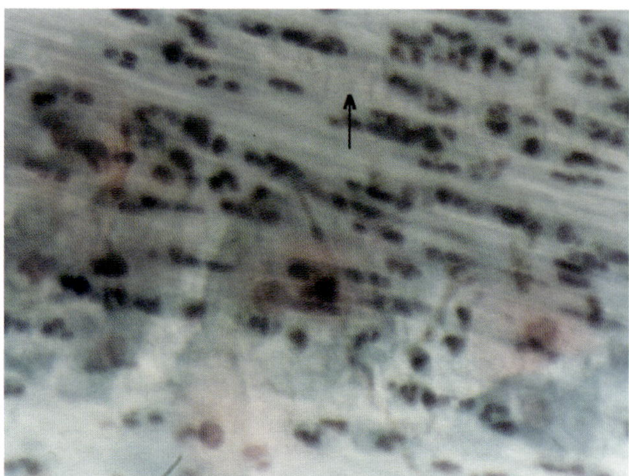

Figure 21.4, Pap Stain x 520

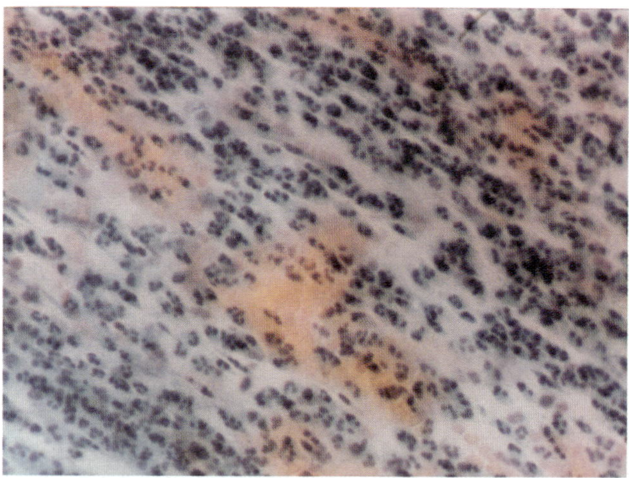

Figure 21.5, Pap Stain x 260

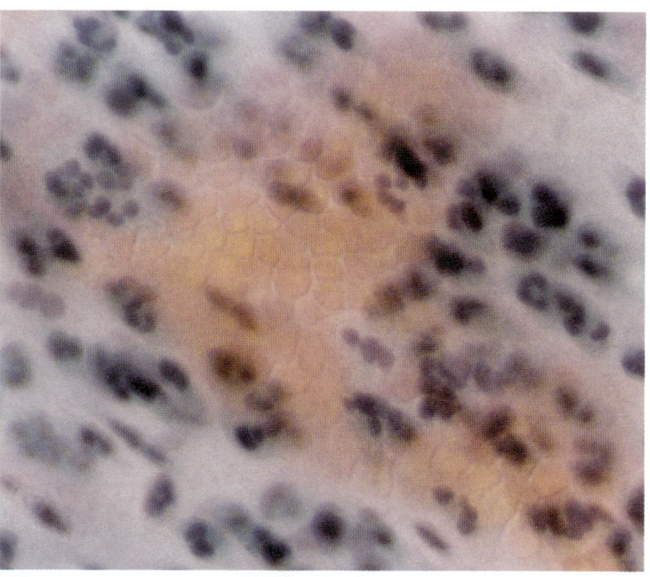

Figure 21.6, Pap Stain x 800

exudate. In figure 21.3, the beginning of development of SN cells starts with fine, long, straight, closely placed, parallel lines passing through the vaginal content. In figure 21.4, the subsequent development of SN cells along these parallel lines is demonstrated. Also, there is the development of uniform, parallel, and linearly placed (red cell sized) vacuoles from the acellular background substance (arrow). In figures 21.5 and 21.6, both the linearly placed SN cells and the intercellular substance are changing into sheets of red cell sized globules. (Figure 21.6 is a magnification of the lower-middle area of figure 21.5.) Figures 21.3 and 21.4 Pap Stain x 520; figure 21.5 Pap Stain x 260; and figure 21.6 Pap Stain x 800

Development of SN cells from vaginal epithelium (fig. 21.7)

Figure 21.7. The development of many SN cells, in sheet form, from dissolving vaginal epithelial cells is shown in this Pap smear. In area (A), the SN cells are in the process of developing and are not yet isolated from epithelial cell substance. Pap Stain x 400

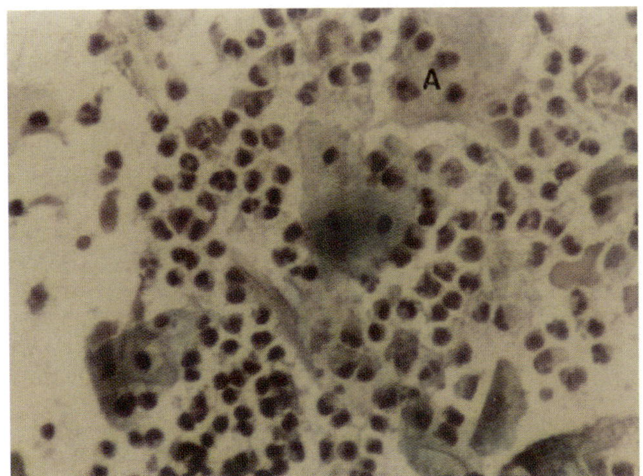

Figure 21.7, Pap Stain x 400

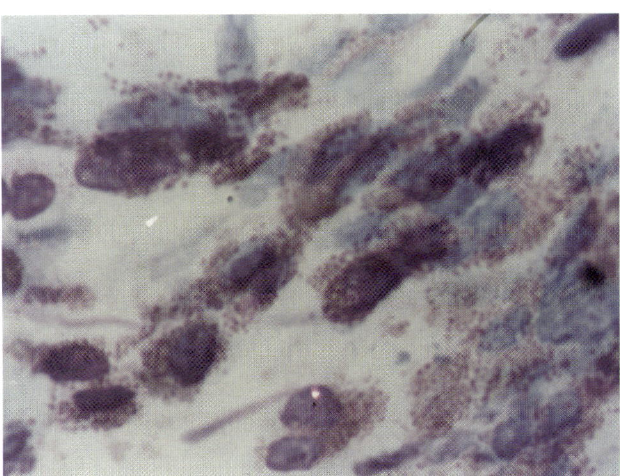

Figure 21.8, Wright x 1300

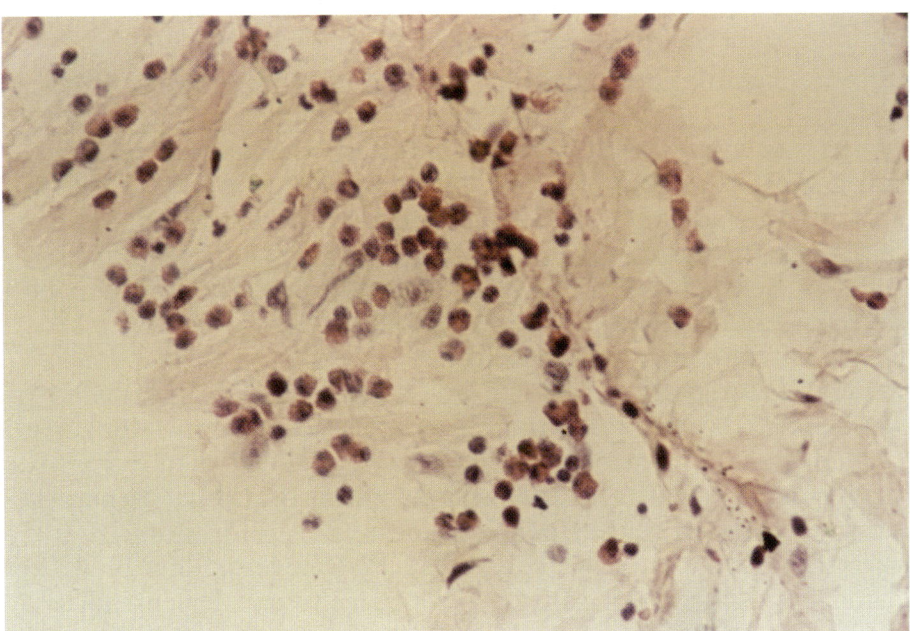

Figure 21.9, H&E x 400

Origin of eosinophils from nasal epithelium (fig. 21.8); and from gastric lining mucosa (fig. 21.9)

Figures 21.8. The origin of eosinophils from desquamated epithelial cells, or those on the verge of desquamation, is shown in this figure and the next figure. In this figure, a large number of eosinophilic granules are present in the nasal epithelial cells and are involved in the development of eosinophils. Many epithelial cell nuclei have not yet transformed into segmented or multilobed nuclei. Wright x 1300

Figure 21.9. In this gastric mucosa, from a surgically excised stomach, many eosinophils are developing in the intact lining as well as in the shedded epithelial cells. H&E x 400

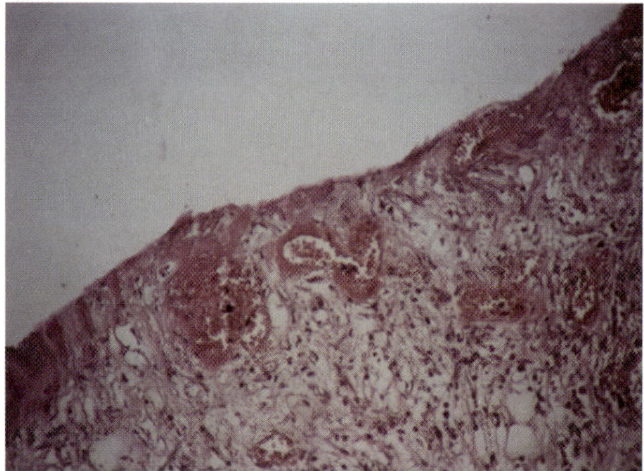

Figure 21.10, H&E x 130

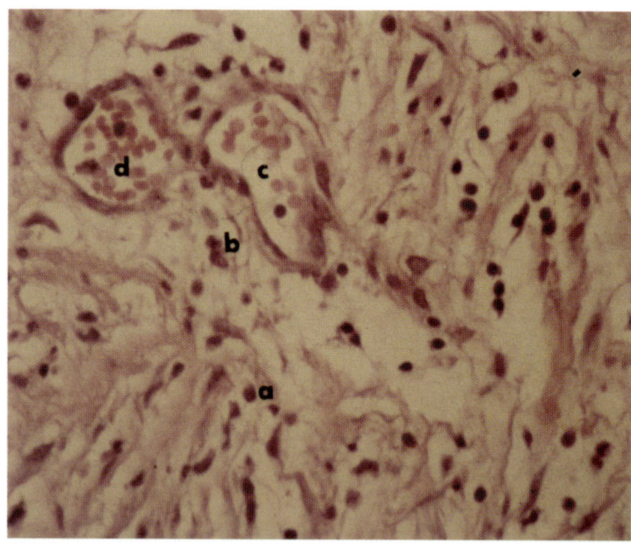

Figure 21.11, H&E x 400

2. Granulation Tissue (figs. 21.10-21.17)
(Blood Formation in Granulation Tissue Consisting of Inflammatory Cells of Local Origin)

In this study of granulation tissue formed during the healing process of skin lesions, with destruction of the epidermis and immediate sub-epidermis tissue, the development of red cells as hemoglobin globules within formative vascular channels is seen. Also, scatterings of inflammatory cells arising locally and with the capacity to form red cells are demonstrated. The inflammatory cells (such as segmented nuclear cells, lymphocytes, and plasma cells) tend to be very active in the production of red cells. Often one can see the formation of red cells in groups from plasma cells (grape cells).

Figures 21.10-21.15 are from a skin lesion (pyogenic granuloma) surgically excised from the back of the neck of an 82-year-old female. Figures 21.16 and 21.17 are from mucocutaneous tissue of the lower lip of a 64-year-old male with baso-squamous cell carcinoma.

Development of red cells and blood capillaries from inflammatory (reactive) cells of local origin (figs. 21.10-21.17)

Figure 21.10. This figure is a low magnification of the superficially ulcerated lesion, with exudates and granulation tissue formation, discussed in figures 21.11-21.15. Collections of red cells are seen close to the surface, and some of these collections are within developing capillaries. Many inflammatory (reactive) cells can be seen throughout this figure. H&E x 130

Figure 21.11. Scattered throughout this edematous, inflamed tissue there are many tiny nuclei with or without definite cytoplasm (a) and (b). These cells can be better seen in higher magnification in figures 21.12-21.15. A few of these locally-developed cells can be categorized as developing segmented nuclear (SN) cells. Irregular and narrow elongated nuclear structures with basophilic pre-collagenous fibrous substances appear in several areas. The local origin of the developing blood capillary (c) is evidenced by the presence of remnants of this liquefying fibrillary substance lying in the vascular lumen. The red cells are developing from the liquefying fibrillary substance. The upper left part of this capillary is formed by local cells which have not yet developed into endothelium. Capillary (d) is further developed than capillary (c); the fibrillary substance is already dissolved and more red cells are present. H&E x 400

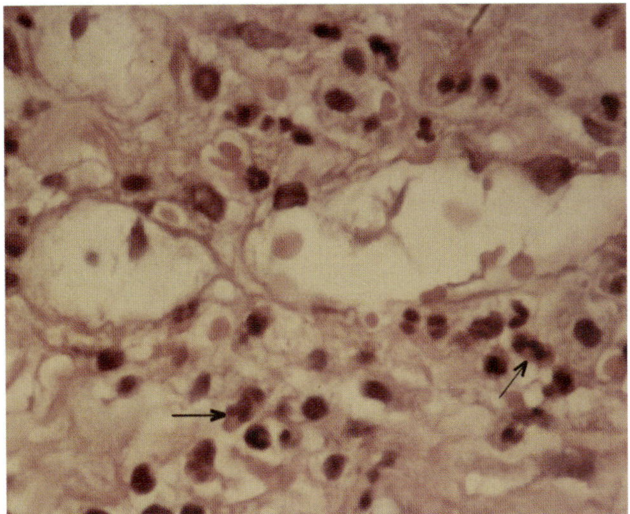

Figure 21.12, H&E x 800

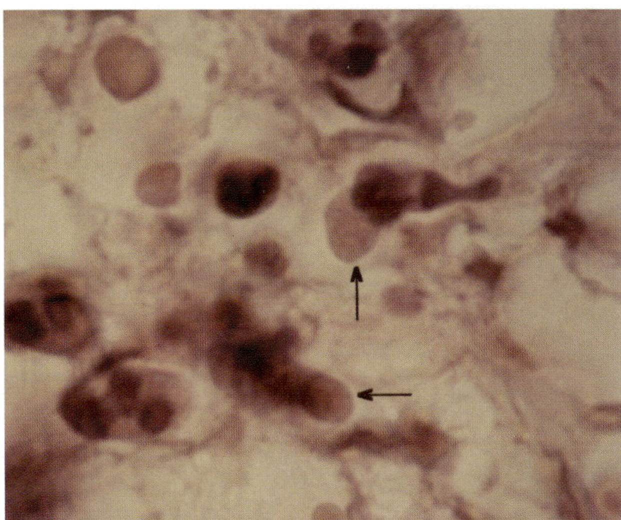

Figure 21.13, H&E x 2000

*Red cell development from SN cells
(figs. 21.12-21.15)*

Figure 21.12. At a slightly higher magnification than figure 21.11, the development of red cells from SN cells (arrows) can be seen. Development of red cells from SN cells is better viewed in figures 21.13-21.15. Note that there are many inflammatory cells, mainly developing SN cells and some cells which cannot be classified, at different stages of formation. The vascular channels in this figure are being developed by the liquefaction of local tissue. The endothelium has not yet developed in these vascular channels. H&E x 800

Figure 21.13. The origin of red cells as hemoglobin globules directly from SN cells (arrows) is seen at a higher magnification. H&E x 2000

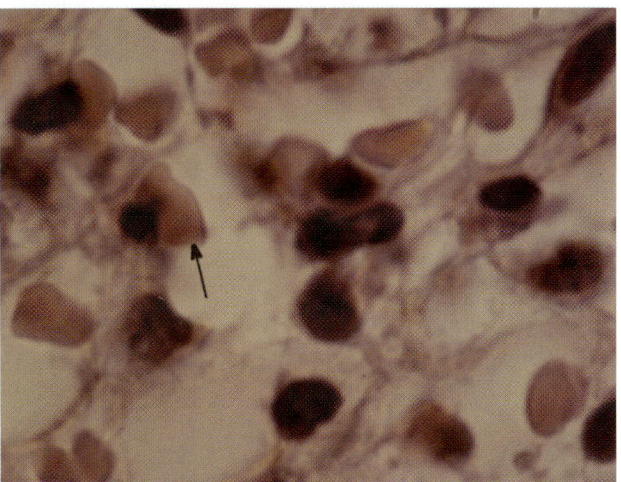

Figure 21.14, H&E x 2000

Figure 21.14. The development of a red cell from a segmented nuclear cell (arrow) is seen in this figure. This process is shown in more detail in SN cells in figure 21.15. H&E x 2000

Figure 21.15. These SN cells (arrows) are developing multiple red blood cells starting as clear globules in the SN cell on the left. The transformation of such clear globules into red cells is suggested in the SN cell on the right.[2] H&E x 2000

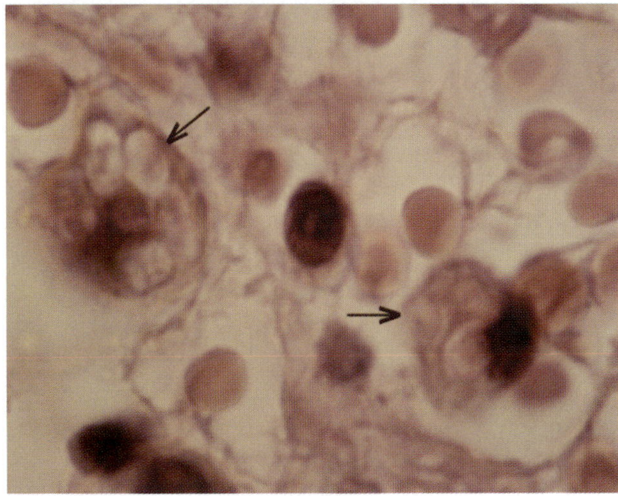

Figure 21.15, H&E x 2000

2. A similar process is described in renal carcinoma in figures 100 and 101 of Volume I (McDonald, 1989).

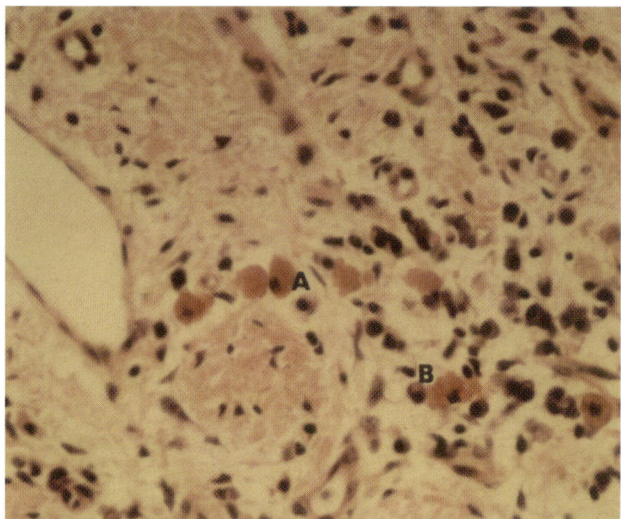

Figure 21.16, H&E x 400

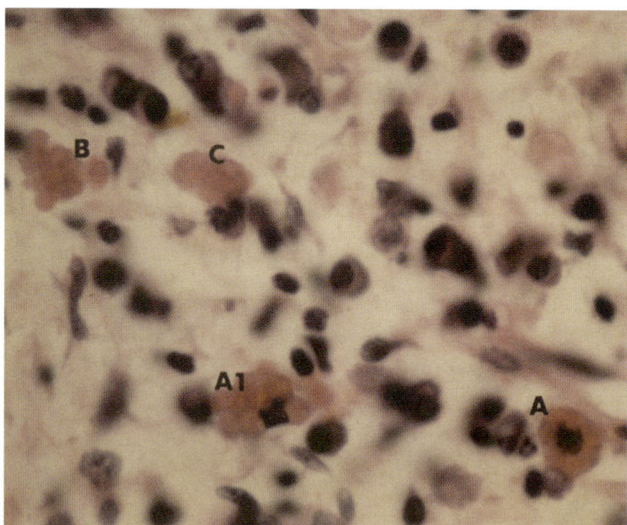

Figure 21.17, H&E x 800

Grape cells (Russell body cells) developing red cells (figs. 21.16 and 21.17)

Figure 21.16. In this figure, multiple grape cells (Russell body cells) with developing red cells are present in a line (A). There is a definite possibility that a vascular channel enclosing these grape cells (with developing hemoglobin globules) will form and subsequently communicate with the already existing channel on the left. The lower right area (B) is shown in higher magnification in the next figure. H&E x 400

Figure 21.17. This figure is a higher magnification of the mid-lower right area of figure 21.16. Multiple plasma cells, lymphocytes, and grape cells (A-C) with developing red cells are better viewed at this magnification. The red cells in grape cell (A1) are more developed than those in (A). In the two grape cells (B) and (C) the nuclei have already dissolved, or are not in view, and the red cells will become part of the formative vascular channel discussed in figure 21.16. The grape cell (C) is composed of a clump of tiny developing red cells. H&E x 800

3. Redefined Vasoformative Cells of Ranvier (RVR cells), or *Fish-Head* Cells, as an Alternate Source for Development of Red Cells, White Blood Cells and Blood Capillaries in Nonmedullary Tissues
(figs. 21.18-21.32)

Both Schwann (1847) and Wedl (1853) described the developmental processes of stellate cells (large cells with cytoplasmic projections in different directions) forming a network of capillaries and producing red cells and plasma in the lumen. In 1874, Ranvier observed distinctive cells participating in the development of independent blood capillaries in the greater omentum of rabbits. Since a variety of other cells, in addition to red cells and endothelial cells (as observed by Ranvier), are found to be generated by these vasoformative cells, I have taken the liberty of renaming these cells as *Redefined Vasoformative Cells of Ranvier* (RVR cells). Because these cells, in a certain posture, have a fish-head-like appearance when stained with Giemsa (see figure 21.22) I am tempted to also name them *"fish-head"* cells.[3]

3.	My findings on RVR cells were originally demonstrated in Volume III, figures 14.11-14.24 (McDonald, 2001).

These vasoformative cells appear as large cells with basophilic angular or dendritic cytoplasm arising from commonly seen background mixoid substance. The nucleus is very pale and usually contains one tiny, hyperchromatic nucleolus. In the process of development, red cells arise (often in columns) from the central part of these vasoformative cells, and the endothelium develops from the peripheral parts. These newly-formed capillaries then join other developing capillaries, or pre-existing ones, by anastomosis. I have also observed such cells in edematous inflamed tissues. Examples are given below from the subserosa of the gastric wall and edematous, inflamed labia majora. (See also figures 3.10 and 3.11 in Chapter 3, Adipose Tissue, for examples of RVR cells in adipose tissue.)

In addition to Ranvier's findings, the multipotential quality of these cells is demonstrated by the development of not only red cells and endothelium in the development of blood capillaries, but also by the development of a large number of SN cells, plasma cells, lymphocytes, eosinophils and fibroblasts. Furthermore, red cells can develop from these second generation inflammatory or reactive cells. These vasoformative cells can also take part in the development of fibroblasts in formation of fibrosis. In some instances, these cells play a large role in the healing processes of edematous, inflammatory conditions. In acute and chronic inflammatory conditions RVR cells often play an important role in the proliferation of SN cells and other reactive cells.

Figures 21.18-21.22 are taken from the edematous subserosa of a surgically resected stomach with a chronic gastric ulcer. Figures 21.23-21.32 are taken from edematous, inflamed labia majora with superficial ulceration. Figures 21.18 and 21.22 are reprinted from Volume III (McDonald, 2001).

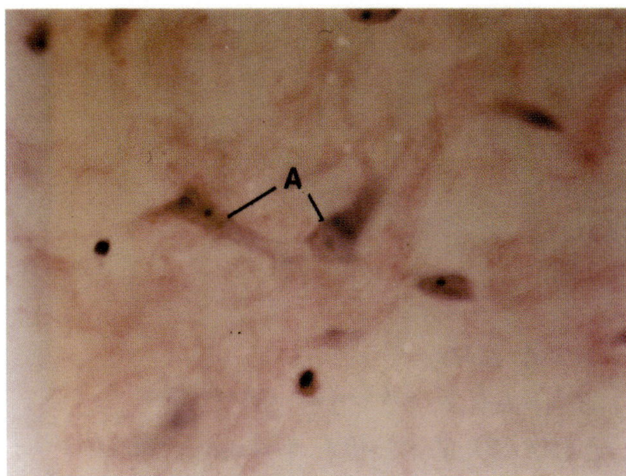

Figure 21.18, Giemsa x 520

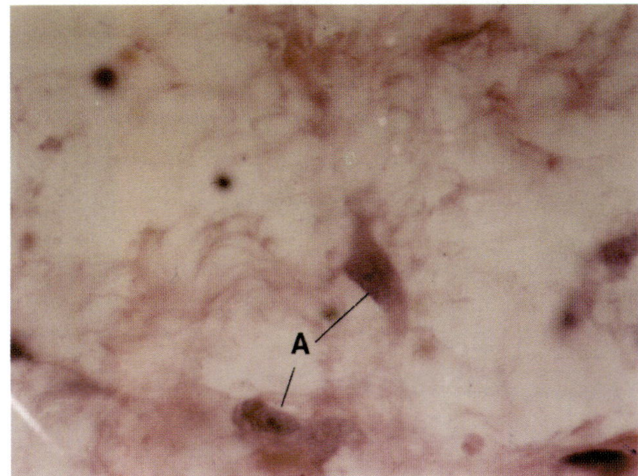

Figure 21.19, H&E x 520

Presence of RVR cells in edematous subserosa of stomach (figs. 21.18-21.22)

Figure 21.18. RVR cells (A) are scattered throughout this edematous gastric serosa. Giemsa x 520

Figure 21.19. This is a photomicrograph of RVR cells (A) taken from a hematoxylin and eosin stained section of the same specimen as seen in figures 21.18-21.22. Because RVR cell nuclei do not stain as distinctly in H&E stain, most of the following figures are from Giemsa stained slides. H&E x 520

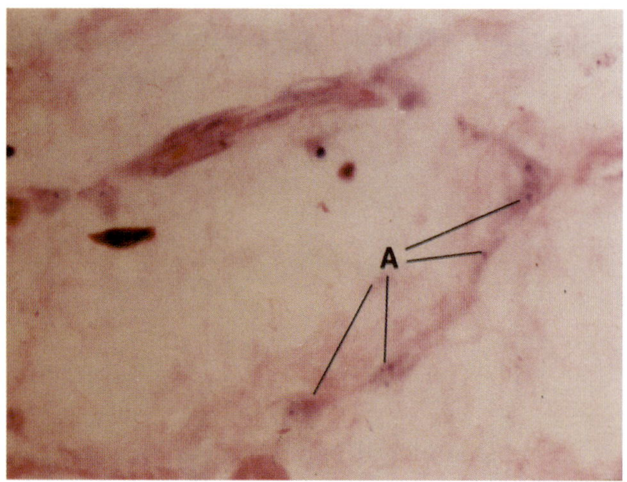

Figure 21.20, Giemsa x 400

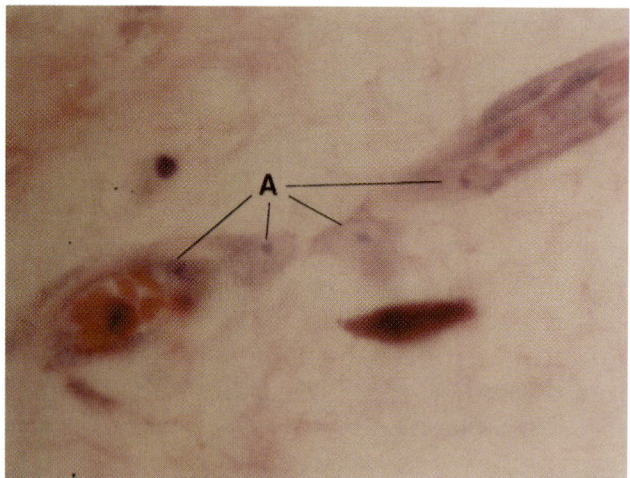

Figure 21.21, Giemsa x 800

Capillary formation and red cell development from RVR cells (figs. 21.20-21.22)

Figure 21.20. In the edematous gastric serosa RVR cells are often found in the development of a network of capillary branches. In this Y-shaped, anastomosing, developing capillary the lower branch is less developed and consists of a single line of RVR cells (A). The upper branch is shown in higher magnification in the next figure. Giemsa x 400

Figure 21.21. On the right, in the capillary lumen there are two red cells developed from RVR cells. In the center, the RVR cells (A) have a beautiful "fish-head" appearance. The capillary on the left is shown in higher magnification in the next figure. Giemsa x 800

Figure 21.22. This figure demonstrates the erythrogenic property of RVR cells and their capacity to develop the endothelial lining of capillaries from their peripheral cytoplasm. It is interesting to note the peculiar fish-head-like appearance of these RVR cells with its nucleus showing pale nucleoplasm and a prominent, tiny, dark nucleolus appearing as the pupil of the eye. In this figure there are two examples of this formation. Red cells and blood plasma arise from the central portion of these cells. Note the development of red cells and an

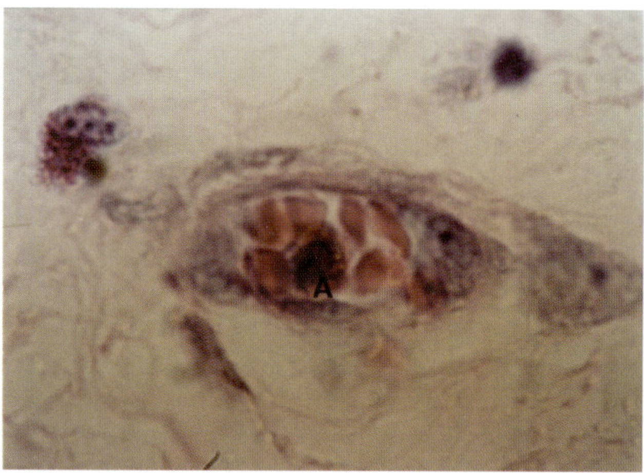

Figure 21.22, Giemsa x 1300

eosinophil (A) from the cytoplasm of RVR cells in the developing capillary lumen. Giemsa x 1300

Presence of RVR cells in edematous labia majora granulation tissue (figs. 21.23-21.32)

Figure 21.23. On the left, scattered in this edematous labia majora granulation tissue there are several prominent individual RVR cells. This stage of RVR cell is commonly and mistakenly classified as a fibroblast. On the right, a group of RVR cells are arranged in a whorl pattern, possibly in preparation for blood vessel development. Similar whorl patterns associated with blood and blood vessel formation are seen in figures 21.24 and 21.25. Giemsa x 800

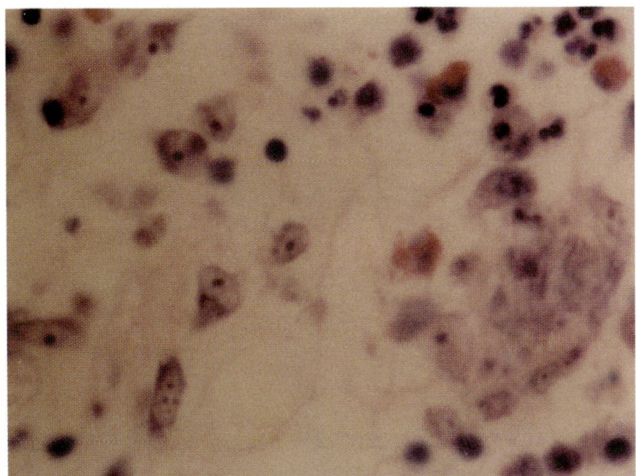

Figure 21.23, Giemsa x 800

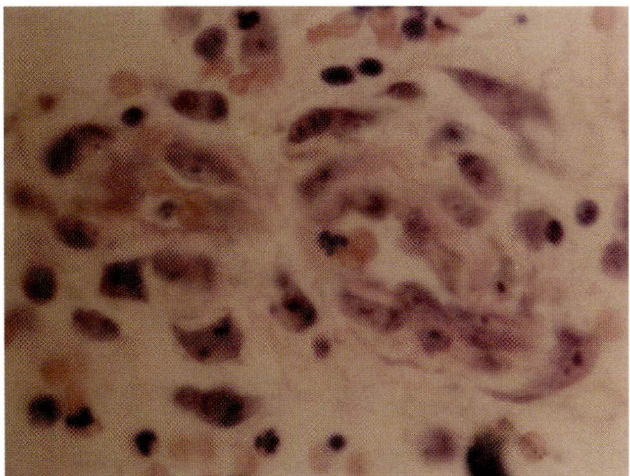

Figure 21.24, Giemsa x 800

*Blood and blood vessel formation from RVR
cells (figs. 21.24 and 21.25)*

Figure 21.24. Two groups of RVR cells
are arranged in whorl patterns in formation of developing blood vessels. In the
lumen of the developing blood vessel on the
left, there are three or four developing red
cells that have not separated and a
nucleated cell. In the lumen of the developing vessel on the right, there is an SN cell
and one red blood cell present. Note that
RVR cells make up the walls of both of these
blood vessels. The inner layer of RVR cells
will eventually transform into endothelial
cells. Giemsa x 800

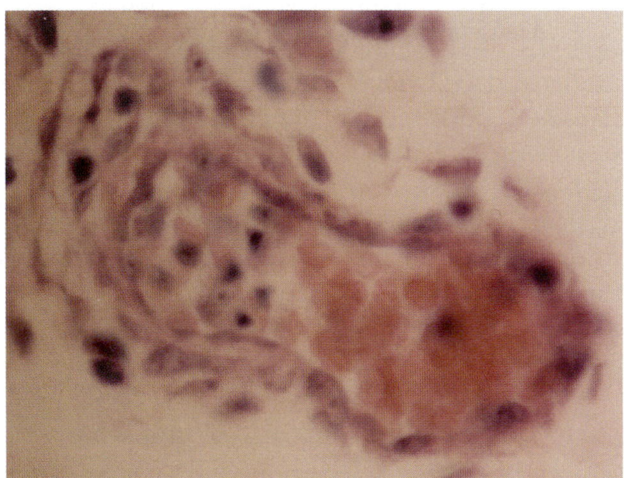

Figure 21.25, Giemsa x 800

Figure 21.25. This is a more organized
blood vessel developed from RVR cells. The
RVR cells on the left are participating in
the widening of this blood vessel to the left.
The nuclei of the RVR cells within the
lumen will eventually dissolve in development of hemoglobin globules (red cells).
Giemsa x 800

*Hemoglobin globule development from RVR
cells (figs. 21.26, 21.27 and 21.30)*

Figures 21.26 and 21.27. Several RVR
cells in these two photomicrographs are
forming hemoglobin globules starting as
multiple red cell sized vacuoles, followed by

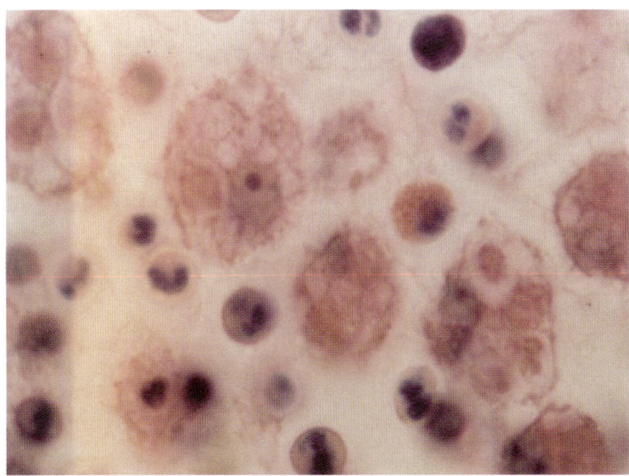

Figure 21.26, Giemsa x 1300

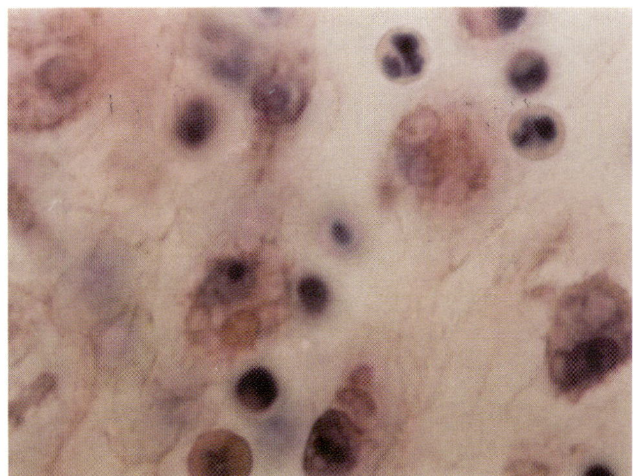

Figure 21.27, Giemsa x 1300

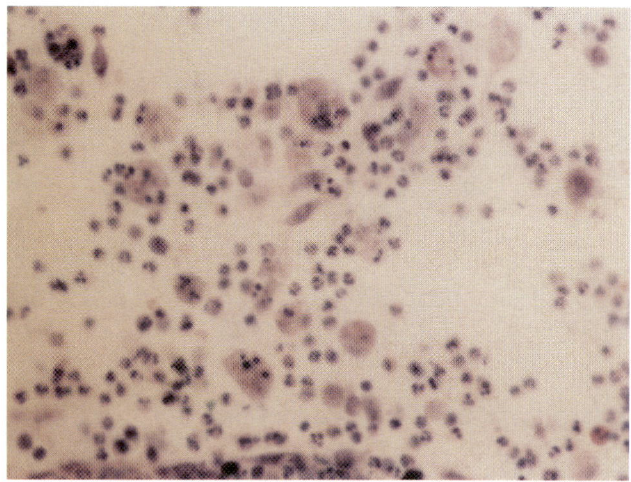

Figure 21.28, Giemsa x 400

hemoglobin globule formation. The development of these red cells is somewhat similar to red cell development from grape cells (Russell body cells), except, in grape cells the hemoglobin globules start out as tiny globules. Also note that these cells have a typical RVR nucleus (fish-head appearance) and grape cells have a plasma cell nucleus (see figures 8.6, 21.16 and 21.17). Giemsa x 1300

Development of SN cells (figs. 21.28-21.30)
and eosinophils (figs. 21.31 and 21.32)
from RVR cells

Figure 21.28. Several RVR (mother) cells are developing a multiple number of nucleated cells, mainly SN cells. A similar RVR cell can be seen in the next figure in higher magnification. Giemsa x 400

Figure 21.29. An RVR cell (A) is developing a multiple number of tiny SN cells. Note that the RVR nucleus is still present in the center of this cell. The release of these tiny SN cells from the mother cell into the surrounding media (B) is suggested. In the upper middle of this figure, there are also several eosinophils probably formed from RVR cells (see figures 21.31 and 21.32). Giemsa x 800

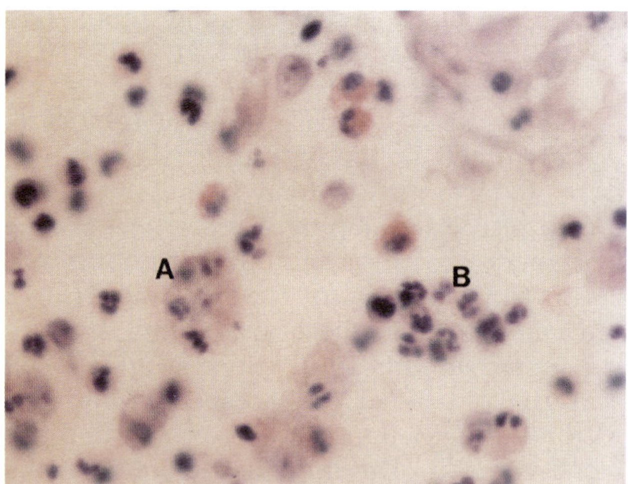

Figure 21.29, Giemsa x 800

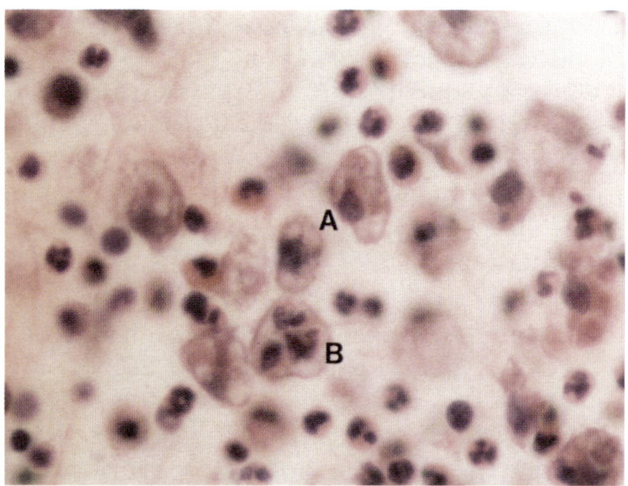

Figure 21.30, H&E x 800

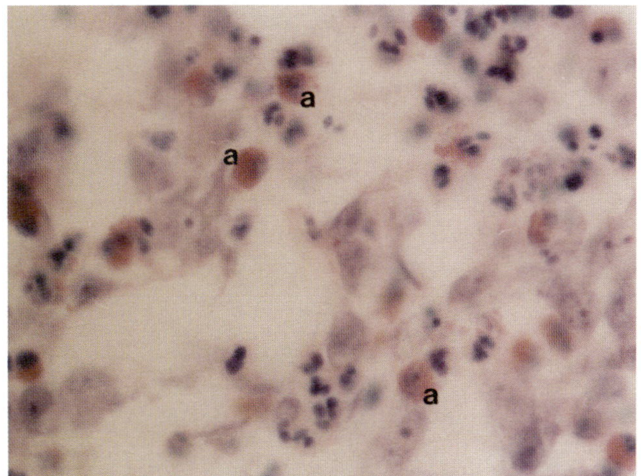

Figure 21.31, Giemsa x 800

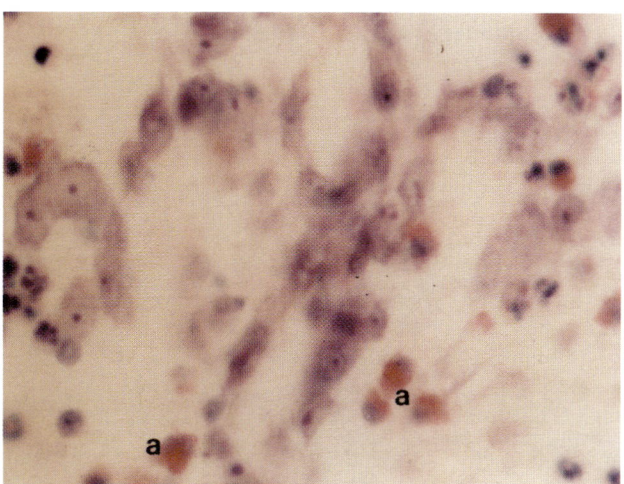

Figure 21.32, Giemsa x 800

Figure 21.30. This is a photomicrograph of RVR cells with vacuole formation (A) and an RVR cell developing SN cells (B) taken from a hematoxylin and eosin stained section of the same specimen as seen in figures 21.26-21.29. Note that these RVR cells are somewhat distorted and their nuclei do not stain as distinctly in H&E stain. H&E x 800

Figure 21.31. RVR cells are forming many eosinophils. The RVR nucleus can still be identified in eosinophils (a). Giemsa x 800

Figure 21.32. On the left, there is a comma shaped line of RVR cells. Many widely scattered eosinophils are developing from RVR cells. The RVR nucleus can still be identified in eosinophils (a). Giemsa x 800

4. Example of Vascularization of a Pericolic Abscess (figs. 21.33-21.34)

Described below is an example of a rarely studied form of blood and blood vessel formation from inflammatory cells of a necrotizing, ulcerative lesion of the colon. The inflammatory cells of this lesion appear to be a heavy concentration of poorly differentiated plasma cells. As usual, these poorly differentiated plasma cells are locally developed without mitosis.

Usually, red cells develop from plasma cells as round regular sized hemoglobin globules (as demonstrated in figures 8.6, 21.16 and 21.17). However, the poorly differentiated plasma cells described below appear to be breaking down into tiny chromatin particles, from which arise irregular clumped hyperhemoglobinized globules (red cells).

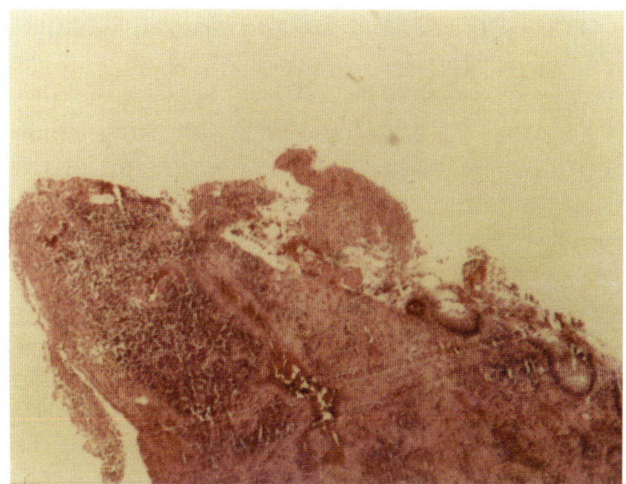

Figure 21.33, H&E x 52

Figures 21.33 and 21.34 are taken from a section of colon surgically removed from a 59-year-old female with advanced,

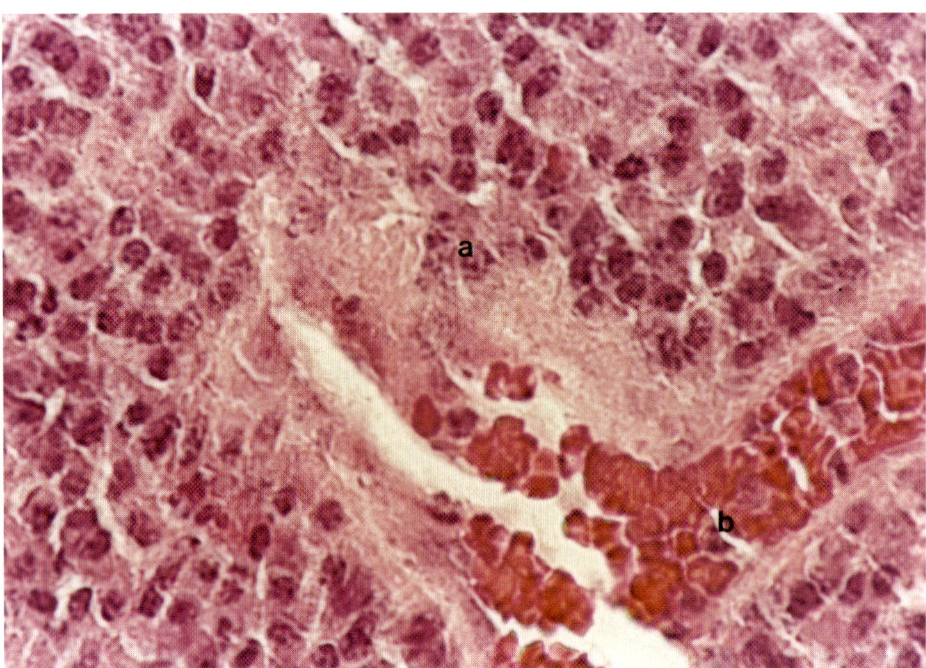

Figure 21.34, H&E x 800

necrotizing, ulcerative colitis. The patient had been on steroid therapy for many years for treatment of idiopathic autoimmune hemolytic anemia.

Red blood cell and blood vessel development from the break down of poorly differentiated plasma cells of an inflammatory lesion (figs. 21.33 and 21.34)

Figure 21.33. This is a low-magnification of a necrotizing, ulcerative lesion of the colon. A higher magnification of this lesion is seen in the next figure. H&E x 52

Figure 21.34. In this area of the necrotizing, ulcerative colitis lesion there is a blood vessel developing from a heavy concentration of locally developed, poorly differentiated plasma cells. The plasma cells are breaking down into a scattering of tiny nuclear chromatin particles (a). Most of the plasma cell nuclei are composed of these chromatin particles instead of the usual smooth nuclear structures. These tiny chromatin particles together with the background substance give rise to the irregular hyperhemoglobinized compressed red cells seen in this developing blood vessel. The red cells are in a wide column surrounded by an irregular, poorly developed wall composed of the broken down plasma cell remains. Unlike the usual red cell development from plasma cells, these red cells are irregularly shaped and clumped together. The tiny nuclear chromatin particles can be seen still attached to the red cells well within this developing vascular column (b). This is a rarely studied form of vascularization of inflammatory (reactive) tissue. H&E x 800

PART TWO

MALIGNANT TUMORS

Chapters 22 - 36

Chapter 22

CANCERS OF THE SKIN (figs. 22.1-22.19)

1. Basal Cell Carcinoma (figs. 22.1-22.6)

Figures 22.1-22.4 are taken from a deeply invading basal cell carcinoma surgically removed from the right side of the face of an adult male. Figures 22.5 and 22.6 are from a specimen surgically excised from the dorsum of the left hand of an 84-year-old male. Cellular changes in benign skin tissues are presented in Chapter 2. Figure 22.6 is reprinted from Volume I (McDonald, 1989).[1]

Deep and widespread ulcerative basal cell carcinoma of the skin (fig. 22.1)

Figure 22.1. This is a low-magnification view of an advanced basal cell carcinoma of the skin showing extensive deep and widespread ulcerative invasion of the skin. Figures 22.2-22.4 are taken from this same tissue section. H&E x 52

Transformation of tumor cells into plasmacytoid cells (figs. 22.2-22.4)

Figure 22.2. Compact tumor cells (A) are transforming (transdifferentiation) into plasmacytoid cells (B). H&E x 400

Red cell development directly from basal cell carcinoma (figs. 22.3 and 22.5)

Figure 22.3. On the right of this figure, tumor cells are transforming into plasmacytoid cells. On the left, there are prominent spindle-shaped tumor cells with hyperchromatic elongated nuclei. In the center of this figure, faintly stained red cells

(A) are developing directly from the cancer cells. H&E x 520

Transformation of plasmacytoid cells into erythrogenic SN cells (fig. 22.4)

Figure 22.4. In area (A), plasmacytoid cells derived from basal cell carcinoma are transforming into SN cells. The erythrogenic capacity of these cells is demonstrated by SN cell (B). Groups of erythrogenic SN cells are developing red cells within formative vascular channels (1, 2 and 3). H&E x 800

Figure 22.5. Development of red cells directly from the tumor cells can be prominently seen at this higher magnification. H&E x 1300

Direct and rapid transformation of spindle-shaped cancer cells into red cells (fig. 22.6)

Figure 22.6. This figure demonstrates, rarely observed, complete hemoglobinization of the spindle-shaped basal cell carcinoma cells. This type of occurrence is probably related to the acute urgency for hemoglobinization where no time is spent acquiring the regular shape of red cells. At the peripheral part of this figure, developmental stages of round red cells with progressive hemoglobinization can also be seen. This occurrence probably took place from the time of surgery to the beginning of fixation in formalin. H&E x 520

---

1. Additional photomicrographs and findings in basal cell carcinoma are presented in Volume I, figures 57-69.

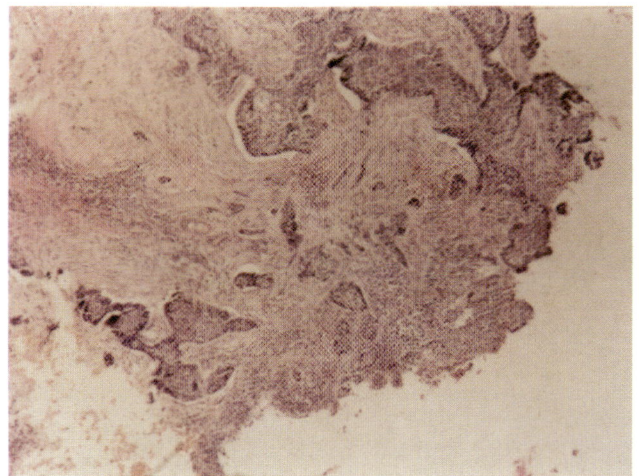

Figure 22.1, H&E x 52

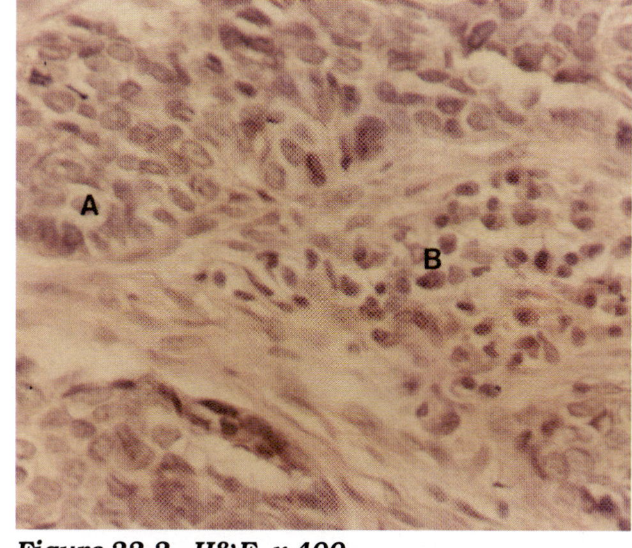

Figure 22.2, H&E x 400

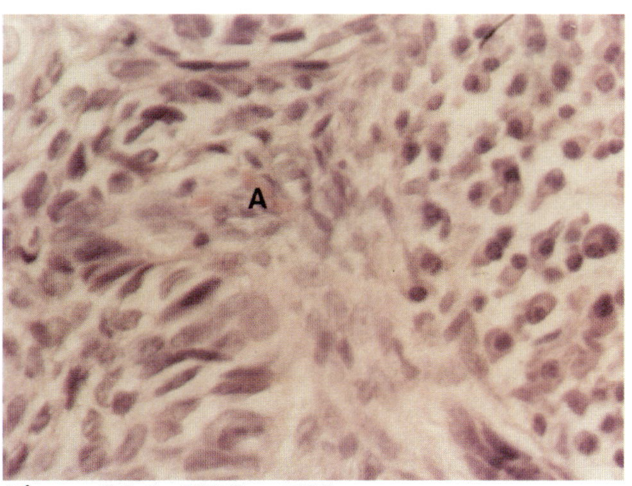

Figure 22.3, H&E x 520

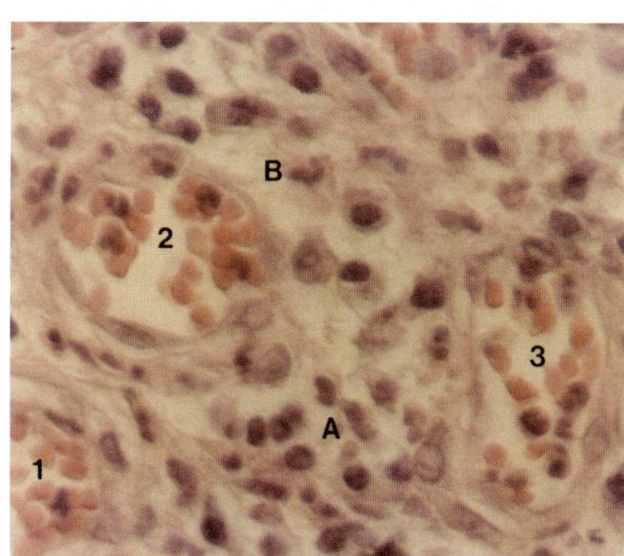

Figure 22.4, H&E x 800

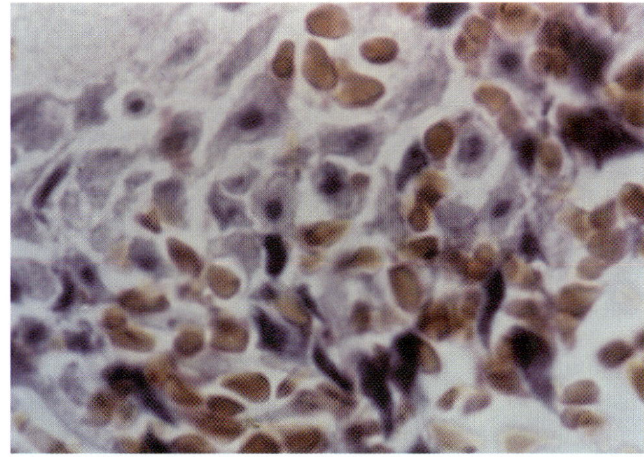

Figure 22.5, H&E x 1300

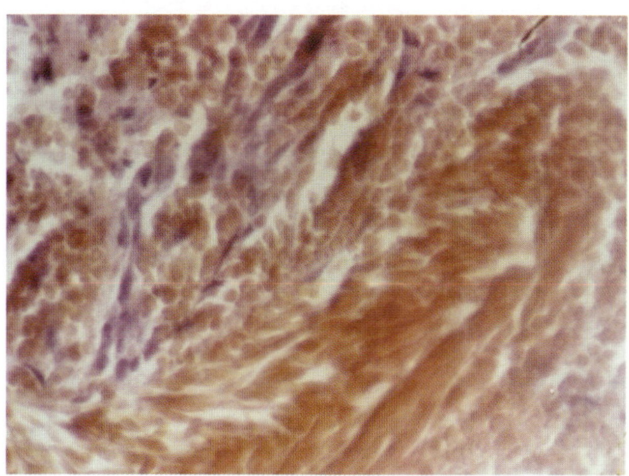

Figure 22.6, H&E x 520

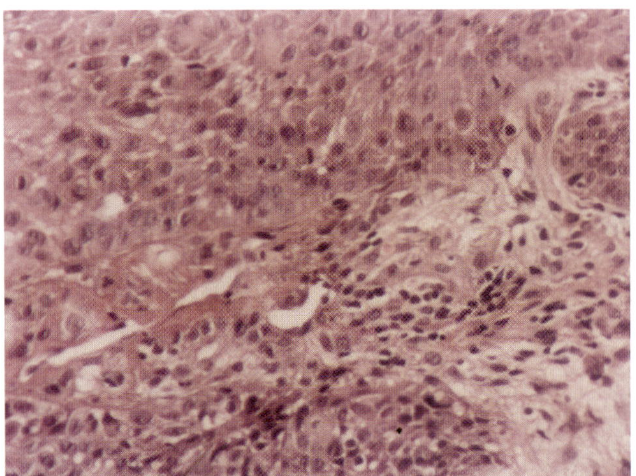

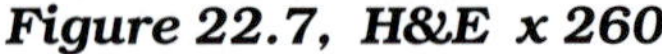

Figure 22.7, H&E x 260

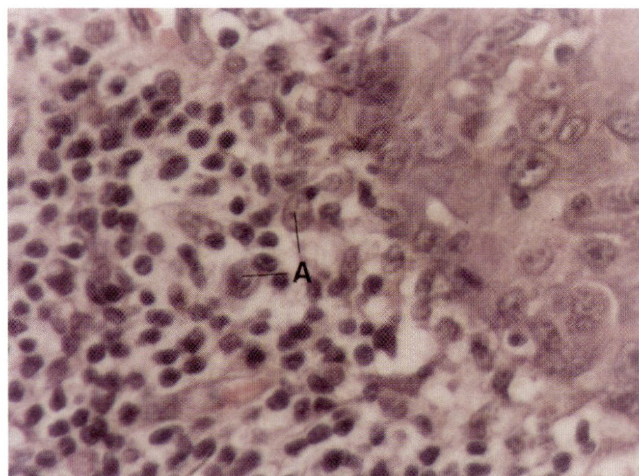

Figure 22.8, H&E x 520

2. Squamous Cell Carcinoma of the Skin and Esophagus (figs. 22.7-22.12)

Figure 22.7 is taken from a squamous cell carcinoma of the right ear surgically excised from a 77-year-old Caucasian male. Figures 22.8 and 22.9 are from a squamous cell carcinoma of the left side of the neck surgically excised from an 84-year-old Caucasian male. Figures 22.10-22.12 are from a surgically excised specimen of squamous cell carcinoma of the esophagus with lymph node metastasis.[2]

Squamous cell carcinoma of the skin transforming into reactive/inflammatory cells (figs. 22.7 and 22.8)

Figure 22.7. In the lower and right side of this figure, adjoining the squamous carcinoma tissue there is prominent inflammatory cell reaction. In this area cancer cells disassociated from the remaining cancer tissue are transforming (transdifferentiation) into reactive cells. H&E x 260

Figure 22.8. In the top right, the lower part of this squamous cell carcinoma shows atypical changes with development of large multinucleated atypical cancer cells. On the left, adjoining the lower border of cancer tissue there are many chronic inflammatory cells in which some remnants of cancer cells can be seen (A). In this area cancer cells disassociated from the remaining cancer tissue are transforming (transdifferentiation) into reactive cells. H&E x 520

Development of red cells from keratinizing squamous cell carcinoma (fig. 22.9)

Figure 22.9. The outer part of this squamous cell carcinoma (A) shows keratin with nuclear remains of saprophytic inflammatory cells. A concentrated area of keratinized tissue is transforming into red cells (B). H&E x 260

---

2. Additional photomicrographs and findings in squamous cell carcinoma are presented in Volume I, figures 70-77.

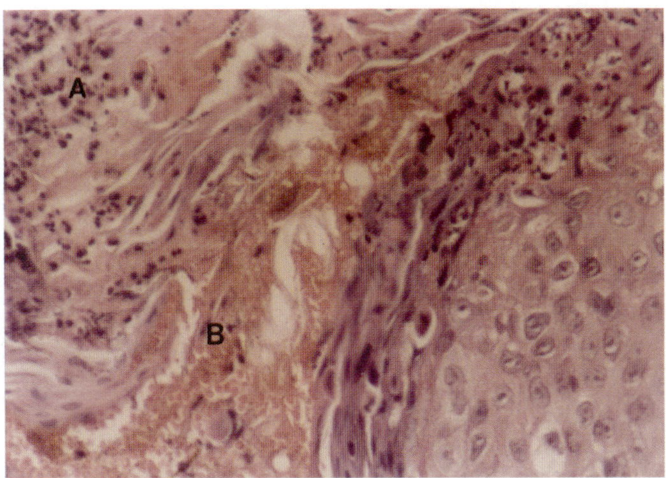

Figure 22.9, H&E x 260

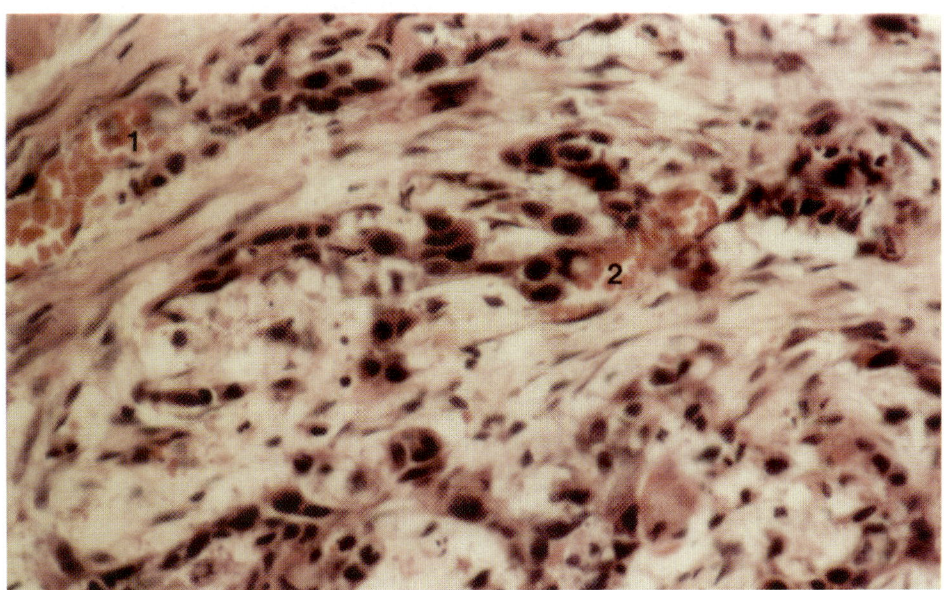

Figure 22.10, H&E x 400

Blood and blood vessel development from squamous cell carcinoma of the esophagus (figs. 22.10 and 22.11)

Figure 22.10. In areas (1) and (2) the squamous carcinoma cells are being replaced by columns of developing red cells, partially lined by developing endothelium. Transformation of cancer cells into red cells can be seen in area (1). H&E x 400

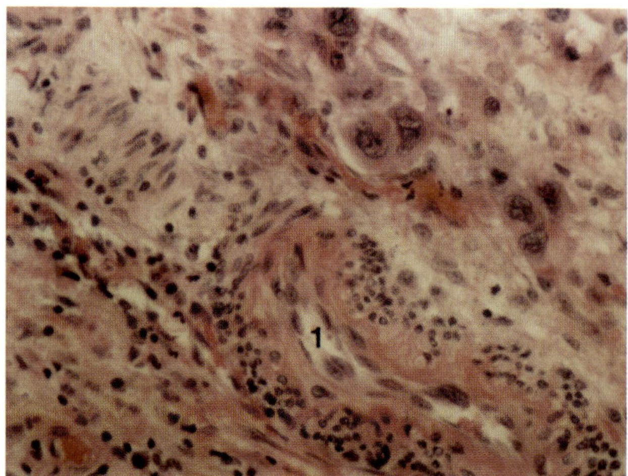

Figure 22.11, H&E x 260

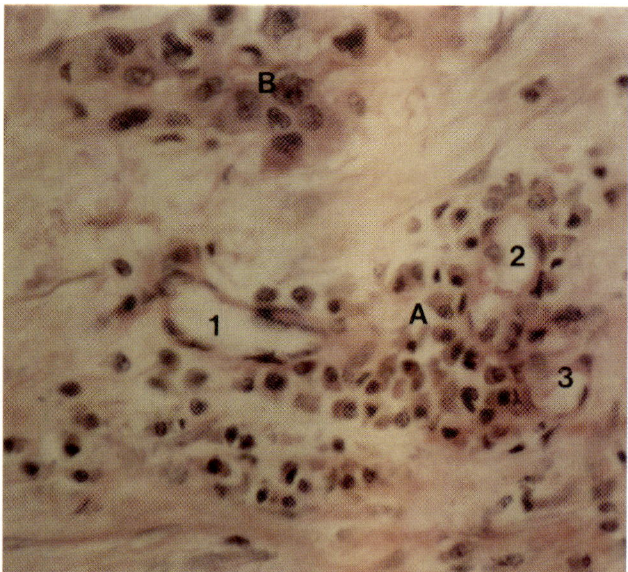

Figure 22.12, H&E x 400

Figure 22.11. In the upper right quadrant of this figure, highly abnormal large multinucleated cancer cells are seen among the regular tumor tissue. Blood vessel (1), with thickened wall, is surrounded by chronic reactive cells. Scattered in the surrounding tumor tissue, there are streaks of hemoglobinization suggesting the beginning of blood and narrow blood vessel formation. H&E x 260

Plasma cell development as a reaction to nearby cancer (fig. 22.12)

Figure 22.12. Proliferation of plasma cells (A) from the esophageal wall is related to stimulation by the nearby cancer cells (B). There is development of vascular channels (1, 2 and 3) from these dissolving plasma cells. H&E x 400

3. **Malignant Melanoma** (figs. 22.13-22.19)

A. *Melanoma of skin* (figs. 22.13-22.16)

Figures 22.13-22.16 are taken from a malignant melanoma of the skin measuring 14 mm in diameter and 1.5 mm in thickness surgically excised from the forearm of a 26-year-old Caucasian male. Figures 22.15 and 22.16 are reprinted from Volume I.[3]

Blood and nucleated cell development from the keratin layer of melanoma of the skin (figs. 22.13 and 22.14)

Figure 22.13. In this tissue section of melanoma of the skin there is extensive inflammation associated with some destruction of the melanoma cells. The tumor tissue extends deeply into the dermal region

---

3. Additional photomicrographs and findings in malignant melanoma are presented in Volume I, figures 78-89.

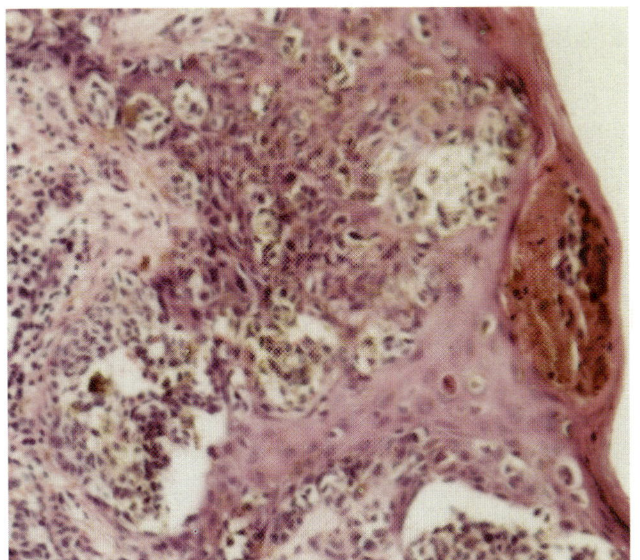

Figure 22.13, H&E x 200

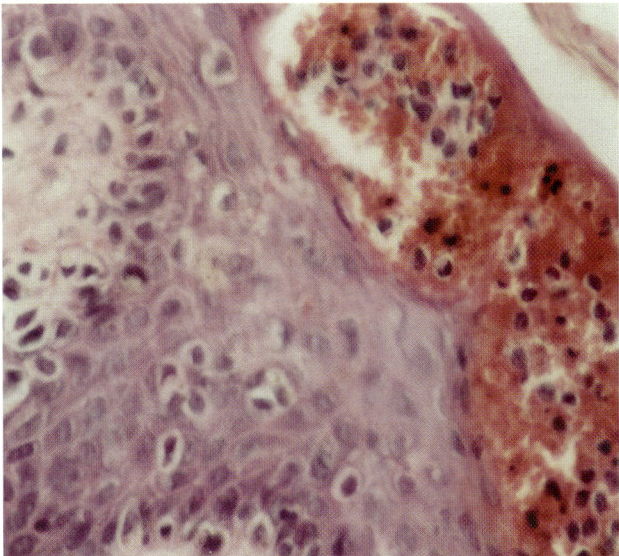

Figure 22.14, H&E x 400

and is transforming into chronic inflammatory cells. Melanin pigment can be seen in many of the cells in the center of this figure where melanoma cells are being infiltrated by locally developed inflammatory cells. On the right, in the keratin layer there is development of blood and pyknotic nuclear fragments (see also the next figure). H&E x 200

Figure 22.14. From the keratin layer of this melanoma lesion there are irregularly shaped red cells and nucleated cells arising from the keratin itself. H&E x 400

*Malignant changes in squamous epithelium
with the development of melanoma cells
producing red cells (fig. 22.15)*

Figure 22.15. There are early malignant changes in the epidermal cells with production of melanin pigment and nuclear changes with prominent nucleoli towards transformation into melanoma cells. An interesting finding in this figure is the highly prominent intercellular bridges (arrows) passing through the cytoplasm as well as partially vanishing cell nuclei. These intercellular bridges of squamous epithelium can be seen in a major portion of this

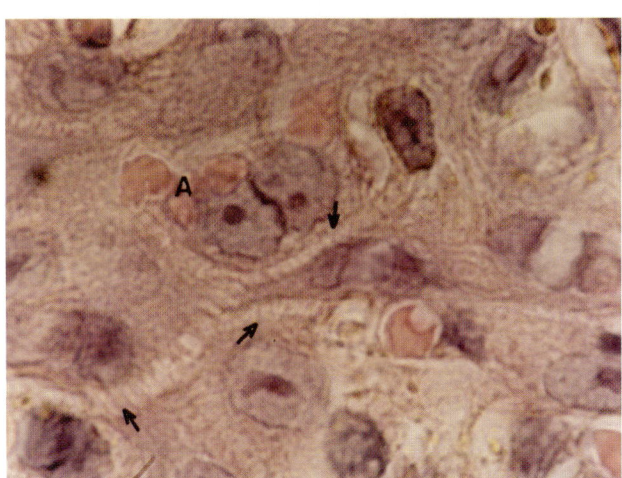

Figure 22.15, H&E x 1300

figure. Granules of melanin pigment can be seen attached to these intercellular bridges. Note also the development of red blood cells (A) with irregular margins from the cytoplasmic and nuclear substance of melanoma cells derived from squamous epithelium. H&E x 1300

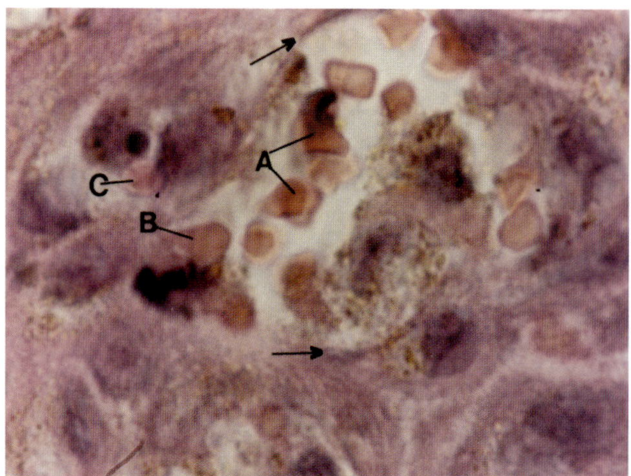

Figure 22.16, H&E x 1300

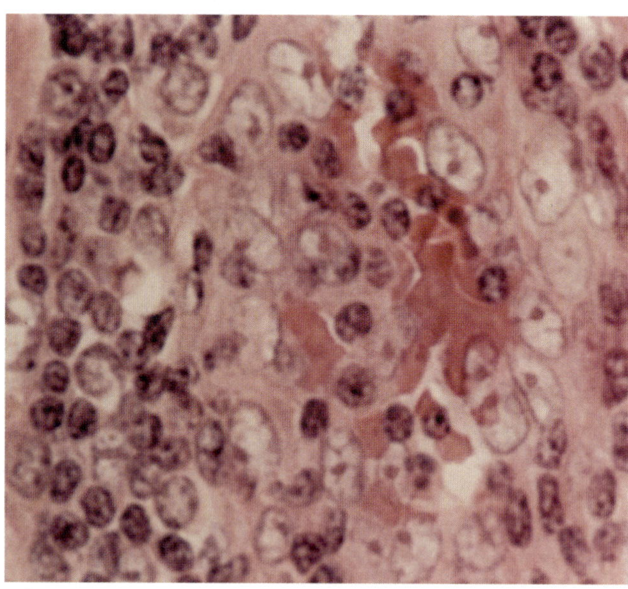

Figure 22.17, H&E x 1040

Blood and blood vessel formation directly from dissolving melanoma cells (fig. 22.16)

Figure 22.16. Within a formative blood vessel there is development of red cells and original plasma from dissolving melanoma cells with melanin pigment. A few red cells still show a brown cast of melanin pigment in them (A). One segmented nuclear cell with a distorted nucleus still has melanin pigment in it and is producing hemoglobin globules (B). Linear markings represent the protoplasmic membrane developed in preparation for the endothelial lining formation (arrows). Red cell (C) is developing from nearby melanoma cells. H&E x 1300

B. Melanoma metastatic to lymph node
(fig. 22.17)

Figure 22.17 is taken from a supraclavicular lymph node with metastatic malignant melanoma removed by excisional biopsy from a 24-year-old male.

Red cell development from metastatic melanoma cells (fig. 22.17)

Figure 22.17. In the center of this figure there are small and large areas of hemoglobinization replacing both metastatic melanoma cells and lymphocytes. There is a suggestion of early development of red cells from this hemoglobinized area. H&E x 1040

C. Melanoma metastatic to lung
(figs. 22.18 and 22.19)

Figures 22.18 and 22.19 are taken from the right middle lobe of a lung with metastatic melanoma removed by lobectomy from a 41-year-old male. The patient had a six year history of skin melanoma of the back.

Red cells developing from metastatic melanoma cells (fig. 22.18)

Figure 22.18. In this figure melanoma cells, mostly in groups, can be seen in the alveolar and interalveolar spaces. A small amount of melanin pigment is noted in most of these tumor cells which have replaced most of the normal lung tissue. There are

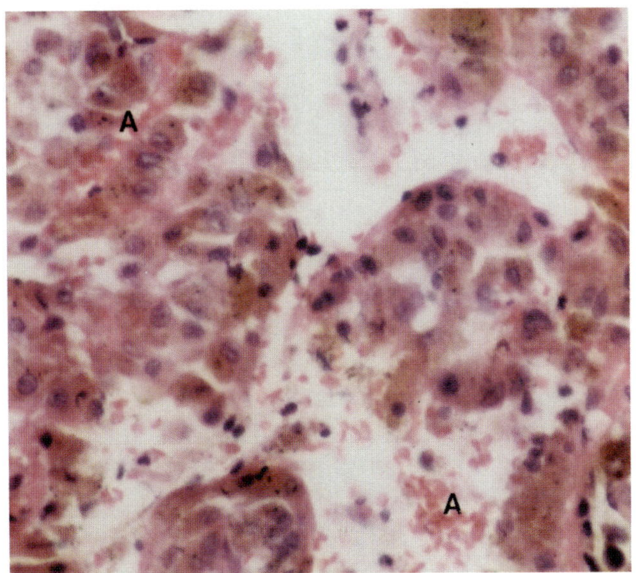

Figure 22.18, H&E x 400

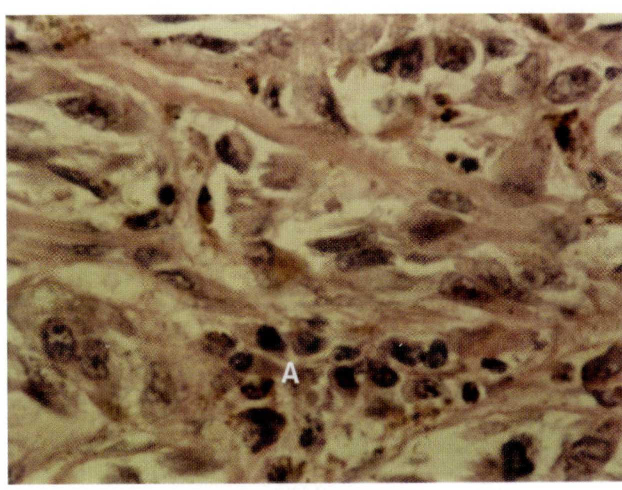

Figure 22.19, H&E x 520

irregular areas developed by the erasement of tumor tissue which contain scattered developing red cells. In areas (A) incompletely formed, not yet separated, red cells are developing from the dissolving melanoma cells. H&E x 400

Plasma cells developing from metastatic melanoma cells (fig. 22.19)

Figure 22.19. In this section of melanoma metastatic to the lung a group of plasma cells (A) are developing from melanoma cells. H&E x 520

Chapter 23

LIPOSARCOMA
(Malignant Tumor of Adipose Tissue) (figs. 23.1-23.3)

Figures 23.1-23.3 are taken from the surgically amputated right ankle of a 67-year-old female with liposarcoma. Cellular changes in benign adipose tissue are demonstrated in Chapter 3.

Blood and blood vessel development from liposarcoma cells (figs. 23.1-23.3)

Figure 23.1. This is a low-magnification view of liposarcoma. The left side of this figure is occupied by liposarcoma cells. On the right, there is adipose tissue and a few widely scattered sarcoma cells. H&E x 200

Figure 23.2. In this figure, there is prominent variation in both tumor cell and nuclear size and shape and nuclear chromatism. Throughout this figure no definite fat cells are identified. There are focal areas of hemoglobinization of liposarcoma cells. In areas (A), there is confluent hemoglobinization of several tumor cells and red cells have not yet begun to develop. At the edge of these areas, there is the appearance of tiny narrow endothelial cells. A regular blood capillary (B) is arising from the tumor tissue. In this blood capillary there is endothelial lining formation surrounding the regular column of red cells. The red cells are not fully rounded and are not yet fully separated from each other. In the surrounding areas, the fat cells are not distinguishable in their alteration toward tumor tissue. H&E x 520

Figure 23.3. In this malignant adipose tissue a blood capillary is arising from and replacing the tumor tissue. H&E x 520

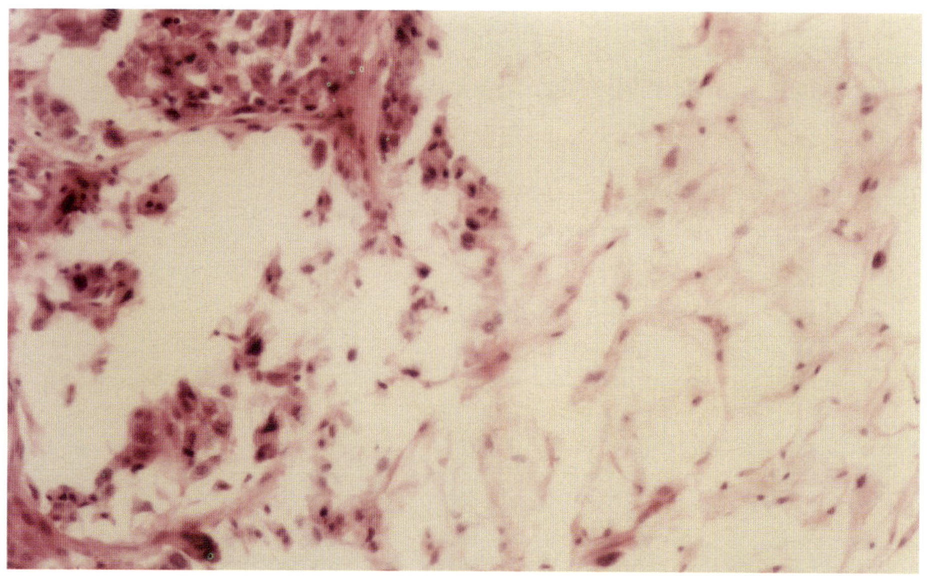

Figure 23.1, H&E x 200

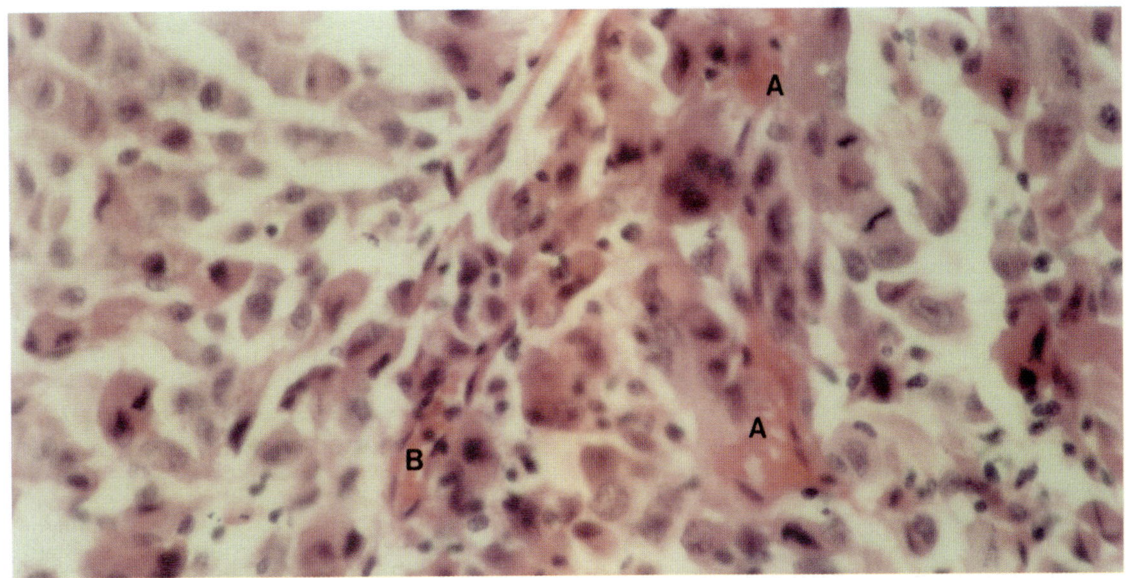

Figure 23.2, H&E x 520

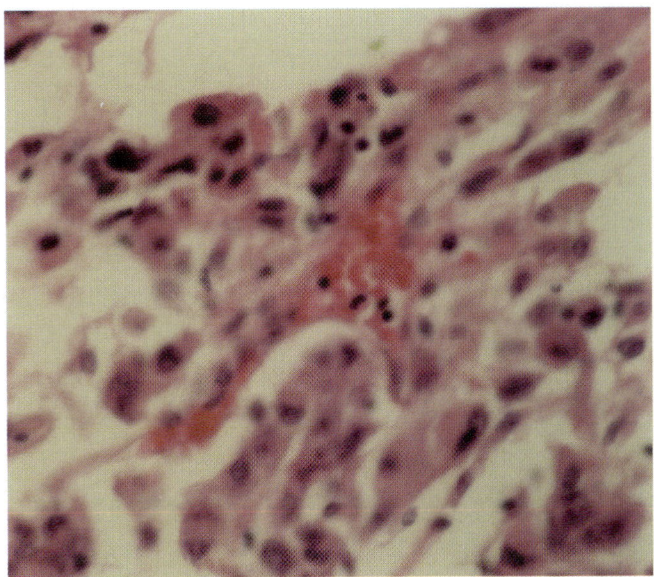

Figure 23.3, H&E x 520

Chapter 24

BREAST CANCER (figs. 24.1-24.17)

1. Mammary Carcinoma in Mice (figs. 24.1-24.12)

Figures 24.1-24.12 are from malignant epithelium undergoing various cellular changes including massive necrosis, cellular lysis, and hyalinization. From malignant epithelium there is development of red cells, blood vessels, blood cysts (so-called hemorrhagic cysts), and stromal tissue.

The photomicrographs presented here are made from autopsy sections of one transplanted (M2391) mammary carcinoma and six spontaneous (M40, M45, M59, M61, M1378 and M2039) mammary carcinomas (adenocarcinomas) in C3H mice: figures 24.1 and 24.2 (M40); figure 24.3 (M45); fig. 24.4 (M1378); figs. 24.5-24.7 and 24.11 (M61); figs. 24.8 and 24.9 (M2391); fig. 24.10 (M59), fig. 24.12 (M2039). These C3H mice were used as part of a small scale experimental chemotherapy program. Figures 24.4-24.12 are reprinted here, from Volume I (McDonald, 1989), for the convenience of the reader.[1]

Mammary carcinoma with prominent, extensive, and rapid blood formation associated with tissue destruction and necrosis (figs. 24.1-24.3)

Figures 24.1-24.3 are taken from fresh autopsy specimens of spontaneous breast carcinomas from mice that died within thirty to ninety minutes after a lethal injection, given inadvertently, of an experimental chemotherapy drug, ethylhydrazine acetate hydrochloride. These mice had small spontaneous mammary carcinomas which were still at a nonlethal stage.[2]

Figure 24.1. In this figure there is rapid development of blood occupying large areas of the tumor in formation of blood cysts with destruction of large areas of tumor tissue. This process is presented in higher magnification in figures 24.2 and 24.3. H&E x 80

Figure 24.2. In this pool of blood the origin of red cells from tumor tissue can be noted by following the remains of necrosed tumor cells within this blood cyst. Some of the nuclear remains have a segmented nuclear appearance. However, these nuclear remains are part of the necrotic tumor tissue transforming into hyper-hemoglobinized red cells. An earlier stage of this process is presented in the next figure. Also, note that the lining epithelial cells at the periphery of the hemorrhagic cystic structure show some evidence of rapid hemoglobinization (A) in formation of hemoglobin globules. H&E x 800

Figure 24.3. In this figure tumor necrosis is more visible in association with rapid blood formation. Nuclear and cellular remnants of tumor tissue can be seen throughout this area of blood formation. H&E x 200

---

1. Additional photomicrographs and findings in mammary carcinoma in mice are presented in Volume I, figures 1 to 33-1.
2. Rapid mobilization of liver tissue in formation of blood in the liver of one of this same group of mice is presented in figure 12.28.

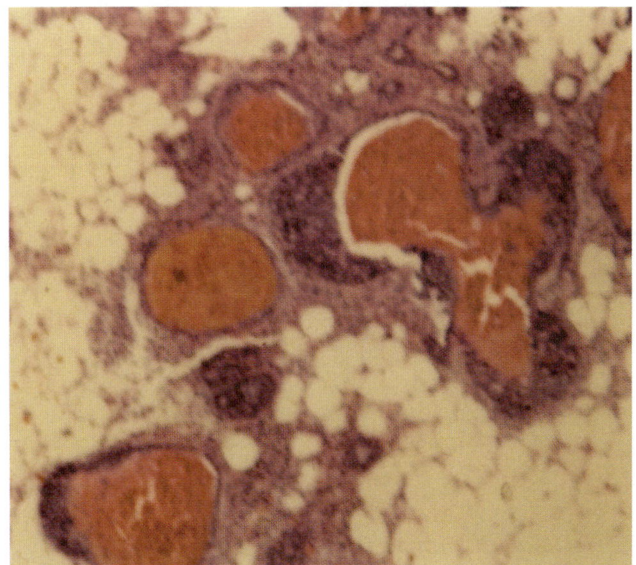

Figure 24.1, H&E x 80

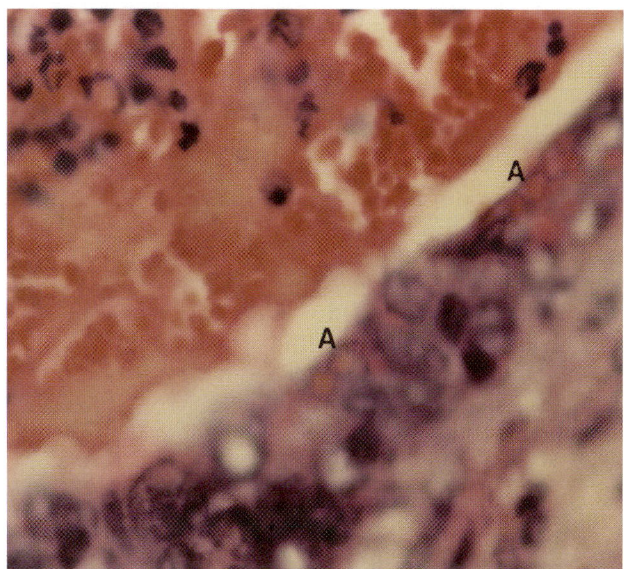

Figure 24.2, H&E x 800

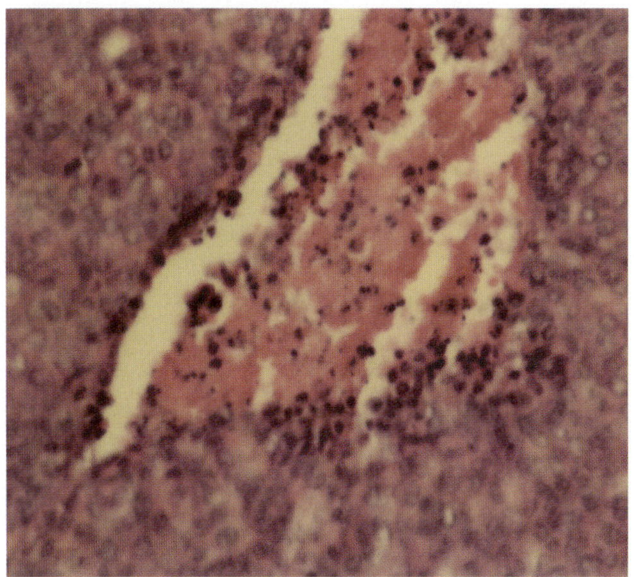

Figure 24.3, H&E x 200

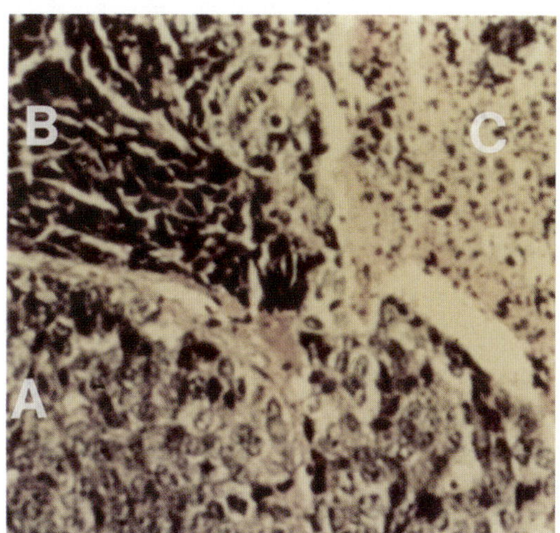

Figure 24.4, H&E x 200

Development of blood and blood cysts (so-called hemorrhage and hemorrhagic cysts common in C3H mice) from necrotic tumor tissue (figs. 24.4-24.7)

Figure 24.4. Wide areas of tumor tissue in compartments (A), (B), and (C) show progressive necrosis to partial dissolution of cancer cells. In area (A), there are scattered small clumps of pyknotic cancer cells. In (B), there is a solid sheet of necrotic tumor tissue with marked pyknosis. In (C), the disintegration of necrotic tumor tissue with scattered remains of tiny, pyknotic fragments can be seen in the background of the light pink areas. H&E x 200

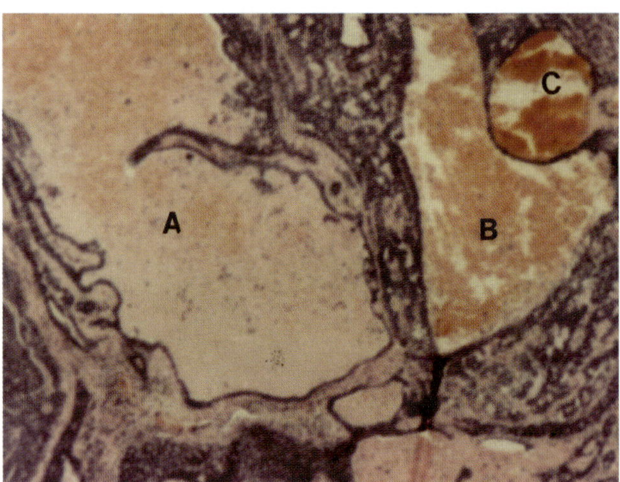

Figure 24.5, H&E x 80

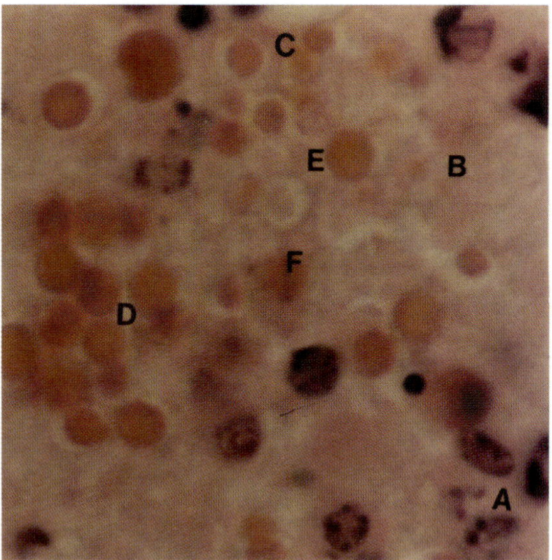

Figure 24.6, H&E x 2000

Figure 24.5. The progressive developmental stages (in three compartments, A, B and C) of blood formation following necrosis from the stage shown in figure 24.4 (C) is seen here. Also note the organization of the peripheral remaining epithelium into smooth epithelial linings in development of blood cysts. H&E x 80

Figure 24.6. Presented here in higher magnification are the progressive stages (A-E) of development of red cells from necrotic tumor tissue as shown in figure 24.5 (A). The stages range from not completely liquefied tumor tissue (A), to fully formed rounded red cells of various sizes and degrees of hemoglobinization. Some red cells (E) are even larger than normal red cells. Note the development of tiny red cells in other areas. (F) denotes a developing red cell with faint remains of necrotic cell substance. This type of erythrogenesis is often categorized as hemorrhagic necrosis. H&E x 2000

Figure 24.7. These multilocular blood cysts are developing from liquefied necrotic tumor tissue, in some places passing through fibrinoid changes. The developing red cells within these cysts are not yet rounded. The interlocular septa consist of tumor cells ranging from a single layer of barely visible cells to cells in multiple lay-

ers. In region (A), red cells are still developing from fibrin-like or fibrinoid material derived from the liquid product (B) of necrosed tumor tissue. The epithelial boundary is beginning to form in the already developed protoplasmic membrane (arrows) as forerunners in which tiny nuclear bodies have appeared in a single line. Instead of epithelial cell production, endothelium may form in the protoplasmic membrane in the formation of multilocular blood sinuses. Vessel formation following necrosis of tumor tissue has been demonstrated earlier (McDonald, 1962). H&E x 200

The simultaneous origin of vascular content, red cells and plasma, and the vessel wall from local cancer tissue (figs. 24.8-24.10)

Figure 24.8. The early stages of blood and blood vessel formation from lysing cancer cells are seen here in the longitudinal plane. Feathery cytoplasmic remnants still adhere to some of the red cells. (A) is a single well-developed endothelial nucleus arising from lysing tumor cell remains, and (B) are the remnants of not yet completely lysed tumor cells lying in the pathway of the formative vascular lumen. One lymphocyte not yet completely rounded, apparently arising from cancer cells, is lying in the vascular lumen. Giemsa x 800

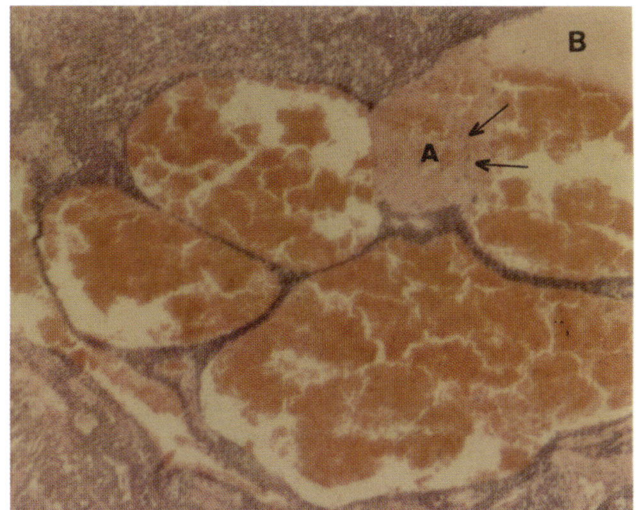

Figure 24.7, H&E x 200

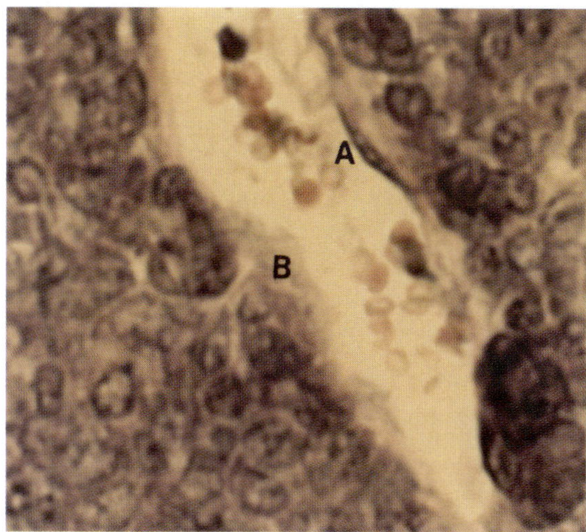

Figure 24.8, Giemsa x 800

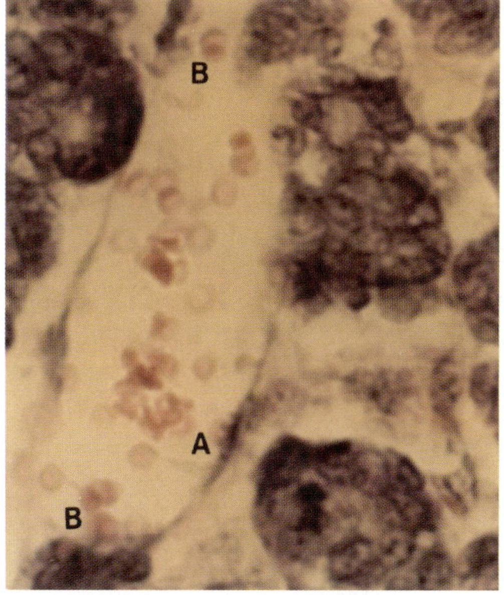

Figure 24.9, Giemsa x 800

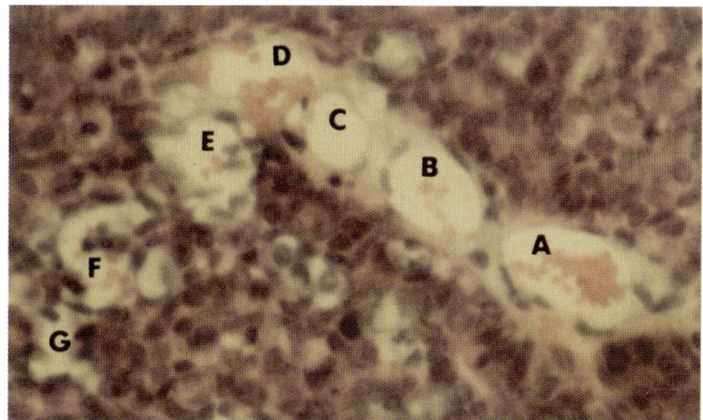

Figure 24.10, H&E x 520

Figure 24.9. The endothelium is beginning to develop as a protoplasmic membrane from lysing cancer cell substance and is projecting through the dissolving tumor cells. (A) marks an early stage of formation of a full-sized endothelial nucleus in this protoplasmic membrane and a developing red blood cell from the same nuclear substance. In the lumen, red blood cells of various degrees of hemoglobinization float in clear plasma, the product of cell lysis. In the periphery, red blood cells are still being developed from a cluster of lysing can-

cer cells (B). On casual observation, these cells (A and B) are usually misinterpreted as erythrophagocytic endothelial cells or erythrophagocytic cancer cells, respectively, instead of erythrogenic cells. Outside the vascular channel, cellular lysis is producing tissue plasma gel or tissue fluid. In this area, an excess amount of cellular lysis results in edema. Giemsa x 800

Figure 24.10. The development of a vascular channel by the unification of adjoining vascular units made of several cells is seen here. In addition to the formation of large capillaries from single columns of dissolving cancer cells as shown in the two previous figures, the possibility of unification of several adjoining multi-cellular vascular units in formation of a

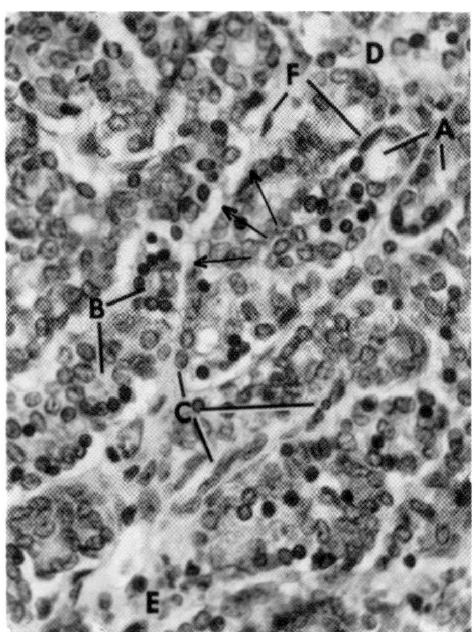

Figure 24.11, H&E x 440

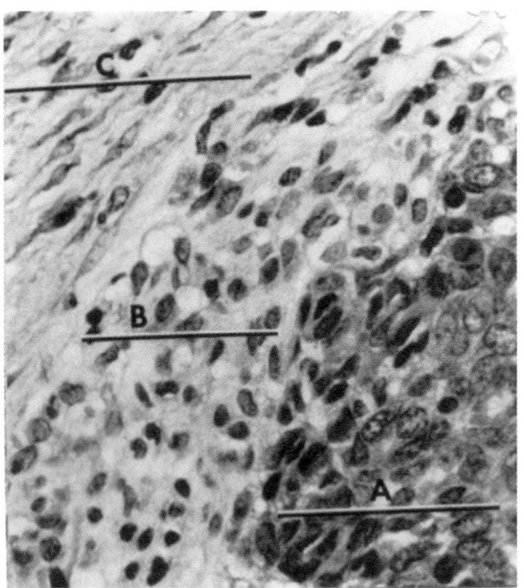

Figure 24.12, H&E x 440

larger capillary is suggested by this figure. The units here (A-G in successive stages of development in reverse order) are composed of several groups of lysing cancer cells with the capacity to form red cells, clear lytic fluid (the original plasma), and the endothelium from the remaining peripheral cells. In unit (F), the origin of red cells can be seen from cancer cells undergoing clear and colorless liquefaction in the process of producing plasma. H&E x 520

The possible transformation of cancer cells into benign cells and the development of collagenous connective tissue stroma (figs. 24.11 and 24.12)

Figure 24.11. Cellular changes in the malignant epithelium of the acinar pattern toward connective tissue stroma formation are demonstrated here. Disruption of the acini by linear hyalinization of the cancer cells is outlined by the arrows. Note the transformation of lining cells into lymphocytes or lymphocyte-like cells in the acinar epithelium (A and B). Linear arrangement of the carcinoma cells and transformed lymphocytes with a tendency toward formation of spindle cells of the stroma is shown in successive stages (C). Acini are becoming hyalinized (D). Irregular cell remnants and a lymphocyte are present in the area of hyalinization (E). Two pyknotic spindle cells are seen at (F), one replacing an acinar epithelial cell and the other in the pathway of linearly hyalinized tumor tissue.[3] H&E x 440

Figure 24.12. From right to left (A-C), the successive stages of connective tissue stroma formation from a solid sheet of cancer cells are seen here. In the lower right zone (A), many tumor cells have become pyknotic and spindle-shaped (pyknotic spindling). A few cells show hyalinization. The cells in the pale middle zone, which may also be called a zone of halo formation (B), show prominent hyalinization and intermediary stages of irregular transformations which no longer fall under the category of malignant cells.[3] H&E x 440

3.   Figures 24.11 and 24.12 are reproductions of photomicrographs which were originally presented in one of my earlier articles (McDonald, 1970a). I greatly appreciate the permission for reproduction given by the editor and publisher of the *Journal of the American Medical Women's Association.*

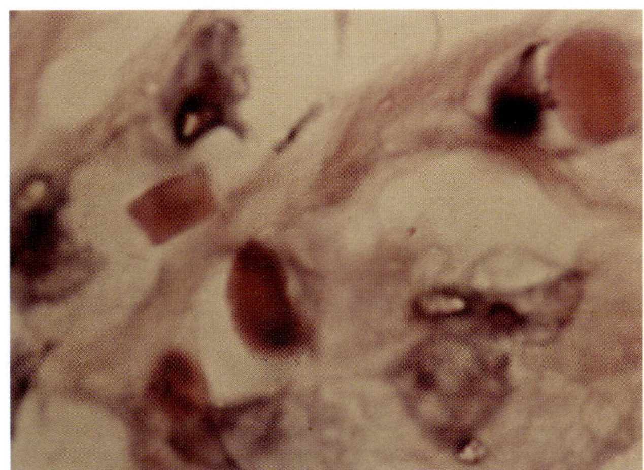

Figure 24.13, H&E x 2000

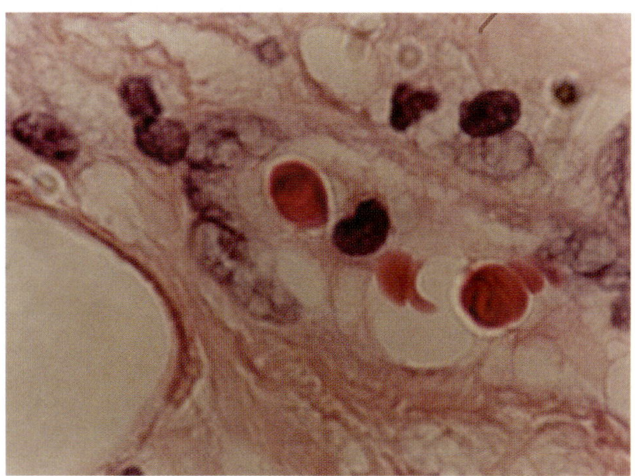

Figure 24.14, H&E x 1300

2. **Mammary Carcinoma in Humans** (figs. 24.13-24.17)

Figure 24.13 is taken from a medullary carcinoma of the breast obtained by radical mastectomy from a 55-year-old female. Figures 24.14, 24.16, and 24.17 are taken from an infiltrating ductal carcinoma of the breast obtained by excisional biopsy from a 73-year-old female, and figure 24.15 is taken from the excisional biopsy specimen of the same tumor obtained eight days before mastectomy. Figures 24.13-24.17 are reprinted from Volume I (McDonald, 1989).[4] For cellular changes in benign human breast tissues see Chapter 18.

The origin of red cells and blood capillaries directly from cancer cells undergoing stromal changes with collagen formation (figs. 24.13 and 24.14)

Figure 24.13. Developing red blood cells from both cytoplasmic and nuclear substance of changing cancer cells is shown in high magnification. Note the appearance of collagen fibers replacing parts of the cancer cells. H&E x 2000

Figure 24.14. The appearance of red blood cells in collagen fibers undergoing watery vacuolar changes (hydropic degeneration) is seen here. Note the marked deformity of the red cells which are crescent-shaped and pressed together like

pressed biscuits. The development of lymphocytes within and outside of this formative vascular channel is believed to be from cancer cells undergoing collagenization. A major part of the cancer tissue in this figure is replaced by collagen. H&E x 1300

The development of red cells and blood capillaries from locally developed plasma cells of cancer cell origin (fig. 24.15)

Figure 24.15. Developing stages of blood and endothelium, within a formative vascular channel, all derived from plasma

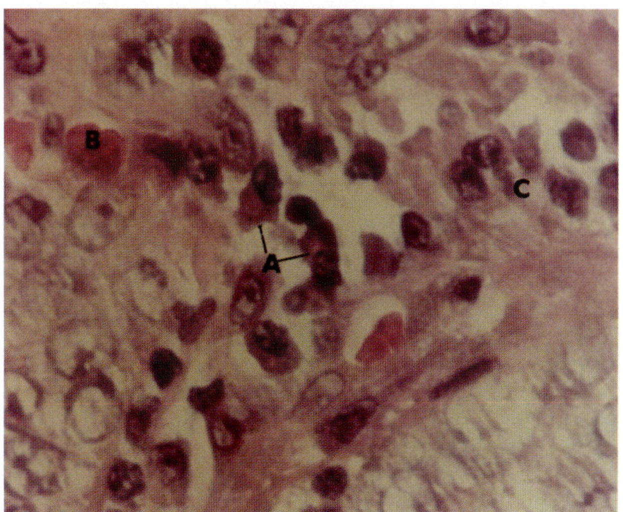

Figure 24.15, H&E x 1040

---

4.   Additional photomicrographs and findings in mammary carcinoma in humans are presented in Volume I, figures 34-56.

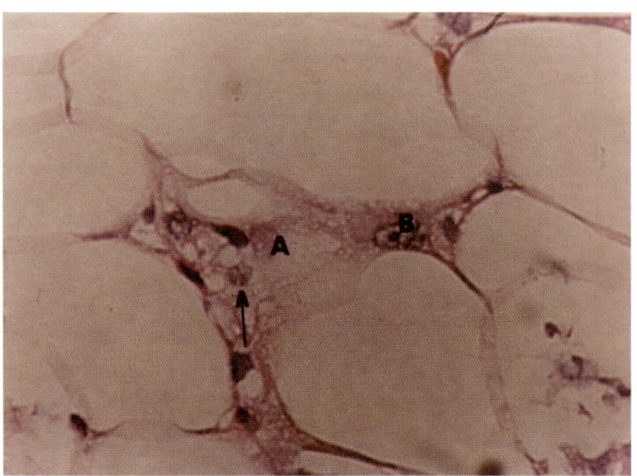

Figure 24.16, H&E x 520

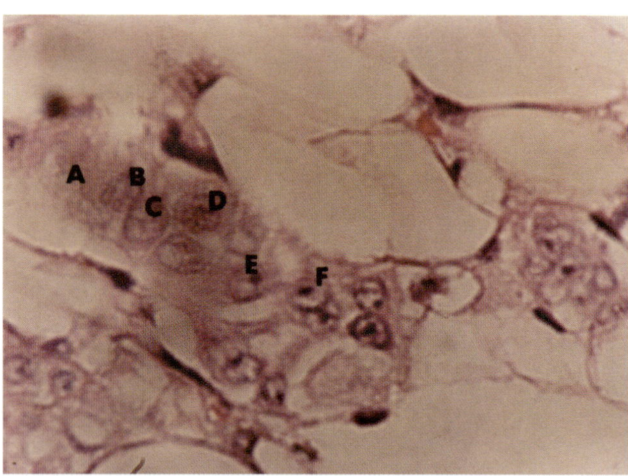

Figure 24.17, H&E x 520

cells in a branching column arising from cancer tissue is demonstrated here. This figure shows the stages of hemoglobinization (A) and red blood cell formation (B) with adherent cellular remnants of plasma cells. The origin of plasma cells from disappearing cancer cells is more apparent in area (C). H&E x 1040

The mechanism of the local spread of mammary carcinoma into adipose tissue demonstrated by the progressive malignant changes in the latter induced in some way by the nearby cancer (figs. 24.16 and 24.17)

Figure 24.16. The early stages of cancer development from adipose tissue are demonstrated here. This is carried out by the development of smoky areas (A) of hyalinization of adipose tissue in which ill-defined, blotchy, lightly stained, basophilic nuclear bodies arise (arrow). In further differentiation from this nuclear state, there appears a fully developed nucleus with a prominent nucleolus (B). H&E x 520

Figure 24.17. Progressive developmental stages of cancer cell nuclei (A to F) from adipose tissue starting with smokey areas of hyalinization are seen here. Note the smoky areas becoming masses of cytoplasm with multiple nuclei like the syncytial layer of chorionic villi. H&E x 520

Chapter 25

THYROID CANCERS   (figs. 25.1-25.4)

Figures 25.1-25.3 are taken from the left lobe of a thyroid surgically removed from a 35-year-old female diagnosed with adenocarcinoma of the thyroid. Figure 25.4 is from the subtotal thyroidectomy of a 62-year-old female with follicular adenocarcinoma. Cellular changes in benign thyroid tissues are presented in Chapter 5, subsection 1.

Development of hemoglobinized colloid and red cells from dissolving cancer tissue (figs. 25.1 and 25.2)

Figure 25.1. In this adenocarcinoma of the thyroid there is highly prominent hemoglobinization of thyroid colloid and the beginning of hyperhemoglobinized red cell development in some areas (arrows) from this product. H&E x 260

Figure 25.2. In the center, cancer tissue is dissolving and forming thick hemoglobinized colloid. Red cell development by crystallization from this thick hyperhemoglobinized erythrogenic fluid is also noted (A). Along the lower edge of this area, some of the surrounding cancer cell nuclei are becoming narrow and pyknotic before dissolving. H&E x 520

Normal thyroid tissue developing hemoglobinized colloid as a reaction to nearby adenocarcinoma (fig. 25.3)

Figure 25.3. Normal thyroid tissue close to adenocarcinoma shows development of crowded masses of hyperhemoglobinized red cells from hemoglobinized colloid, which is produced by the liquefied product of thyroid tissue (see also figs. 5.9-5.11). Surrounding the hemoglobinized colloid the lining epithelial cells are small and flattened and the nuclei are tiny and hyperchromatic. H&E x 260

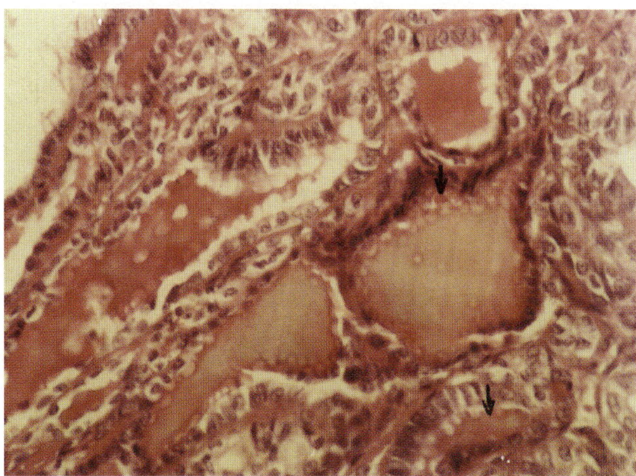

Figure 25.1, H&E x 260

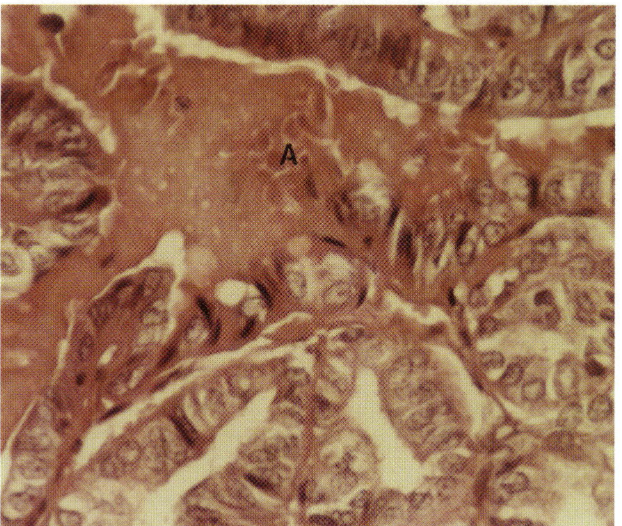

Figure 25.2, H&E x 520

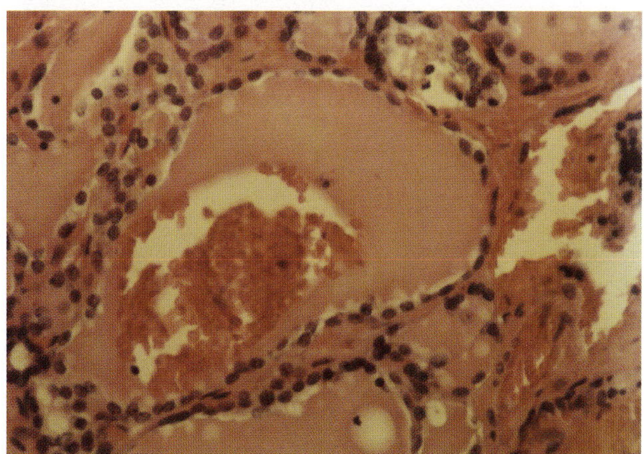

Figure 25.3, H&E x 260

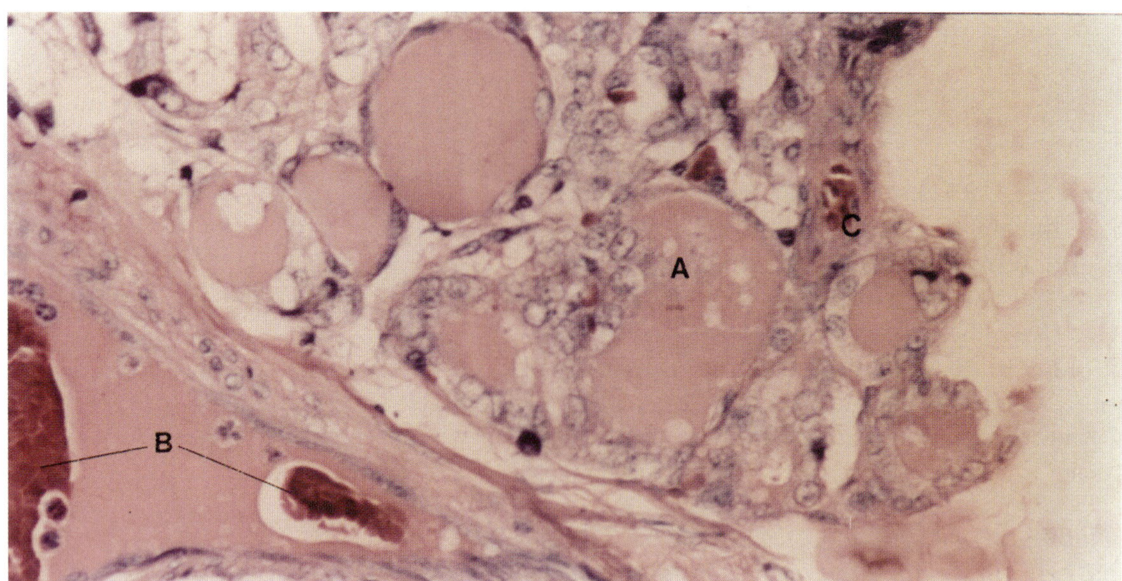

Figure 25.4, Giemsa x 520

Blood and blood vessel development from follicular carcinoma of the thyroid (fig. 25.4)

Figure 25.4. In this figure, follicular carcinoma of the thyroid is stained with Giemsa stain. The pink staining fluid is thyroid secretion with vacuoles. Cancer cells are forming the vacuoles in connection with the developing fluid. Faint remnants of dissolving cancer cells (A) can be seen in the pinkish fluid. Note the irregular atypical crowding of cancer cells lining the border of this follicle and adjoining follicles. These follicles are lined by atypical more or less cuboidal epithelium with lightly stained vacuolated nuclei. In the lower left quadrant of this figure, there are two irregular areas of dark red hemoglobinization (B) with indistinctly noted development of compact hyperhemoglobinized red cells within a large follicle filled with light pink colloid. The lining epithelium of this follicle is indistinct and flattened. Near the right edge of the section, (C) points to a small blood vessel developing in the stromal tissue. Faintly visible remnants of nuclear substance can be seen attached to the developing red cells within this capillary. Giemsa x 520

Chapter 26

ADENOCARCINOMA OF SALIVARY GLAND
AND REACTIONS IN NEARBY NORMAL GLANDULAR TISSUE
(figs. 26.1-26.6)

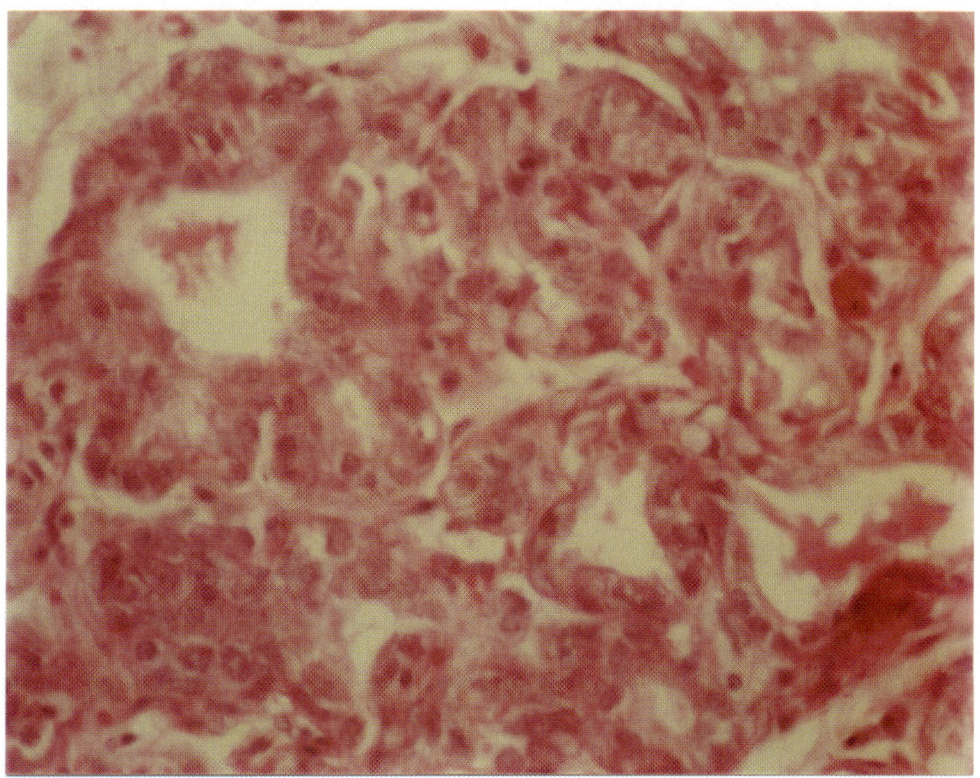

Figure 26.1, H&E x 520

The salivary gland tissue in figures 26.1-26.6 is part of a surgical specimen excised from a patient with poorly differentiated papillary adenocarcinoma of the salivary gland. Cellular reactions in benign glandular tissues are demonstrated in Chapter 5.

Poorly differentiated papillary adenocarcinoma of the salivary gland (fig. 26.1)

Figure 26.1. In this section of salivary gland with poorly differentiated papillary adenocarcinoma, malignant changes with marked disorganization of salivary cell pattern and nuclear and cellular variation can be seen. Nearby, normal salivary gland tissue with reactions to this approaching cancer is presented in figures 21.2-21.6. H&E x 520

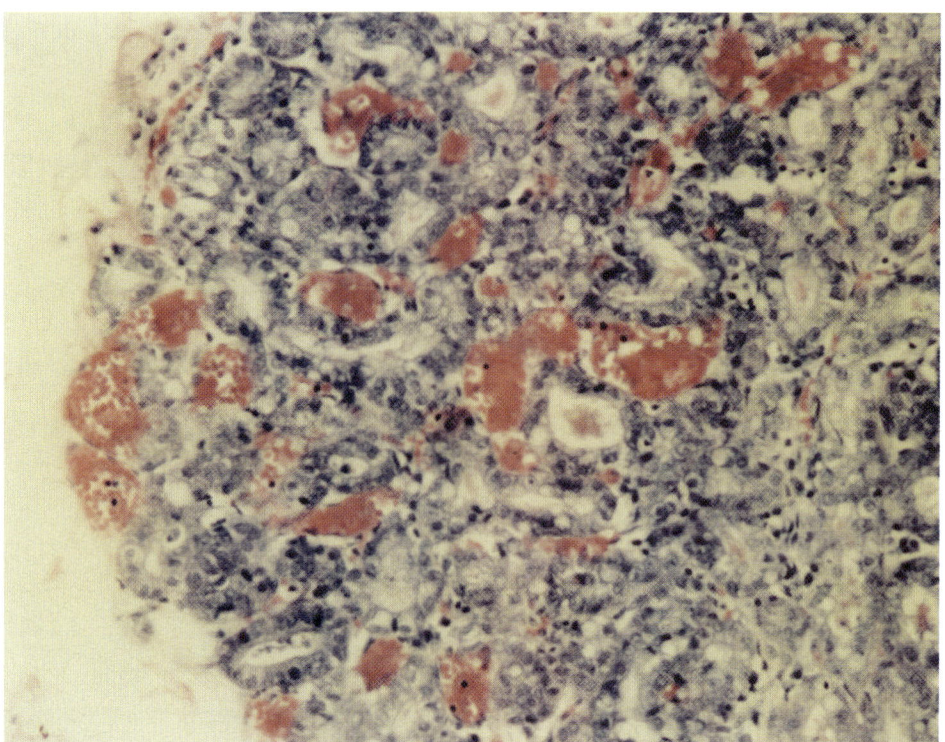

Figure 26.2, Giemsa x 260

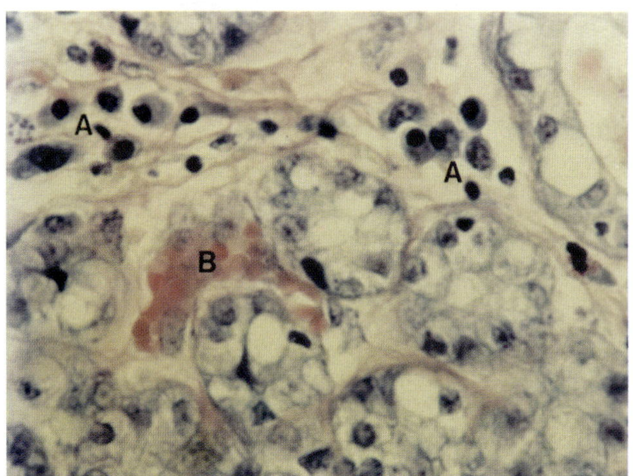

Figure 26.3, Giemsa x 520

Blood and blood vessel development from glandular epithelium as a reaction to nearby poorly differentiated papillary adenocarcinoma (figs. 26.2-26.6)

Figure 26.2. This is a section of normal salivary gland tissue near adenocarcinoma showing marked development of blood within the glandular structures as a reaction to the nearby cancer. Giemsa x 260

Transformation of glandular tissue into plasma cells as a reaction to nearby cancer (figs. 26.3-26.6)

Figure 26.3. In the upper third of this figure there are inflammatory cells, mainly plasma cells and small lymphocytes. The lymphocytes are transforming into plasma cells (areas A). In area (B) the glandular epithelium is transforming into hemoglobin and red cells. Giemsa x 520

Figure 26.4. In this figure many plasma cells of various sizes are developing with erasement of the glandular tissue. Crossing this figure transversely, there is a large developing blood vessel filled with red cells. The endothelium of this vessel is only partially formed. There is some irregularity in the glandular epithelium indicating early malignant changes. On close examination one can see some atypical changes in the glandular epithelium (arrows) with irregular multiplication of epithelial cells and many small hyperchromatic nuclei. Giemsa x 520

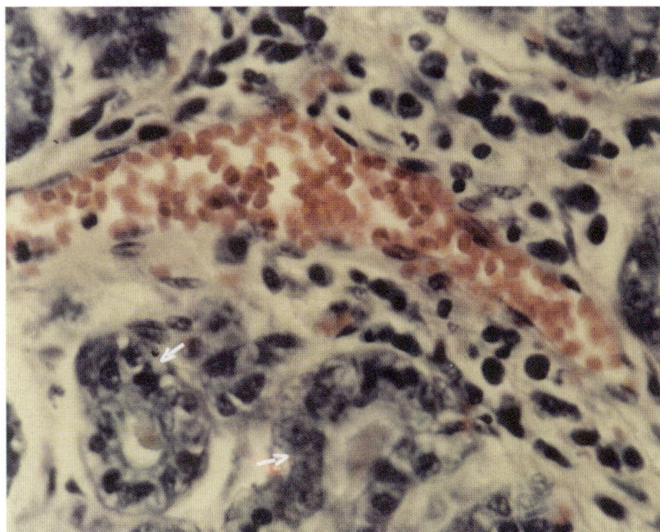

Figure 26.4, Giemsa x 520

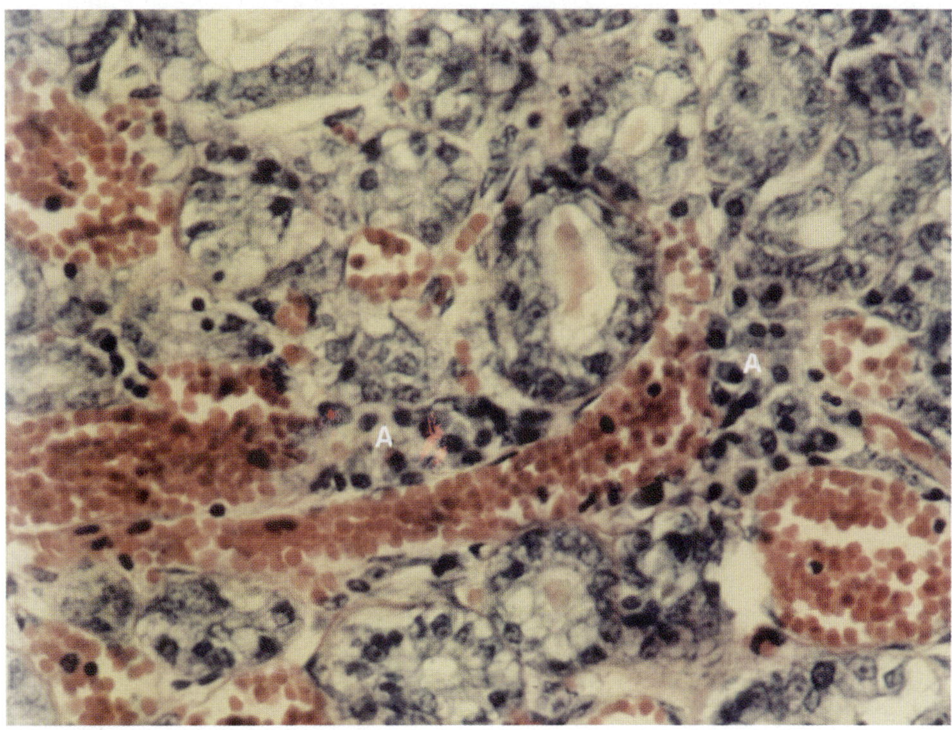

Figure 26.5, Giemsa x 520

Figure 26.5. Salivary gland showing prominent blood and blood vessel development from and replacing the glandular tissue is seen here. In some of these vessels the endothelium is barely developed from the remaining peripheral tissue. In the center (areas A), the glandular tissue is also transforming into plasma cells and lymphocytes. Giemsa x 520

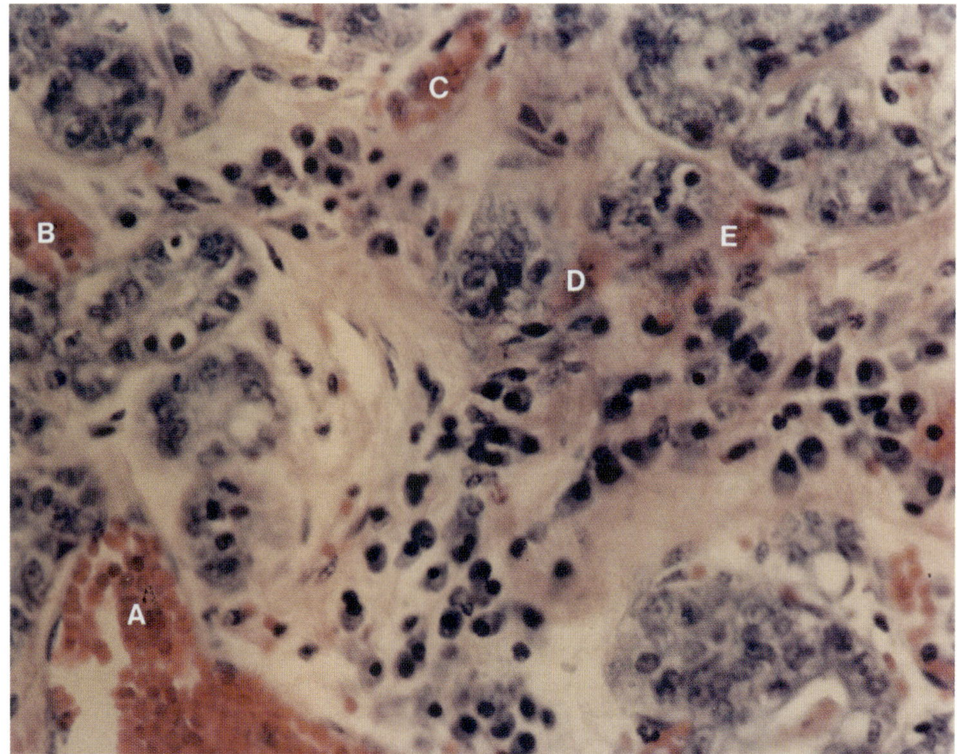

Figure 26.6, Giemsa x 520

Figure 26.6. Stimulated by nearby cancer, the salivary gland epithelium is transforming into plasma cells preliminary to development of stromal tissue. There are also areas of replacement of the glandular tissue by columns of developing red cells (A), (B), and (C). An early stage of developing red cells from glandular tissue is seen at (D and E). Remnants of nuclear material can still be identified in each of these areas of blood formation. Giemsa x 520

Chapter 27

LUNG CANCERS (figs. 27.1-27.7)

1. Bronchogenic Squamous Cell Carcinoma of the Lung (figs. 27.1-27.2)

Figures 27.1 and 27.2 are taken from a surgical specimen from a patient with bronchogenic squamous cell carcinoma of the right lung. For cellular changes in benign lung tissue see Chapter 11.

Development of inflammatory cells, blood and blood vessels from tumor cells (figs. 27.1 and 27.2)

Figure 27.1. In this section of bronchogenic squamous cell carcinoma of the lung sheets of cancer cells can be seen. Note the characteristic squamous cell nature of the tumor cells. Also note the irregular cancer cells with atypical nuclei. In the upper left quadrant, there are many SN cells in various stages of development from necrosed tumor cells. There is a vascular channel (A) developing from cancer tissue and inflammatory cells, which have also developed from cancer tissue. Blood vessel (B) is developing from disappearing cancer tissue. A portion of the vessel is filled with developing red cells. The lining endothelium is not well developed. H&E x 520

Figure 27.2. A large area of this tumor is dissolving in development of red cells and a large vascular space occupying almost half of this figure. Within the lumen, remnants of partially dissolved necrosed cancer cells associated with developing red cells can still be identified. The developing red cells are not fully formed and have not yet separated from each other. Along the upper border, of this vessel many small endothelial nuclei can be seen. Along the lower border of this vessel, the delicate endothelial membrane

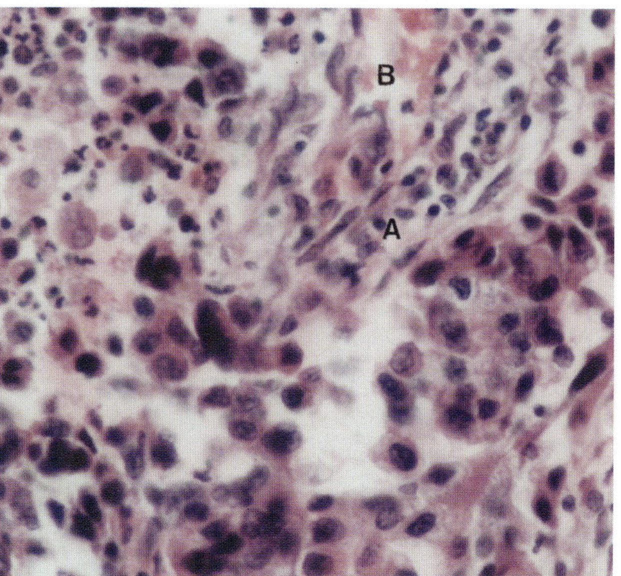

Figure 27.1, H&E x 520

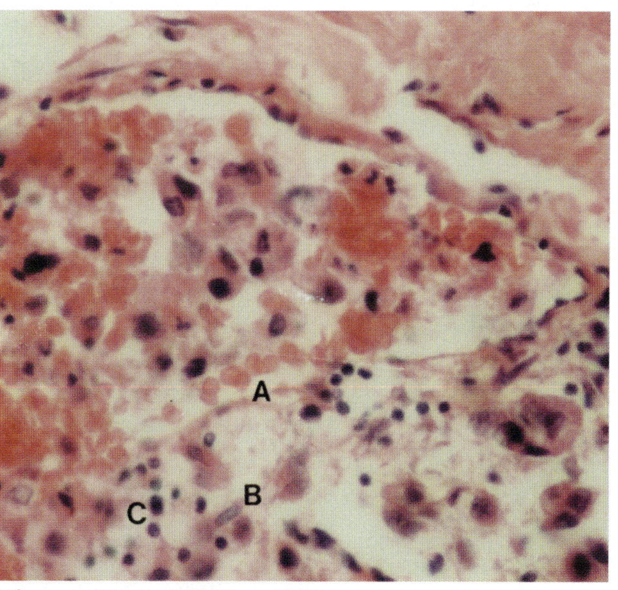

Figure 27.2, H&E x 520

(A) is just beginning to form. At the same time there is an interrupted line of new endothelial cells (B) forming in an attempt to extend the lumen. Within the extended lumen area (C), there are scattered tiny inflammatory cell nuclear remnants and necrosed cancer cells. In the lower right corner of this figure, only a few isolated tumor cells can still be recognized. H&E x 520

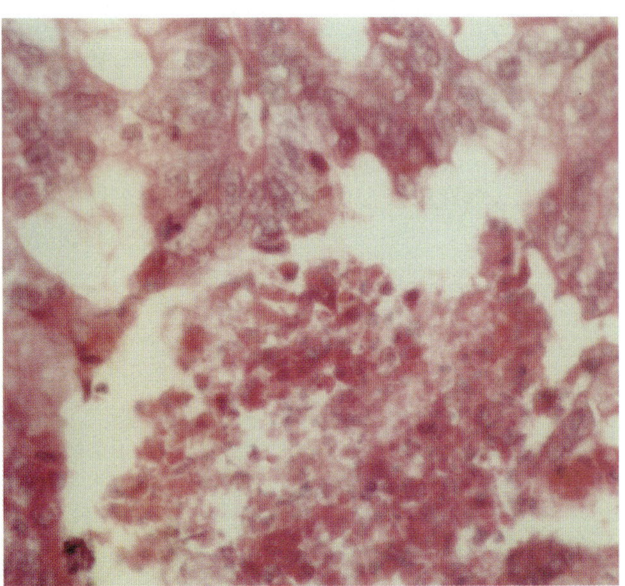

Figure 27.3, H&E x 400

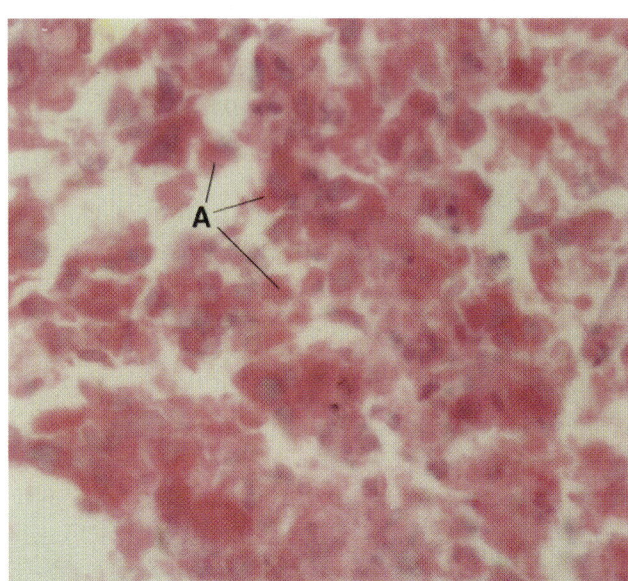

Figure 27.4, H&E x 800

2. Adenocarcinoma (figs. 27.3-27.4)

Figures 27.3 and 27.4 are taken from a section of the left upper lobe of a lung with adenocarcinoma.

Plasma and red cell development from liquefying hemorrhagic necrosed tumor tissue (figs. 27.3 and 27.4)

Figure 27.3. In the lower portion of this figure, liquefaction of the tumor tissue results in a large cystic space. This space is filled with clear plasma and the remains of hemorrhagic necrosed tumor tissue. From the necrosed tumor tissue there is develop-ment of red cells (better demonstrated in the next figure). Malignant epithelial cells can be recognized lining the cystic space. In the upper left of this figure, the acinar structure of the malignant epithelium can still be recognized. H&E x 400

Figure 27.4. In this section of adenocar-cinoma of the lung, extensive hemorrhagic necrosis and liquefaction of the tumor tissue is seen. There is red cell develop-ment (A) from the necrosed tumor tissue. H&E x 800

3. **Mesothelioma** (figs. 27.5-27.7)

Figures 27.5-27.7 are from the surgical lobectomy of the left lower lobe of a 50-year-old male with mesothelioma. On gross examination the tumor measured 3 x 2.5 x 2.5 cm.

Fibrinoid changes in mesothelioma with development of red cells and vessel formation (figs. 27.5-27.7)

Figure 27.5. This is an atypical solid type of mesothelioma. There are two areas (A and B) where the spindle-shaped tumor cells are undergoing reddish fibrinoid changes towards channel formation containing clear fluid. H&E x 520

Figure 27.6. In the mid-lower part of this figure there is an irregular area of red cells developing directly from liquefied mesothelioma cells. In the lower part of this same area, there are pyknotic changes in the tumor cells undergoing hemoglobinization and red cell formation. H&E x 1040

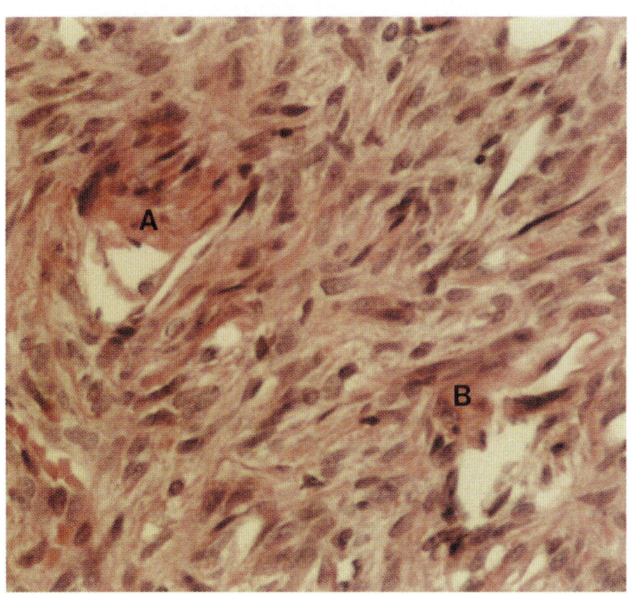

Figure 27.5, H&E x 520

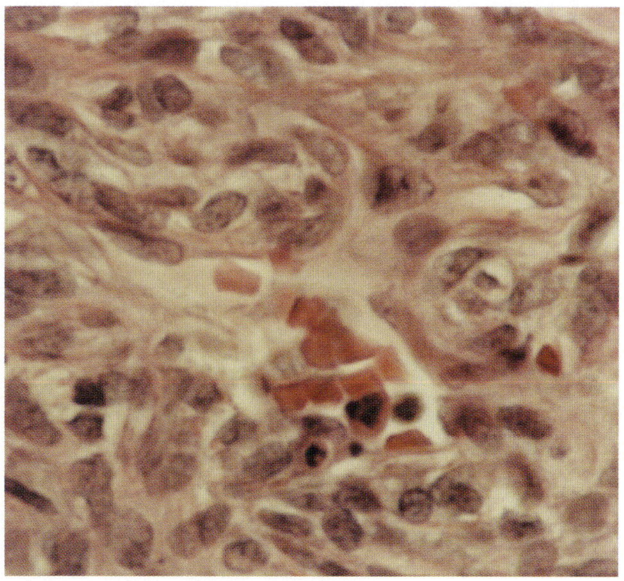

Figure 27.6, H&E x 1040

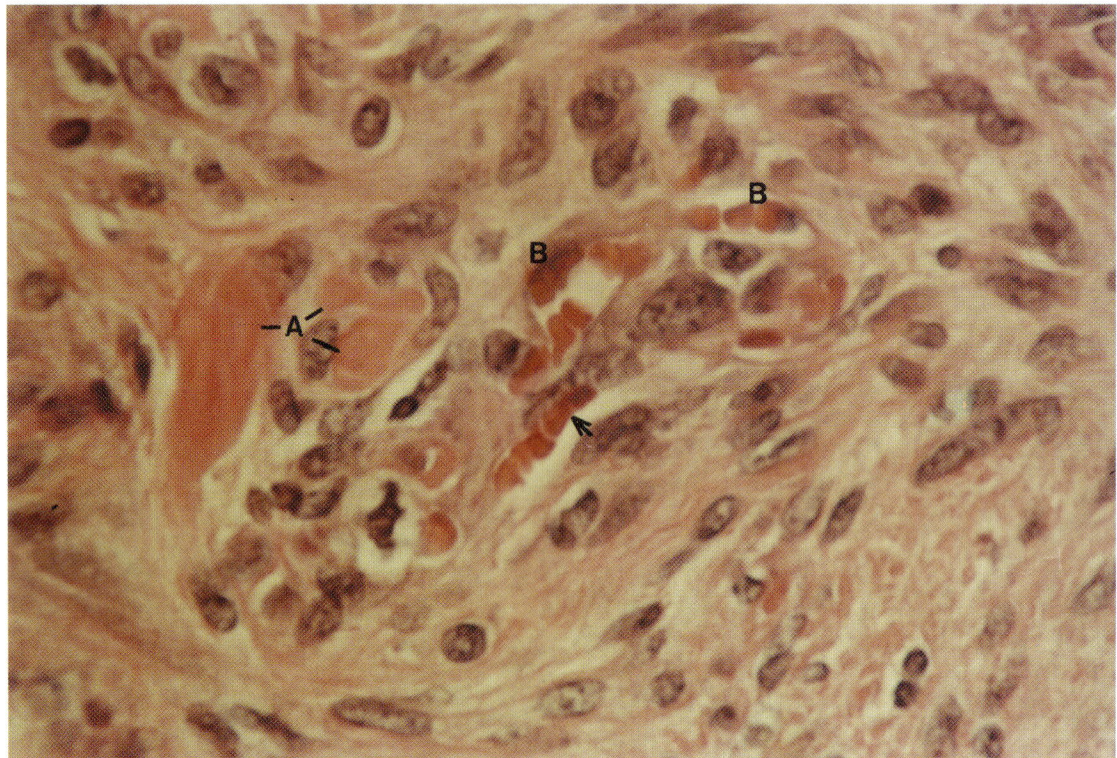

Figure 27.7, H&E x 1040

Figure 27.7. In this figure, there are reddish fibrinoid changes (A). In the middle third of this figure, there is an irregular area of red cells developing directly from tumor cells with early changes towards blood capillary formation. Remnants of mesothelioma cells can be seen still attached to red cells (B). In this irregular area there is a narrow single column of pressed-biscuit-like red cells (arrow) developing from the mesothelioma cells lining a narrow vascular channel. H&E x 1040

Chapter 28

PANCREATIC CANCERS (figs. 28.1-28.8)

1. Adenocarcinoma of Pancreas (figs. 28.1-28.4)

Figures 28.1-28.4 are taken from a 64-year-old male with adenocarcinoma of the pancreas surgically excised by Whipple operation. Cellular changes in benign pancreatic tissues are presented in Chapter 5, subsection 5.

Adenocarcinoma of the pancreas showing hyalinization with marked erasement of pancreatic tissue structures (figs. 28.1-28.4)

Figure 28.1. This is an adenocarcinoma of the pancreas in which only a few of the acinar and ductal structures can be identified. There is prominent hyalinization, disorganization, and disappearance of the pancreatic tissue structures. No mitosis is seen here. H&E x 800

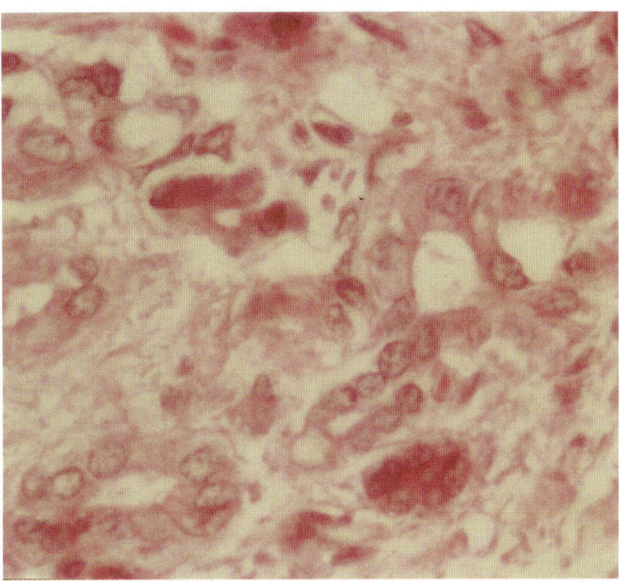

Figure 28.1, H&E x 800

Formation of inflammatory cells and spindle cells (fig. 28.2) and red cells (figs. 28.2-28.4) from pancreatic tissue

Figure 28.2. In some areas of this figure, epithelial cell nuclei are becoming pyknotic and transforming into inflammatory/reactive cells. There is erasement of the damaged pancreatic tissue with partial replacement by spindle cells towards stroma formation. (A) points to two acinar cells becoming hemoglobinized preliminary to red cell formation. Acinar cells (B) are transforming into reactive cells with lightly hemoglobinized cytoplasm. In the upper right corner of this figure, red cells are developing from similar hemoglobinized cells. H&E x 400

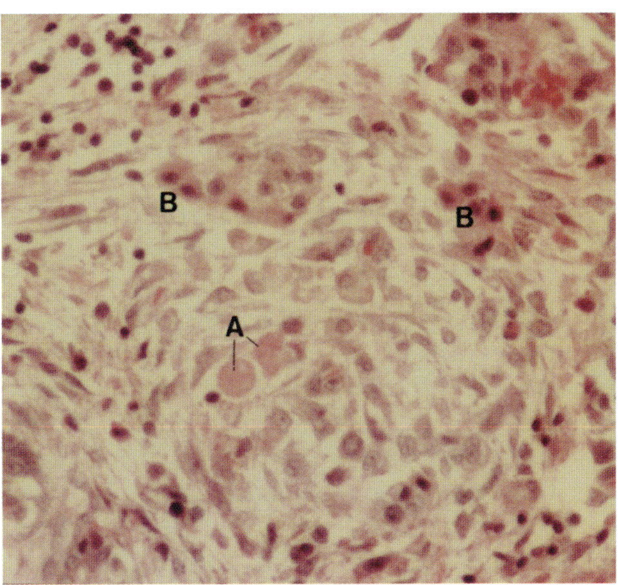

Figure 28.2, H&E x 400

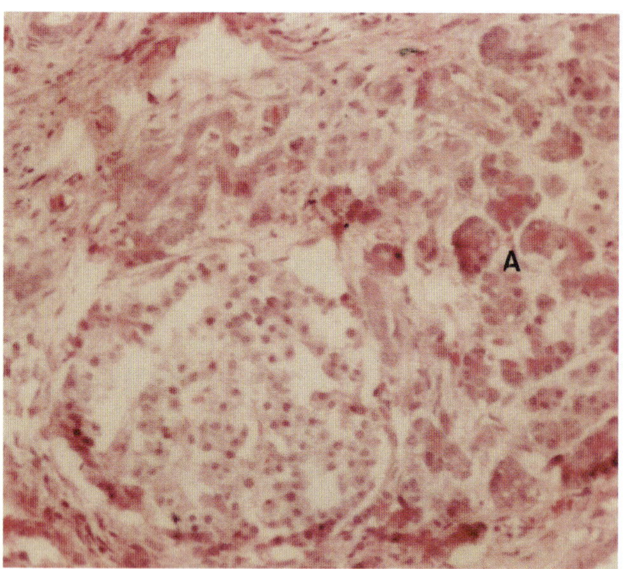

Figure 28.3, H&E x 200

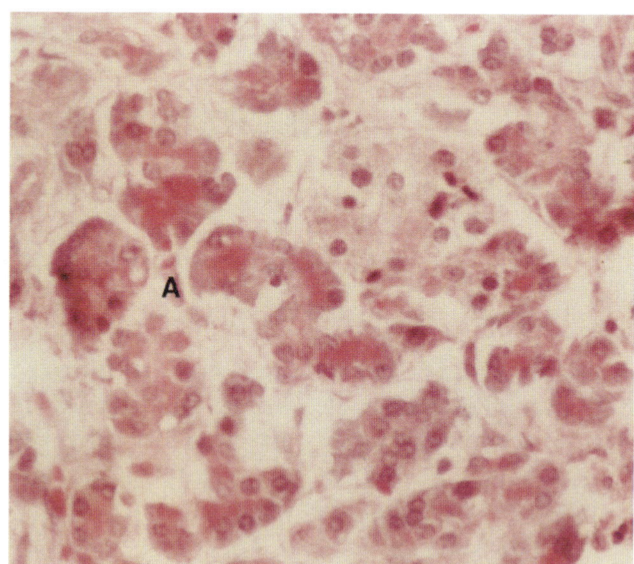

Figure 28.4, H&E x 400

Figures 28.3 and 28.4. Figure 28.3, in this section of pancreas the Islet of Langerhans is pretty much preserved. Surrounding the islet cells, the pancreatic tissue is distorted and erased and there are abnormal cellular changes with replacement of normal acinar structures. On the right, a few acinar structures (A) show the beginning of hemoglobinization and red cell development from pancreatic tissue. This area is seen in higher magnification in figure 28.4. Red cells are not yet formed and no attempt at vessel formation is noted. Figure 28.3 H&E x 200; and figure 28.4 H&E x 400.

2. Islet Cell Tumor of Pancreas (figs. 28.5-28.6)

Figures 28.5 and 28.6 are from the autopsy of a 66-year-old male with malignant islet cell tumor of the pancreas metastatic to liver and lymph nodes.

Blood development from necrosed tumor tissue (figs. 28.5 and 28.6)

Figures 28.5 and 28.6. Figure 28.5, in this section of islet cell tumor of the pancreas, hyperplastic islet cells can be clearly seen in the lower part of this figure. Hemorrhagic tumor tissue can be seen in the upper middle. This hemorrhagic area is probably an area of blood developing from necrosed Islet cell tumor tissue. The red cells are not yet fully rounded. Area (1) is shown in higher magnification in figure

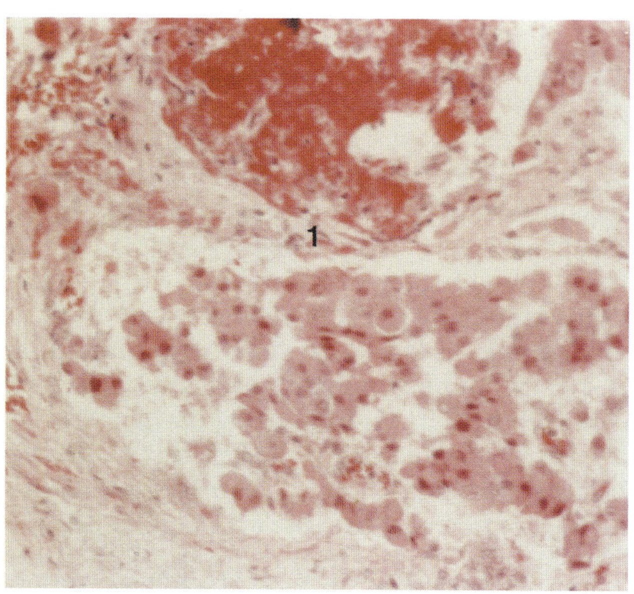

Figure 28.5, H&E x 260

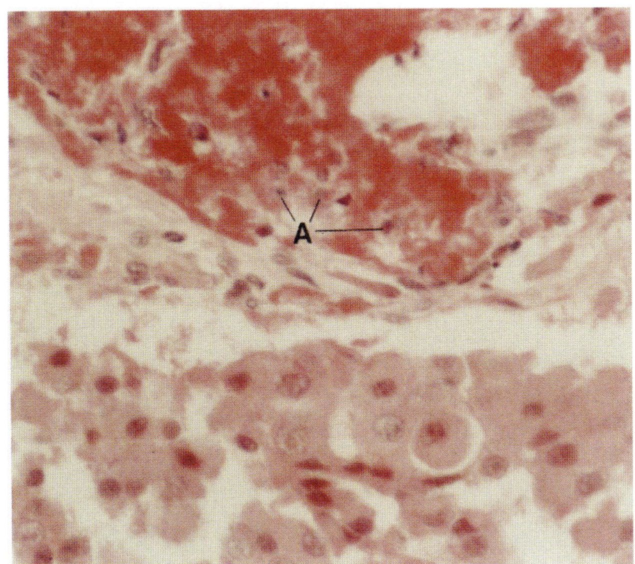

Figure 28.6, H&E x 520

28.6. A few remaining necrosed islet cells (A) can be seen in the hemorrhagic area. Figure 28.5   H&E x 260;   and figure 28.6   H&E x 520

3. Anaplastic Carcinoma Head of Pancreas Metastatic to Liver (figs. 28.7-28.8)

Figures 28.7 and 28.8 are from a surgical specimen of liver taken from a patient with anaplastic carcinoma of the head of the pancreas with metastasis to liver.

Development of blood from necrosed liver tissue as a reaction to nearby carcinoma (figs. 28.7 and 28.8)

Figure 28.7. On the left, metastatic anaplastic carcinoma of pancreas can be seen. On the right, the remains of liver tissue can still be identified. H&E x 260

Figure 28.8. Near the metastatic anaplastic carcinoma blood is forming from necrosed liver tissue stimulated by the nearby cancer. Red cell development starts as clumps of hemoglobinized material from which the individual red cells form and become separated. H&E x 520

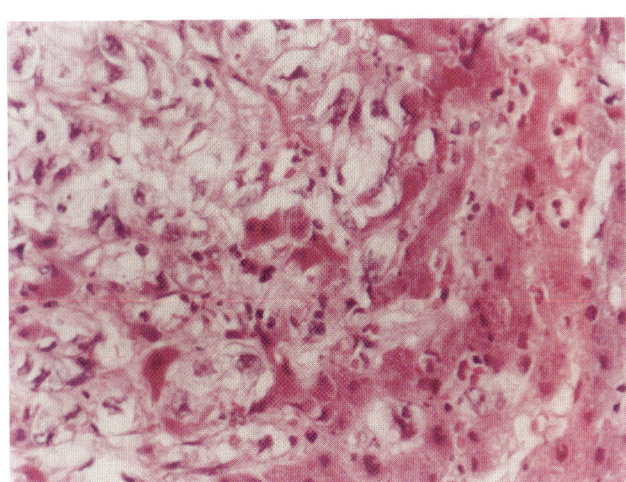

Figure 28.7, H&E x 260

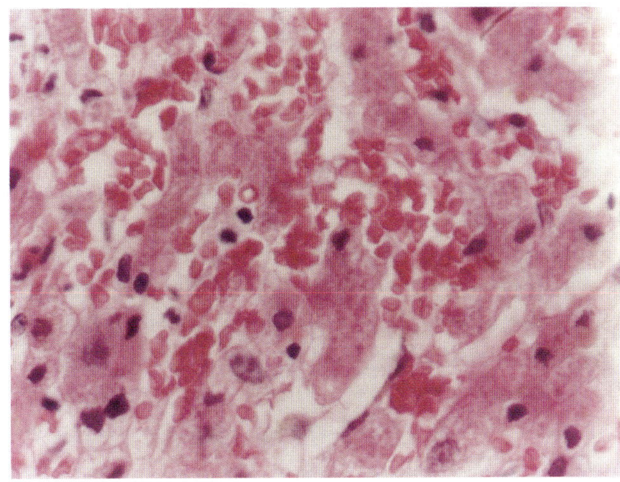

Figure 28.8, H&E x 520

Chapter 29

ADENOCARCINOMA OF LIVER (figs. 29.1-29.5)

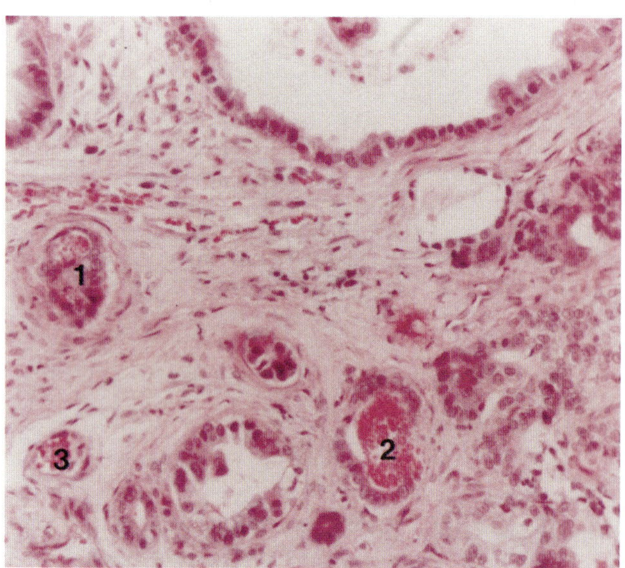

Figure 29.1, H&E x 200

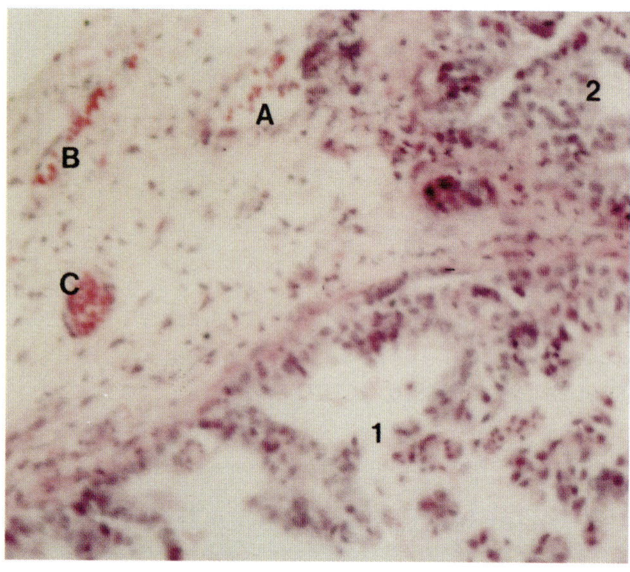

Figure 29.2, H&E x 260

Figures 29.1-29.5 are taken from the surgical specimen of a liver from a patient with adenocarcinoma. Cellular changes in benign liver tissue are demonstrated in Chapter 12.

Red cell and blood vessel development from malignant epithelium (figs. 29.1 and 29.2)

Figure 29.1. This is a low-magnification view of adenocarcinoma of the liver showing the glandular and stromal structures of the tumor. The lining of the glandular structures is composed of various sizes of malignant epithelial cells with irregular nuclear chromatism. There are two malignant acinar structures (1 and 2) partially filled with red hemorrhagic substance in which developing red cells can be identified. This red hemorrhagic substance is formed from the liquefaction of tumor tissue. Another glandular structure (3) has lining cells which compare more with

endothelial cells and in the lumen there are developing red cells. H&E x 200

Figure 29.2. Proliferating delicate malignant epithelial structures form a lace-like pattern within the two papillary cystic areas (1 and 2). Tiny inflammatory cells, including SN cells, can be seen in some of the lacy structures. In the upper left of this figure, in the loose stromal tissue there are also minute scattered inflammatory cells, mainly SN cells. There is a dilated vascular space (A) with red cells, clear plasma, and indistinct (barely visible) remnants of necrosed tumor tissue. This vascular channel is lined by indistinct endothelium and partially necrosed tumor tissue. In the upper left corner there is a vascular channel (B), not well lined, with more developed red cells. In vascular channel (C), the red cells are loosely clumped together and the lumen is partially lined by endothelial cells. H&E x 260

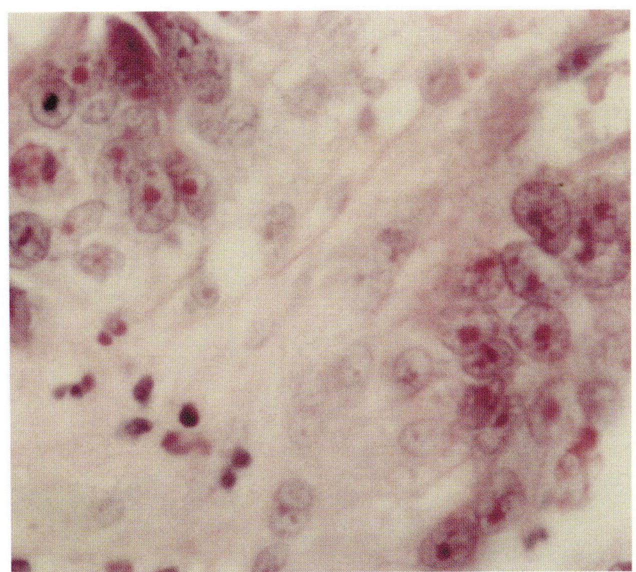

Figure 29.3, H&E x 1040

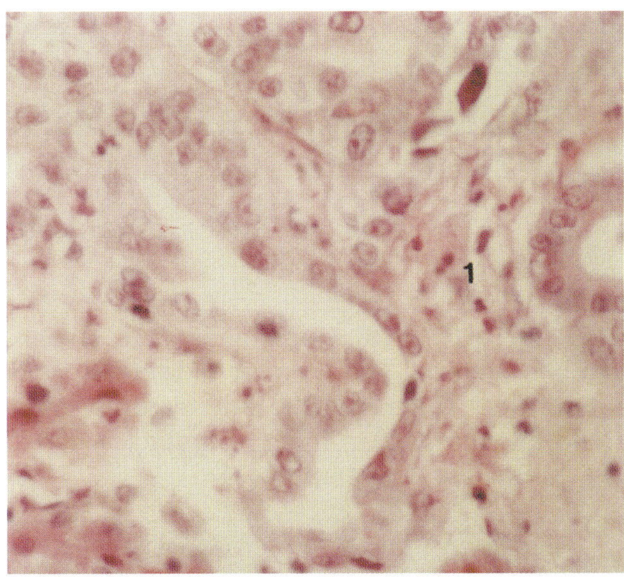

Figure 29.4, H&E x 520

*Stroma development from cancer
cells (fig. 29.3)*

Figure 29.3. In this higher magnified
view, two groups of rather atypical tumor
cells are clearly seen (upper left and lower
right). Most of these cancer cells have a
non-visible cytoplasmic border, a prominent
nucleus with somewhat clear nucleoplasm,
and a central single distinct nucleolus. In
between these two groups of cancer cells,
lying diagonally across the figure, the
cancer cells are lightly stained and vanishing towards development of stromal tissue.
In the lower left, the cancer cells have
almost vanished leaving only a few nuclear
remnants. H&E x 1040

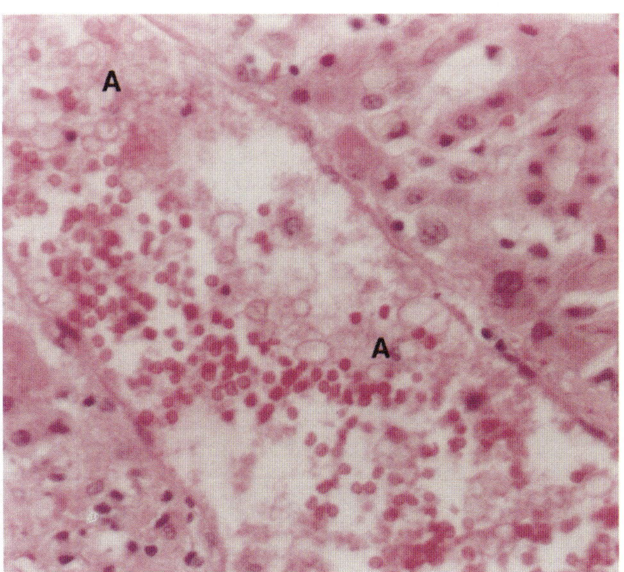

Figure 29.5, H&E x 400

*Invasion of normal liver tissue by
cancer cells (fig. 29.4)*

Figure 29.4. In this figure, the adenocarcinoma is invading the normal liver
tissue with replacement of the latter. There
is sparse inflammatory reaction in the liver
tissue (area 1) with the presence of
scattered tiny SN cells. In the lower left,
remnants of normal liver tissue can still be
identified. H&E x 520

*Normal liver tissue near the approaching
cancer developing blood and a blood
vessel (fig. 29.5)*

Figure 29.5. Near the approaching
cancer, there is a large blood vessel developing from liver tissue. Remains of
vacuolated liver cells (A) are noted within
the blood vessel. Many red cells are
developing from the dissolving liver tissue.
H&E x 400

Chapter 30

TUMORS OF THE GASTROINTESTINAL TRACT   (figs. 30.1-30.21)

1. Adenocarcinoma of Stomach   (figs. 30.1-30.4)

Figures 30.1-30.4 are taken from a partial gastrectomy of a 69-year-old female with moderately differentiated adenocarcinoma.[1] Cellular changes in benign stomach tissues are presented in Chapter 14.

Blood and blood vessel development through locally developed erythrogenic inflammatory cells of local origin (figs. 30.1-30.4)

Figure 30.1. This is a low-magnification view of adenocarcinoma invading the gastric fundus, showing many inflammatory/reactive cells and several irregular dilated vascular channels. In the upper part of this figure, there are many inflammatory cells and dilated newly formed blood capillaries of gastric wall origin. The capillaries are not yet properly endothelialized. In the lower part of this figure, the architecture of the adenocarcinoma is nicely preserved. H&E x 130

Figure 30.2. This is a higher magnification of another area of this adenocarcinoma where malignant glandular structures, with disorganization and distortion of lining epithelial cells, are better seen. There are locally developed reactive cells sprinkled between the glandular structures and more prominent on the left border of this figure. Many of these reactive cells are SN cells, including erythrogenic SN cells. (A) points to a blood vessel partially filled with red cells derived from erythrogenic SN cells (see also next figure). Occupying the middle and

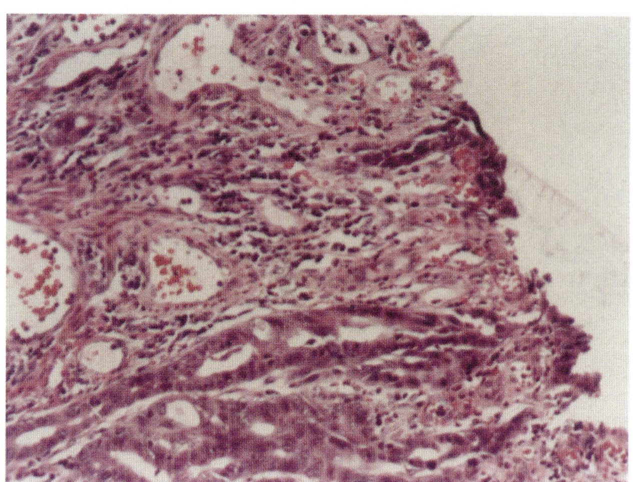

Figure 30.1, H&E x 130

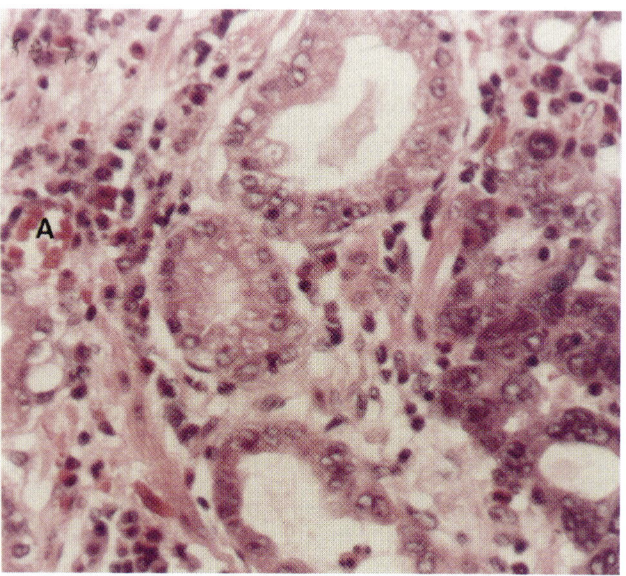

Figure 30.2, H&E x 400

---

1.   A large blood vessel developing from the smooth muscle is described in figure 4.35. Blood and blood vessel formation from reactive cells of smooth muscle origin is described in figure 4.36.

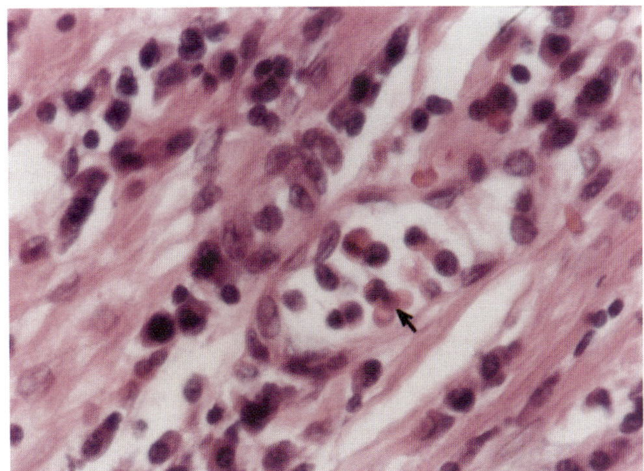

Figure 30.3, H&E x 520

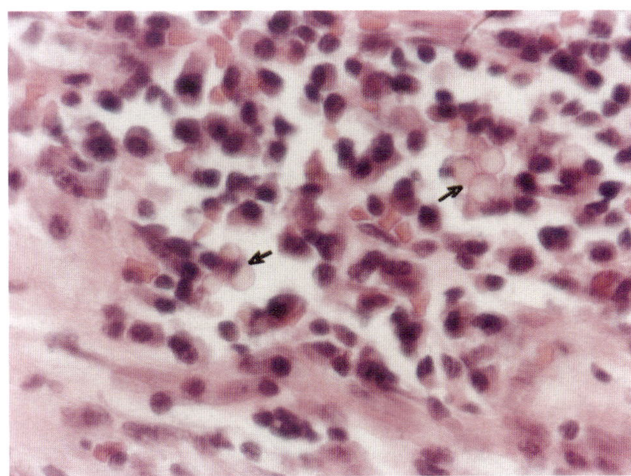

Figure 30.4, H&E x 520

right of this figure, there is well-differentiated adenocarcinoma of gastric origin. On
the right, the cancer tissue is more hyperchromatic and disorganized. H&E x 400

Figure 30.3. In the center, chronic and
subacute erythrogenic reactive cells are
seen forming red cells within a developing
vascular lumen. One erythrogenic SN cell
is forming multiple red cells (arrow). These
erythrogenic reactive cells are developed locally, from the smooth muscle fibers, as a

reaction to the nearby cancer. Many of these
reactive cells are developing along the
smooth muscle fibers. H&E x 520

Figure 30.4. Initiated by the nearby adenocarcinoma, reactive cells (including SN
cells and plasma cells) are developing locally from the smooth muscle coat. Many
hemoglobin globules (arrows) and red cells
are developing from the reactive cells. These
hemoglobin globules are of various sizes.
H&E x 520

2. Reticulum Cell Sarcoma of Stomach (figs. 30.5-30.9)

Figures 30.5-30.9 are taken from the
autopsy specimen of a 66-year-old male with
reticulum cell sarcoma of the stomach superimposed with amoebic ulceration and
metastatic to perigastric lymph nodes.

*Reticulum cell sarcoma with blood and blood
vessel formation (figs. 30.5-30.7)*

Figure 30.5. This is a low-magnification
view of the reticulum cell sarcoma of the
stomach, with widespread necrosis superimposed with amebiasis, presented in figures 30.6-30.9. H&E x 260

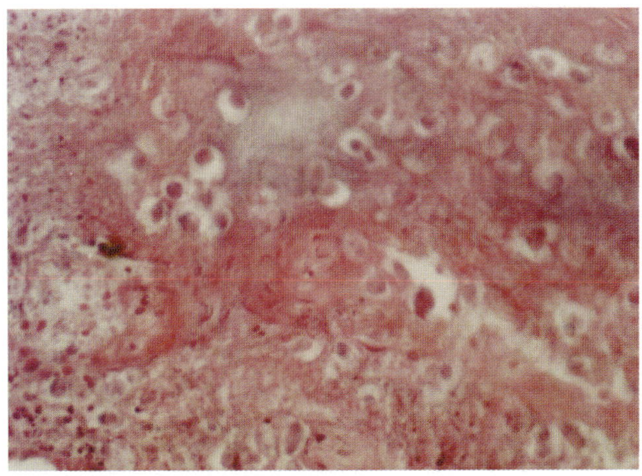

Figure 30.5, H&E x 260

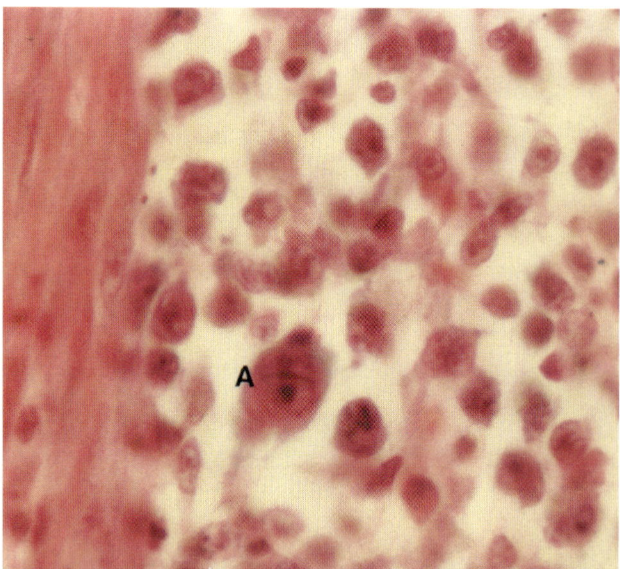

Figure 30.6, H&E x 800

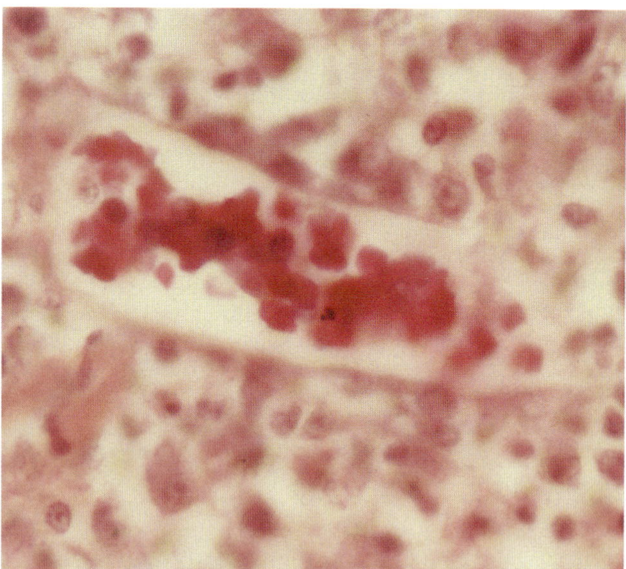

Figure 30.7, H&E x 800

Figure 30.6. This is a higher magnification of this reticulum cell sarcoma of the stomach. A highly abnormal large multinucleated cell (A) can be seen among the smaller surrounding, dissolving tumor tissue. H&E x 800

Figure 30.7. In this figure, a blood vessel is developing from an inflamed area of reticulum cell sarcoma of stomach. Necrosed blood forming tissue can be seen in the lumen of this vessel. H&E x 800

Reticulum cell sarcoma of stomach with amebiasis (figs. 30.8 and 30.9)

Figure 30.8. This is a low-magnification view of an Iron Stained section of reticulum cell sarcoma of the stomach superimposed with amebiasis.[2] See the next figure for a magnified view of this tissue section. Iron Stain x 104

Figure 30.9. Trophozoite forms of Entamoeba histolytica (amebiasis) and reticulum cell sarcoma of stomach are seen here. Vegetative forms of Entamoeba

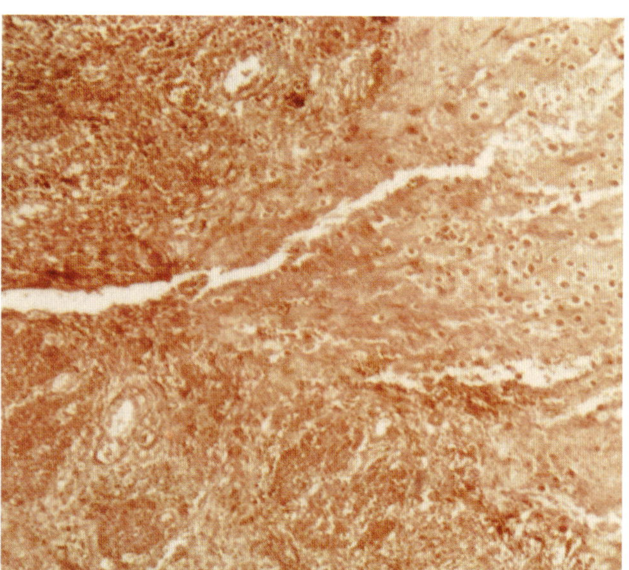

Figure 30.8, Iron Stain x 104

histolytica in this specimen are identified by their large round to oval forms averaging 20 to 25 microns in diameter, and by a single nucleus (small in comparison to the abundant granular cytoplasm) with only one nucleolus. Iron Stain x 1040

---

2. This case was originally presented with black and white photomicrographs in *JAMA* (McDonald and Moore, 1965).

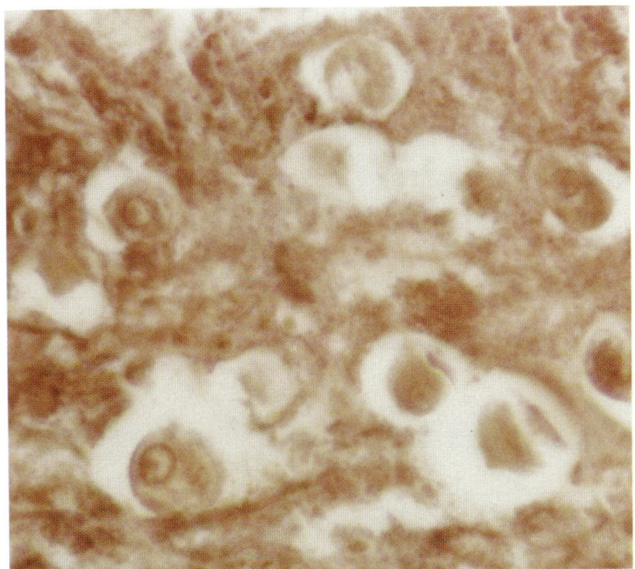

Figure 30.9, Iron Stain x 1040

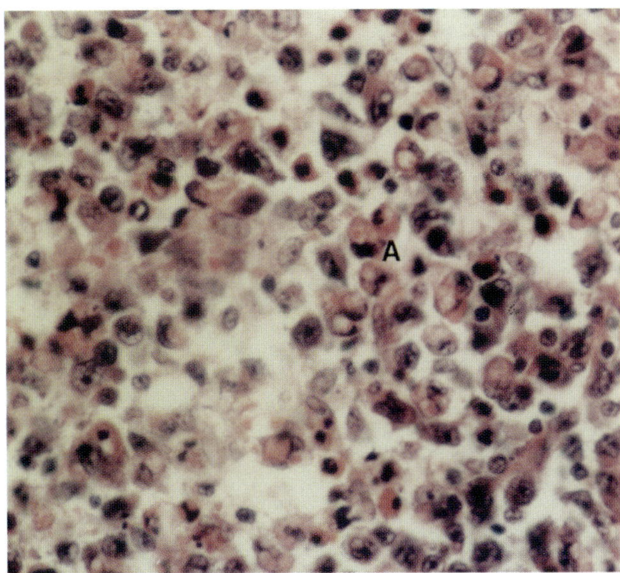

Figure 30.10, H&E x 520

A. Reticulum cell sarcoma of stomach metastatic to perigastric lymph nodes
(figs. 30.10-30.11)

The perigastric lymphatic tissue in figures 30.10 and 30.11 is taken from the autopsy of a 66-year-old male with reticulum cell sarcoma of the stomach (metastatic to perigastric lymph nodes) superimposed with amoebic ulceration.

Red cell development from metastatic reticulum cell sarcoma (figs. 30.10 and 30.11)

Figure 30.10. This autopsy specimen of reticulum cell sarcoma of the stomach metastatic to lymph nodes shows many tumor cells forming vacuolation, area (A), preliminary to red cell formation (see next figure). H&E x 520

Figure 30.11. Here the vacuolated sarcoma cells (A) are becoming hemoglobinized and developing multiple red cells. Red cells are also developing directly from cancer cells without vacuoles (B). H&E x 1040

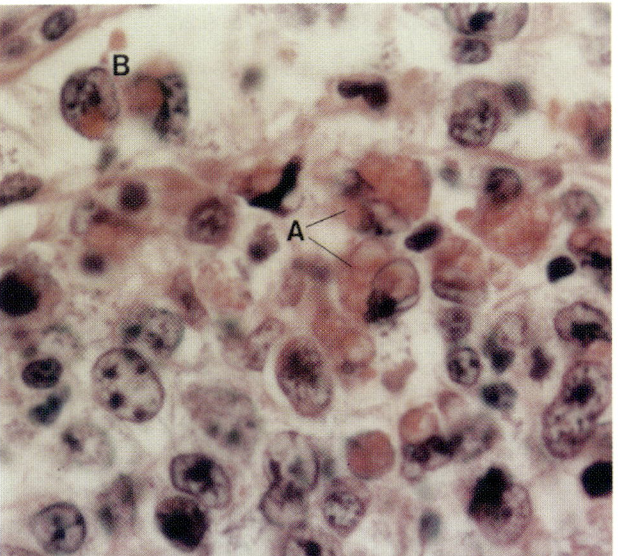

Figure 30.11, H&E x 1040

B. Perigastric adipose tissue near reticulum cell sarcoma of stomach
(fig. 30.12)

The perigastric adipose tissue in figure 30.12 is taken from the autopsy of a 66-year-old male with reticulum cell sarcoma of the stomach superimposed with amoebic ulceration presented in figures 30.5-30.9.

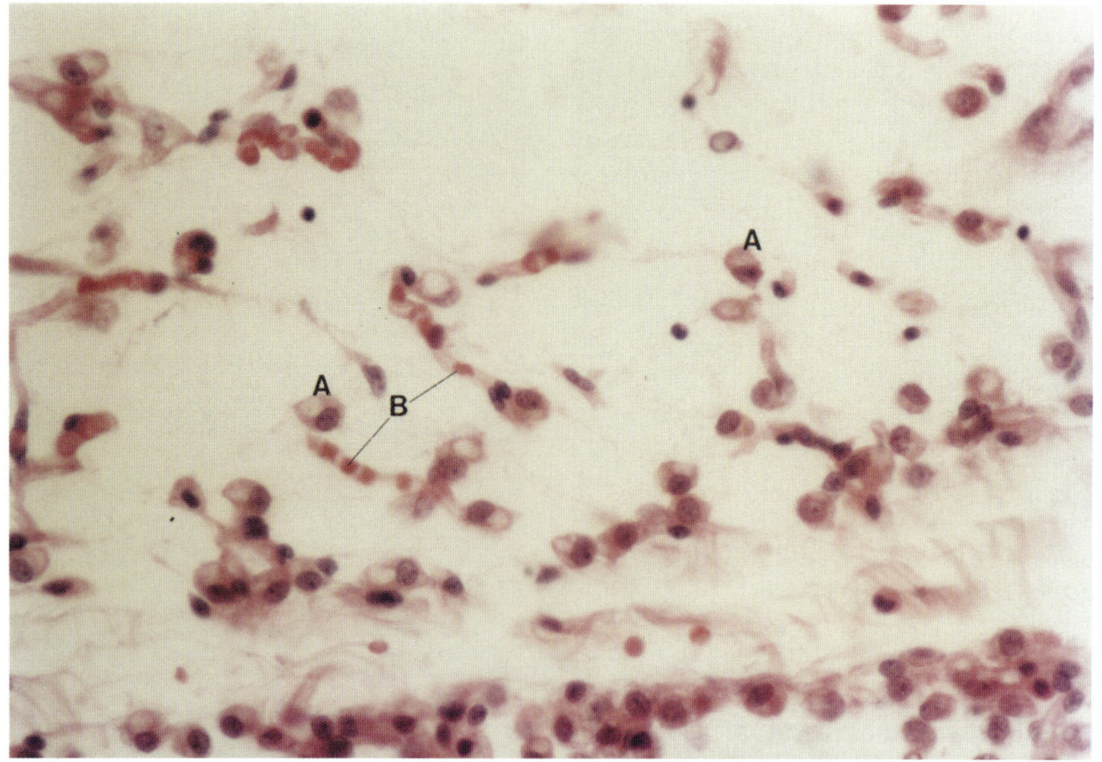

Figure 30.12, H&E x 520

Blood and blood vessel development from perigastric adipose tissue as a reaction to approaching cancer (fig. 30.12)

Figure 30.12. Excited by nearby reticulum cell sarcoma, the normal perigastric adipose tissue transforms and often develops into the small vacuolated cells (A) seen here. There is evidence of blood capillary formation with single columns of red cells (B) developing within the tubular (hyphae-like) structures of these transformed adipose cells. H&E x 520

3. Colon Carcinoma (figs. 30.13-30.21)

Figures 30.13-30.20 are taken from a surgically removed colon carcinoma from an 84-year-old male. Figure 30.21 is taken from a surgically removed adenocarcinoma of the sigmoid colon taken from a 59-year-old male.[3]

Blood and prominent blood vessel development from highly necrotic tumor tissue (figs. 30.13-30.15)

Figure 30.13. This is a low-magnification view of the ulcerative necrotizing adenocarcinoma of the colon presented in figures 30.14-30.20. There is marked pro-

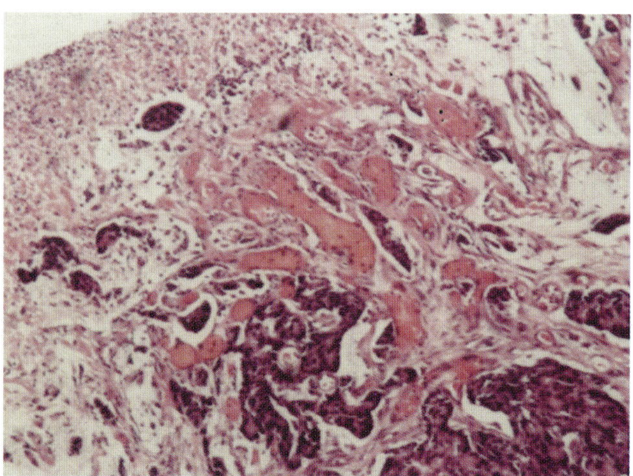

Figure 30.13, H&E x 104

---

3. Additional photomicrographs and findings in colon carcinoma are presented in Volume I, figures 90-95.

liferation of developing blood vessels from the necrotic tumor tissue. These areas are commonly called dilated thrombotic vessels but are actually areas of transformation of local necrosed tumor tissue into developing blood and blood vessels. The central area is seen in higher magnification in the next figure.[4] H&E x 104

Figure 30.14. This whole area of colon cancer is becoming vascularized through the necrosis of the tumor tissue. Cancer cells (A) are transforming into erythrogenic SN cells (B) towards development of blood vessels (C) filled with fine hemoglobinized particles which have not yet formed into red cells. There are a few remaining nuclei within the lumens of these two blood capillaries. Also, even in the sparsely cellular necrotic area (D) note the appearance of very fine, almost invisible, hemoglobin particles. These fine particles will enlarge to form an area similar to (D1). Development of red cells from these areas of fine hemoglobinized particles is demonstrated in the next figure. H&E x 400

Figure 30.15. In this figure, there is one prominent dilated engorged blood capillary which is not yet fully developed from necrotic tumor tissue. The upper part of this developing blood capillary is filled with fine hemoglobinized particles derived from necrotic tumor tissue. In the center and lower end of this capillary, red cells (A) are developing from the fine hemoglobinized particles. The red cells are not yet fully rounded. There are a few remaining nuclei within the lumen of this blood capillary. On the left, there is marked necrosis of tumor tissue which is producing erythrogenic inflammatory cells. The origin of this blood vessel from these necrotic erythrogenic inflammatory cells can be visualized when one examines the upper left corner where the two elements are contiguous. The

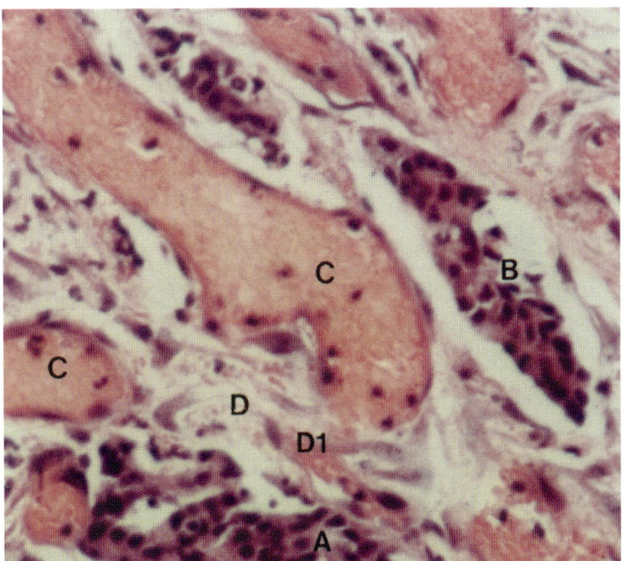

Figure 30.14, H&E x 400

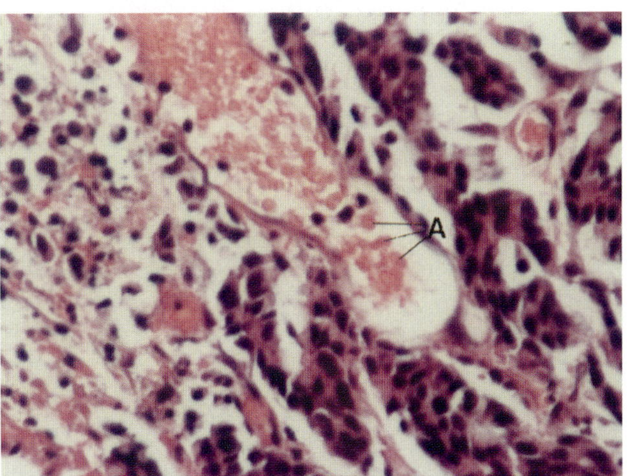

Figure 30.15, H&E x 400

endothelium of this blood vessel is formed from the remaining peripheral necrotic tumor tissue. Hemoglobinization of necrotic tumor tissue can also be seen in the lower left corner. In the right and lower portion of this figure, the cancer cells are more preserved. H&E x 400

---

4. Origin of erythrogenic SN cells and blood and blood capillaries from the pericolic adipose tissue of this lesion are presented in figures 3.7-3.9.

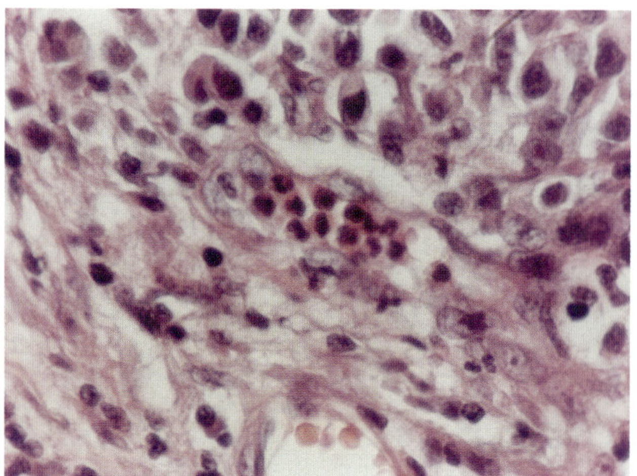

Figure 30.16, H&E x 520

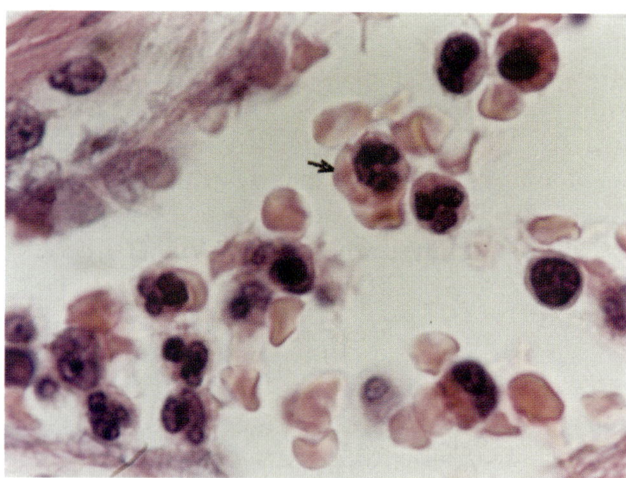

Figure 30.17, H&E x 1300

*Red cell development from SN cells
of necrotic tumor tissue origin
(figs. 30.16 and 30.17)*

Figures 30.16 and 30.17. The developing
blood capillary in the center of figure 30.16
is filled with erythrogenic SN cells. In
figure 30.17, several erythrogenic SN cells
are developing red cells within a vascular
lumen. The SN cell (arrow) is developing
multiple red cells. Figure 30.16 H&E
x 520; and figure 30.17 H&E x 1300

*Development of mucin from this tumor
(fig. 30.18)*

Figure 30.18. This figure demonstrates
the extensive production of mucin by the
cancer cells. Remaining cancer cells can
still be identified in the lower left corner.
H&E x 260

*Connective tissue stroma development
from epithelial tumor tissue
(figs. 30.19 and 30.20)*

Figures 30.19 and 30.20. These two figures demonstrate connective tissue stroma
developing from a non-necrotic area of this
colon cancer. In figure 30.19 crossing the

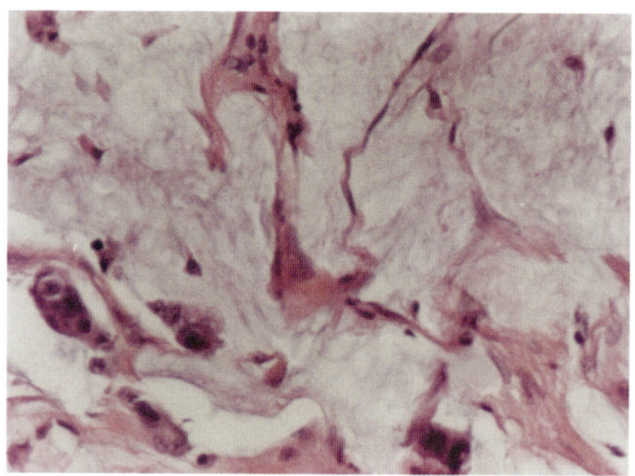

Figure 30.18, H&E x 260

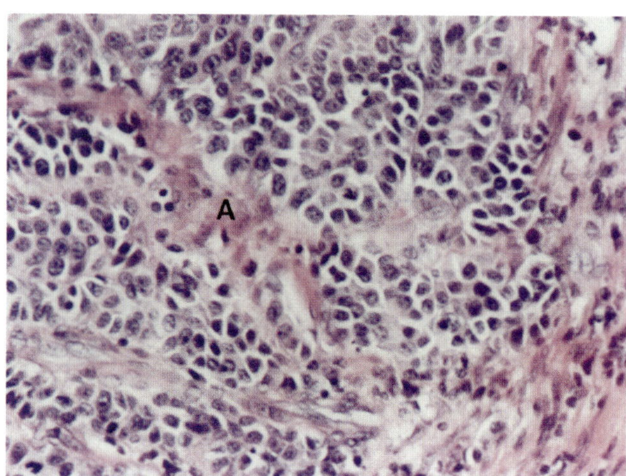

Figure 30.19, H&E x 260

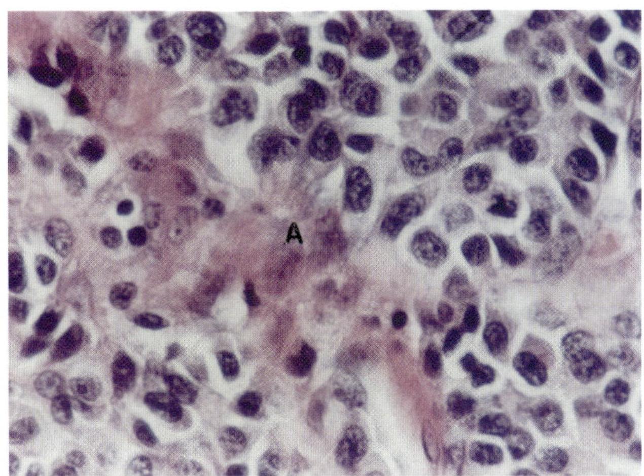

Figure 30.20, H&E x 520

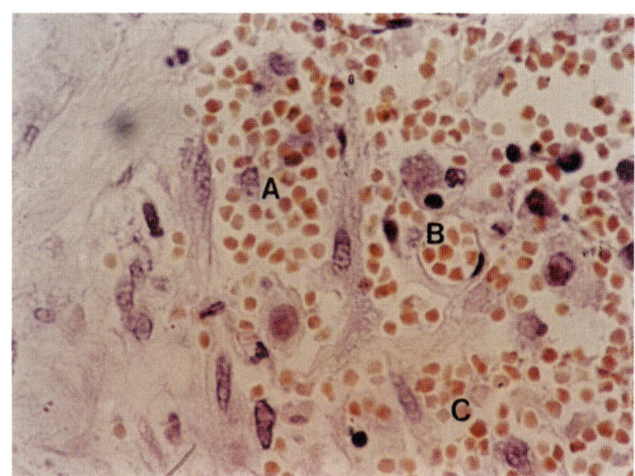

Figure 30.21, H&E x 520

tumor tissue, there are several linear tracts of fibrous stroma formation. In the lower right corner, more developed collagen fibers can be seen in the formation of connective tissue stroma. Area (A) is seen in higher magnification in the next figure. In figure 30.20, the cancer cells are becoming hyalinized (A) towards the development of collagenous stroma. Also, along this tract of stroma formation a few lymphocyte-like cells are arising from cancer cells. Figure 30.19 H&E x 260; figure 30.20 H&E x 520

Red cell and blood vessel development directly from cancer tissue (fig. 30.21)

Figure 30.21. Developing blood and dilated capillaries filled with red cells occupy two-thirds of this figure. Capillary (A) is shown in longitudinal plane and capillary (B) is in transverse plane. The lumen of these blood vessels are partially filled with red cells directly developed from tumor tissue. Remnants of tumor tissue can still be seen in vessel (A). In area (C), red cells which are directly developing from vanishing cancer cells still have a bluish tinge. H&E x 520

Chapter 31

CARCINOID TUMORS (figs. 31.1-31.42)

1. Primary Carcinoid of the Ileum (figs. 31.1-31.16)

Figures 31.1-31.7 are taken from an adult patient with primary carcinoid of the ileum removed by surgical resection. Figures 31.8-31.16 are taken from the small bowel removed from a 58-year-old female during exploratory laparotomy. Two small tumors of the distal ileum measuring approximately 1-2 cm, removed by wedge resection, were felt likely to be the primary carcinoid site with metastasis to the mesenteric lymph nodes and liver. Metastasis to mesenteric lymph node of this same patient is presented in figures 31.21-31.23. Biopsy sections of the anterior edge of the right lobe of the liver from this same patient are presented in figures 31.24-31.33. Cellular reactions in nearby uninvolved intestinal wall of this patient are presented in figures 31.34-31.42.

Typical carcinoid tumor (fig. 31.1)

Figure 31.1. This figure shows a typical carcinoid tumor with compact monotonous (similar appearing) cells with small dark stained nuclei, inconspicuous cytoplasm, and ill-defined cell margins. As usual, mitotic cells are not seen here. The carcinoid cells are grouped in solid clusters which appear as small and large islands of tumor tissue separated by a small amount of fibrous stroma. Figures 31.2-31.7 are taken from the same tissue section of this tumor. H&E x 200

Hemorrhagic necrosis of tumor tissue in production of red hemoglobin-like substance (figs. 31.2 and 31.3)

Figure 31.2. In the center of this figure, there is prominent hemorrhagic necrosis[1] of a nest of carcinoid tissue leaving a layer

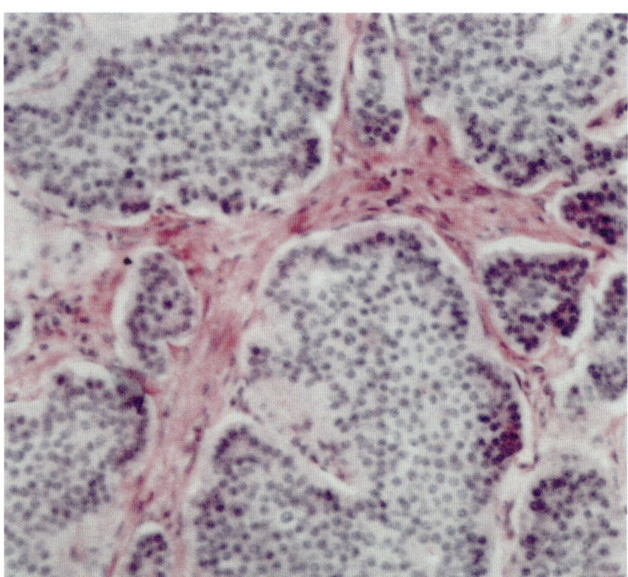

Figure 31.1, H&E x 200

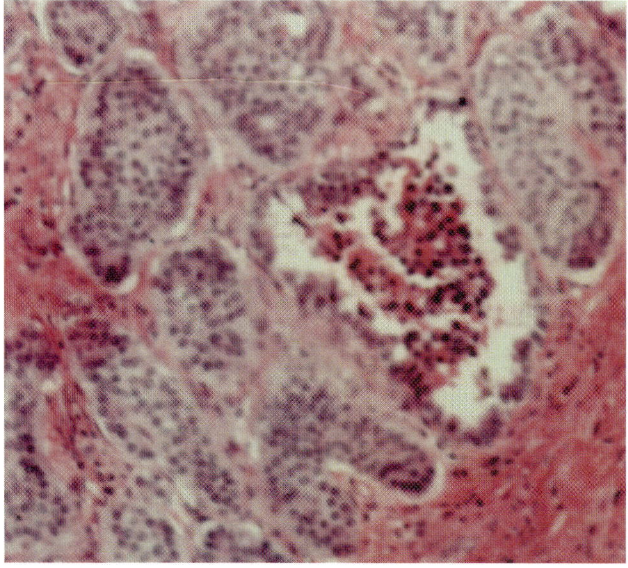

Figure 31.2, H&E x 200

of carcinoid cells at the margin. The dissolution of the necrosed carcinoid cells forms

---

1. A common term for the presence of blood (red cells) in necrosed tissue. The development of red cells through the process of necrosis of local tissues is often mistaken as hemorrhage.

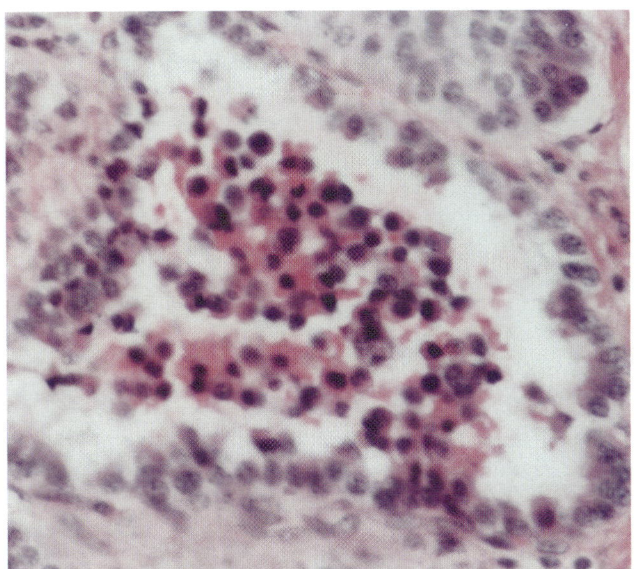

Figure 31.3, H&E x 400

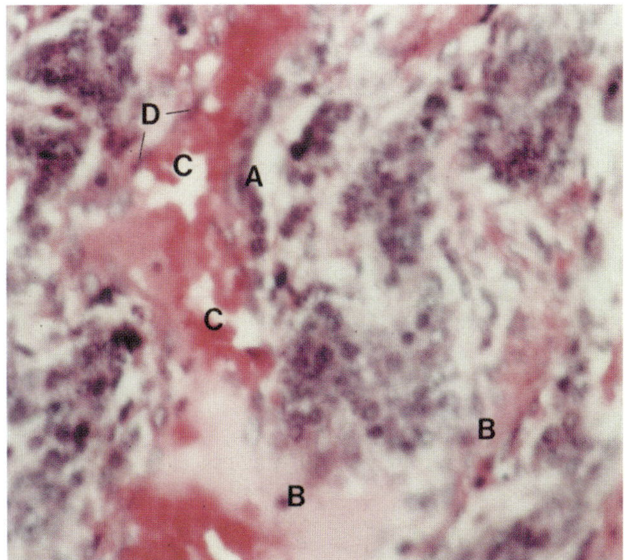

Figure 31.4, H&E x 400

somewhat clear fluid (plasma) around the edge. In the center, the carcinoid cells are undergoing changes in formation of crimson-red hemoglobin-like substance. This area is seen in higher magnification in the next figure. H&E x 200

Figure 31.3. Hemorrhagic necrosis of carcinoid cells is forming red hemoglobin-like substance which gives rise to a few red cells floating in the peripheral clear fluid. H&E x 400

Hyalinization and liquefaction of tumor tissue in development of red hemoglobinized substance, red cells and blood vessels (figs. 31.4 and 31.5)

Figure 31.4. Here the carcinoid tissue is undergoing hyalinization and liquefaction and developing into a crimson-red hemoglobinized substance in development of blood and a blood vessel. This process has left a linear arrangement of carcinoid cells along the edge (A). Carcinoid cells are becoming hyalinized and disappearing (B), leaving a light pink substance with a fine ground glass-like appearance. This light pink substance develops into the crimson-red hemoglobin-like substance and red cells (C). Early stages of faintly visible endothelial lining cells are seen at (D). This

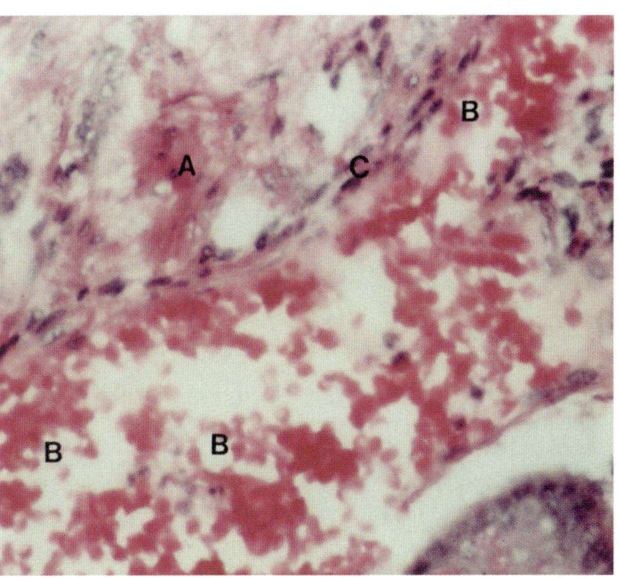

Figure 31.5, H&E x 400

endothelial lining will probably disappear with the further imminent expansion of this vascular channel. H&E x 400

Figure 31.5. There is hyalinization and liquefaction of tumor tissue in development of crimson-red hemoglobin-like substance (A). In this area only a few fragments of vanishing carcinoid tissue can be identified. A large blood vessel with developing red cells (B) is forming from the crimson-red hemoglobin-like substance derived from

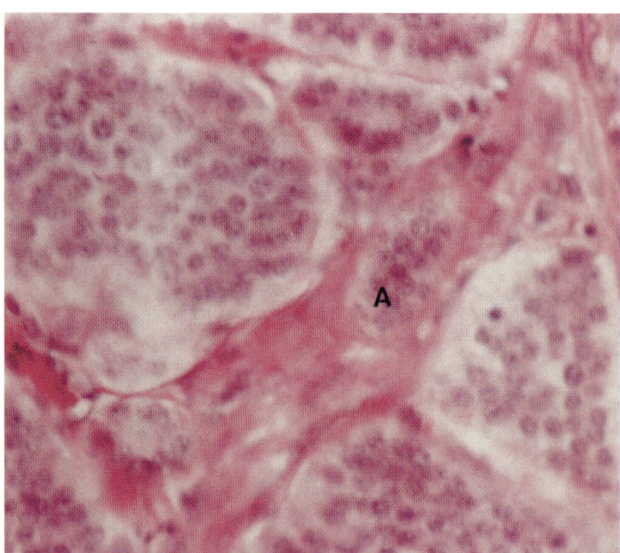

Figure 31.6, H&E x 400

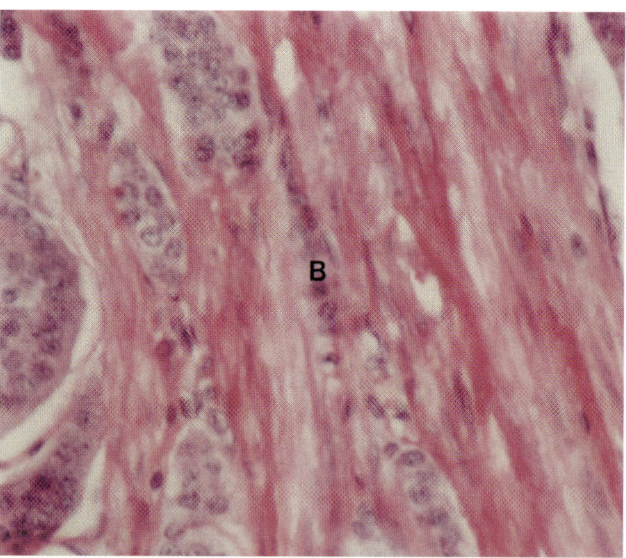

Figure 31.7, H&E x 400

dissolving carcinoid tissue. Hyper-hemoglobinized red cells are mostly seen in clumps and have not yet separated. There seems to be an attempt for endothelial lining formation with two parallel lines of tiny endothelial cells (C). Much of the inner lining is becoming hyalinized and disappearing in the process of red cell formation. Only a small portion of intact carcinoid tissue can be seen in the lower right. H&E x 400

Transformation of carcinoid cells into fibrous stroma (figs. 31.6 and 31.7)

Figures 31.6 and 31.7. In these two figures, groups of carcinoid cells are transforming into developing fibrous stromal tissue. In figure 31.6, in transverse plane, there is a group of remaining carcinoid cells (A) surrounded by developing stroma. In figure 31.7, more extensive fibrosis in longitudinal plane of the collagen fibers is seen. A streak of remaining carcinoid cells (B) can still be identified among the developing stromal tissue. H&E x 400

Development of fibrous connective tissue from hyalinized carcinoid tissue (figs. 31.8-31.10)

Figure 31.8. This figure shows solid groups of carcinoid tissue occupying the mucosal and submucosal tissue of the il-

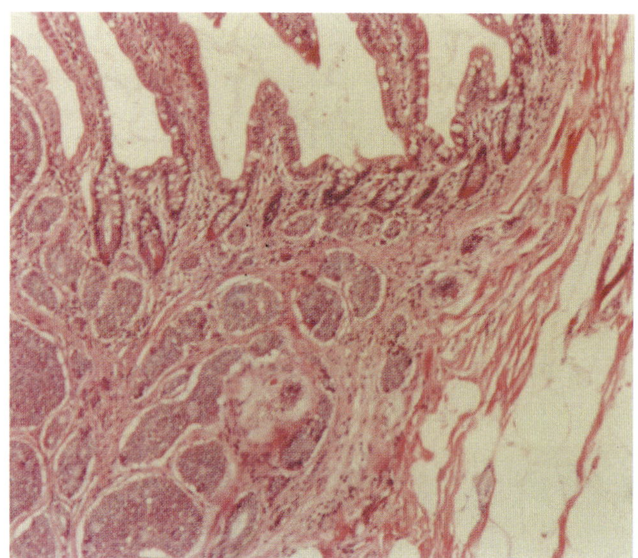

Figure 31.8, H&E x 80

eum. The nests of carcinoid tissue are separated by connective tissue stroma. Figures 31.9-31.16 are taken from the same tissue section as this figure. H&E x 80

Figure 31.9. In this figure the carcinoid cells are arranged in both solid and glandular patterns. Distinct glandular formation (A) is not commonly seen in carcinoid tumors. In the solid compact areas (B), the tumor cells have distinct nuclei but ill defined borders and some of the nuclei are vacuolated. In the central

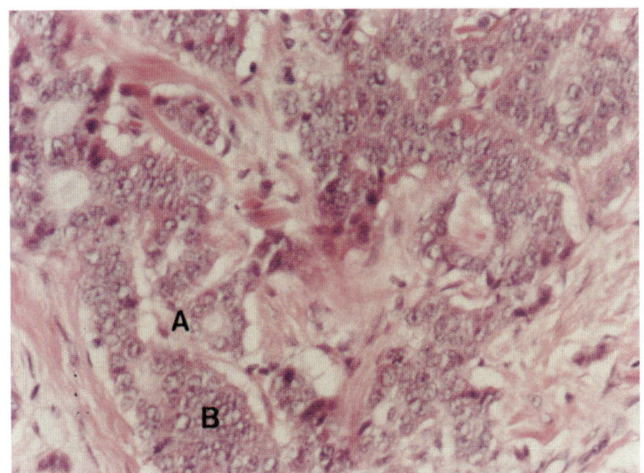

Figure 31.9, H&E x 260

area of this figure, there is hyalinization of carcinoid cells with formation of fibrous stroma. H&E x 260

Figure 31.10. Prominent erasement of the carcinoid cells with development of fibrous stroma is seen here. Remnants of individual carcinoid cells (A) can be seen among the developing stroma. There is development of spindle-shaped nuclei (B) among the collagen fibers. H&E x 400

Red cell development directly from carcinoid tissue (fig. 31.11)

Figure 31.11. Bright crimson-red hyperhemoglobinized erythrocytes are arising directly from carcinoid tumor tissue. Scattered throughout the figure, a few of the carcinoid nuclei show vacuole formation (A). H&E x 400

Extension of carcinoid into smooth muscle (figs. 31.12 and 31.13)

Figure 31.12. This is a section of the smooth muscle coat of the ileum. Here we find the appearance of tiny dedifferentiated carcinoid nuclei,[2] some with vacuole formation (A). These nuclei gradually increase in size (A1), and eventually develop into

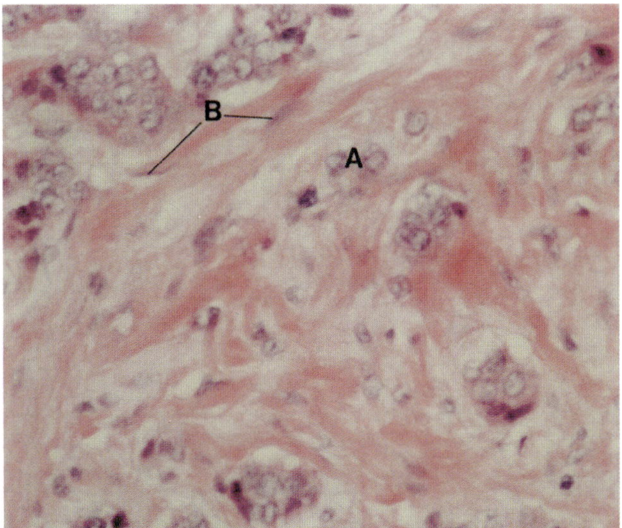

Figure 31.10, H&E x 400

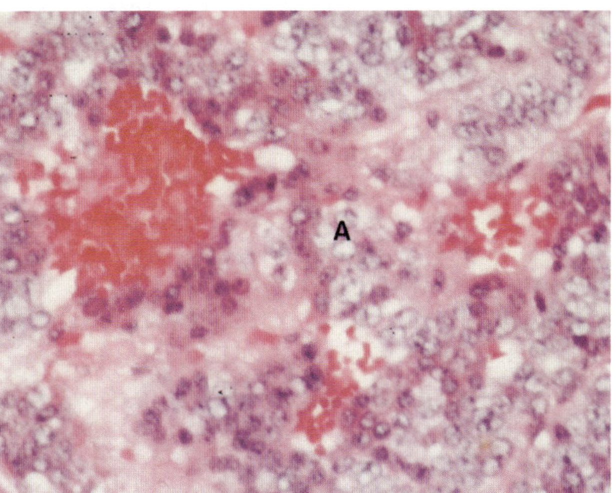

Figure 31.11, H&E x 400

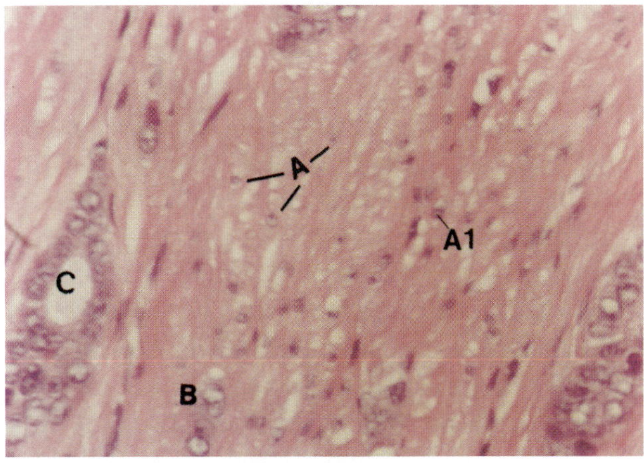

Figure 31.12, H&E x 400

---

2. Regeneration of myocardial fibers through tiny dedifferentiated 'seed state' nuclei arising from lysing cardiac muscle fibers was described by the author in an original paper published in the *Journal of Clinical Pathology* (McDonald, 1975).

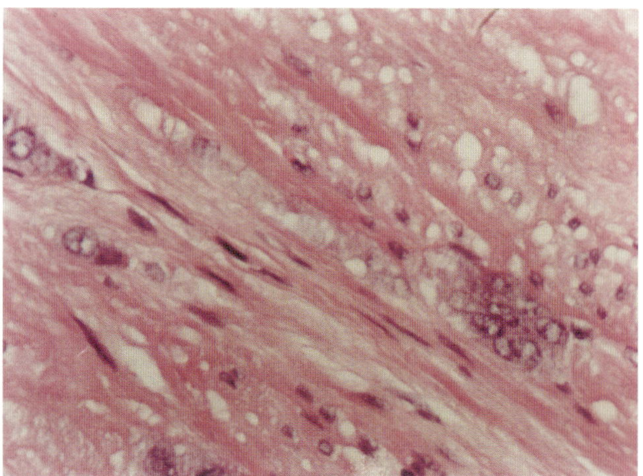

Figure 31.13, H&E x 520

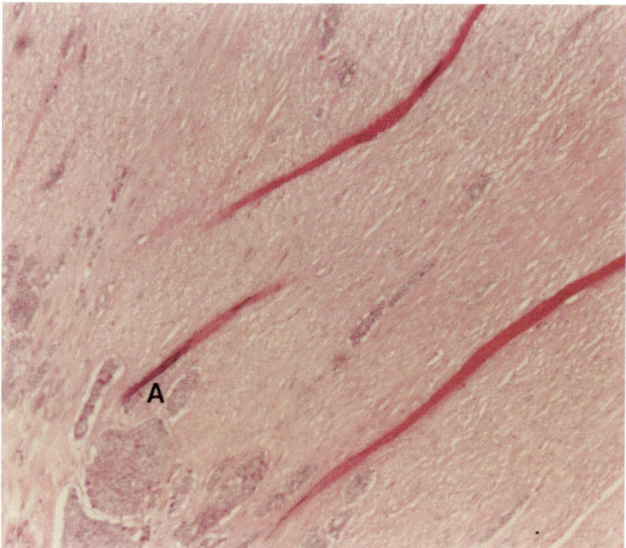

Figure 31.14, H&E x 80

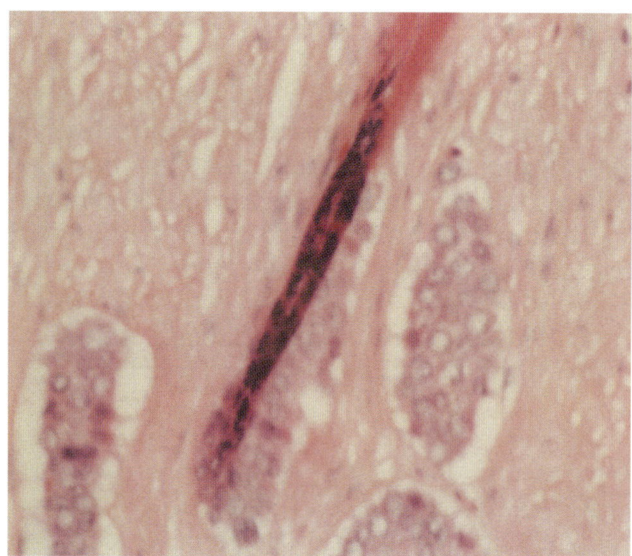

Figure 31.15, H&E x 400

mature carcinoid cell nuclei, some with vacuole formation (B). On the left, there are carcinoid cells with liquefaction forming a small cystic pattern (C). H&E x 400

Figure 31.13. Throughout this section of frothy smooth muscle coat, there are many various sized vacuoles. Some of the vacuoles are associated with the minute lightly stained nuclei of the carcinoid cells infiltrating the muscle fibers. A small cluster of carcinoid cells can be seen in the lower right area and a few more near the left border; some of these nuclei are vacuolated. H&E x 520

Development of calcifications by carcinoid tissue (figs. 31.14-31.16)

Figures 31.14 and 31.15. There are three long, narrow, linear, slightly-curved, bright red, thread-like structures crossing the carcinoid tissue and the muscular coat of the ileum. These structures are the beginning of calcifications which, when seen by x-ray, are suppose to be a diagnostic feature of carcinoid. Area (A) is seen in higher magnification in figure 31.15 which shows the red streak filled with pyknotic nuclei where it crosses a nest of carcinoid cells. Similar structures developing in carcinoid tissue metastatic to the liver of

this same patient are presented in figures 31.31-31.33; and similar structures developing in uninvolved muscle coat and mucosal tissue are presented in figures 31.40-31.42. Figure 31.14 H&E x 80; and figure 31.15 H&E x 400

Figure 31.16. This figure shows prominent calcification of a structure similar to the ones described in figure 31.14. H&E x 80

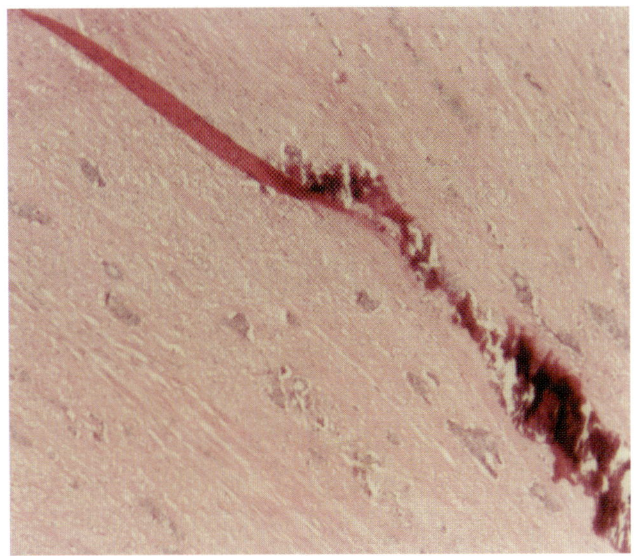

Figure 31.16, H&E x 80

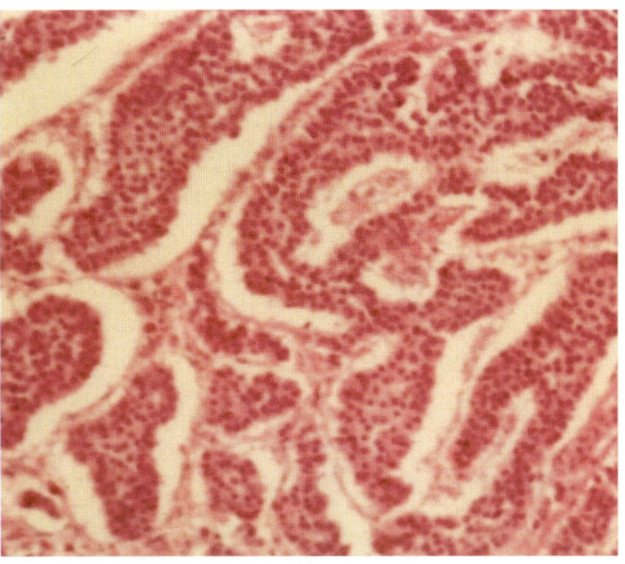

Figure 31.17, H&E x 200

2. **Primary Carcinoid of the Colon** (figs. 31.17-31.20)

Figures 31.17-31.20 are from a primary carcinoid of the large bowel near the rectum removed by abdominoperineal resection from a 48-year-old female.

Lacy pattern of carcinoid of colon
(fig. 31.17)

Figure 31.17. This primary carcinoid of the colon shows a lacy pattern formed by the dissolution of carcinoid cells leaving slit-like spaces between the preserved carcinoid tissues. H&E x 200

Development of red cells (fig. 31.18),
plasmacytoid cells (fig. 31.19), and stromal
tissue (figs. 31.19 and 31.20) from primary
carcinoid of the colon

Figure 31.18. In this figure, there is dissolution of carcinoid cells in association with the development of red cells and clear fluid. There are remnants of carcinoid cells associated with developing red cells (A). Giemsa x 800

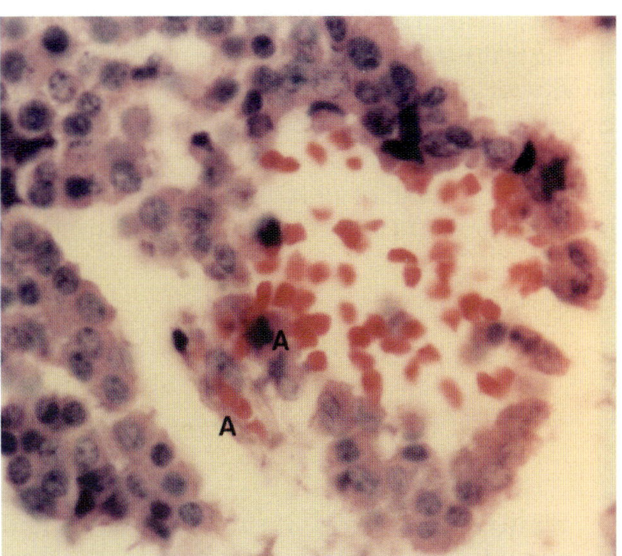

Figure 31.18, Giemsa x 800

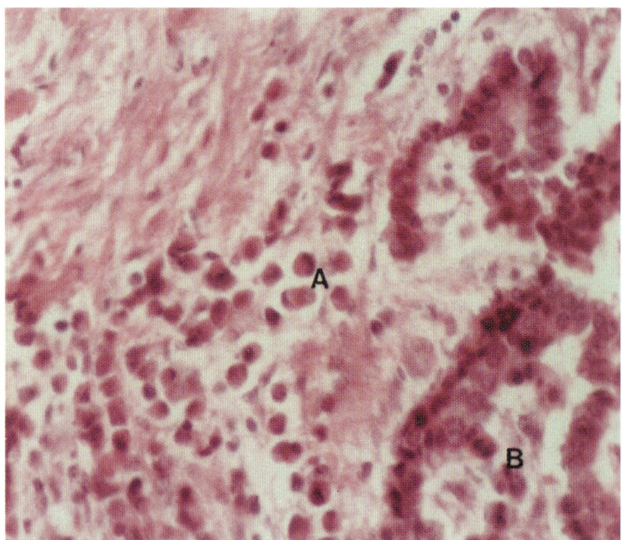

Figure 31.19, H&E x 400

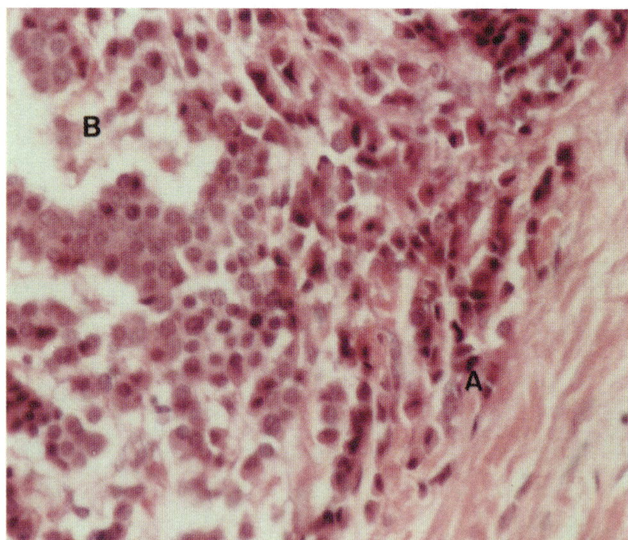

Figure 31.20, H&E x 400

Figure 31.19. There is a large area of liquefaction and cellular changes in the carcinoid tissue resulting in the development of plasmacytoid cells, including erythrogenic plasma cells of various sizes (A) with light crimson-red cytoplasm. These cells have peripheral nuclei and develop hemoglobin-like cytoplasm in the formation of red cells. In the mid-upper left, transformation of the locally developed plasmacytoid cells into fibrous stroma with spindle-shaped nuclei can be seen. On the lower right, in the center of a nest of carcinoid tissue there is liquefaction and dissolution of the carcinoid cells with development of a few SN cell-like nuclei (B). H&E x 400

Figure 31.20. In the center there is a large area of carcinoid tissue undergoing cellular changes in preparation for development of fibrous tissue stroma. In the lower right, the carcinoid cells are undergoing pyknotic changes (A) in the direct development of loose fibrous tissue stroma. On the left, the carcinoid cells are dissolving and transforming into clear fluid with remnants of carcinoid cell substance (B) floating in the clear fluid. H&E x 400

3. Carcinoid Metastatic to Lymph node (figs. 31.21-31.23)

Figures 31.21-31.23 show the carcinoid tumor of the ileum presented in figures 31.8-31.16 metastatic to a mesenteric lymph node.

Carcinoid of the ileum metastatic to mesenteric lymph node (fig. 31.21)

Figure 31.21. On the left half of this figure, there is normal lymph node tissue. On the right, there are several nests of carcinoid tissue metastatic from the ileum presented in figures 31.8-31.16. These nodules of carcinoid tissue are separated by developing stroma. The carcinoid cells are not very distinct as is typical of the primary tumor. Figures 31.22 and 31.23 are taken from the same tissue section as this figure. H&E x 200

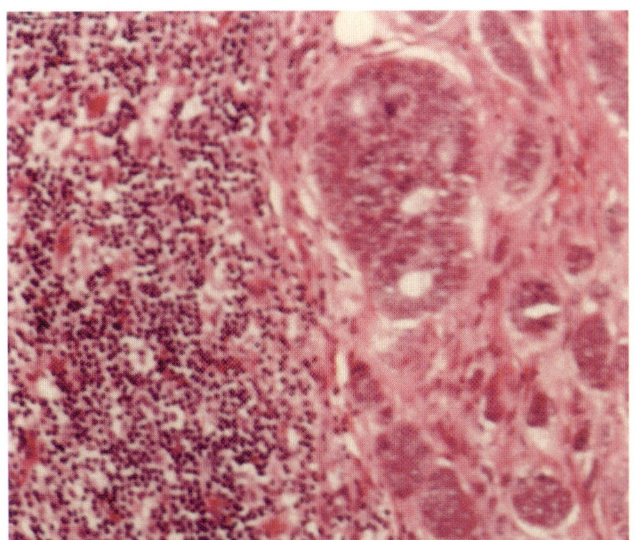

Figure 31.21, H&E x 200

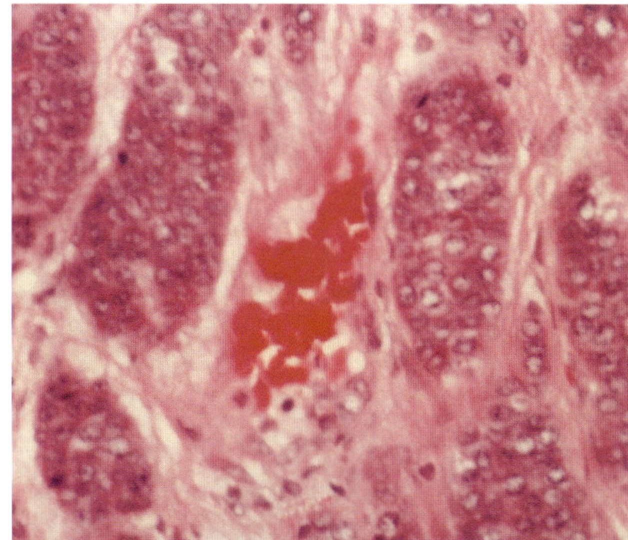

Figure 31.22, H&E x 400

*Red cell development from metastatic
carcinoid (figs. 31.22)*

Figure 31.22. In the center of this
figure, one of the metastatic carcinoid
nodules is replaced by bright crimson-red
blood cells originating from and replacing
the carcinoid tissue. H&E x 400

*Stroma development from
metastatic carcinoid (fig. 31.23)*

Figure 31.23. In the center of this figure
there is hyalinization of metastatic carcinoid tissue in development of fibrous stroma.
The vanishing carcinoid cells can still be
identified in the developing fibrous tissue.
H&E x 520

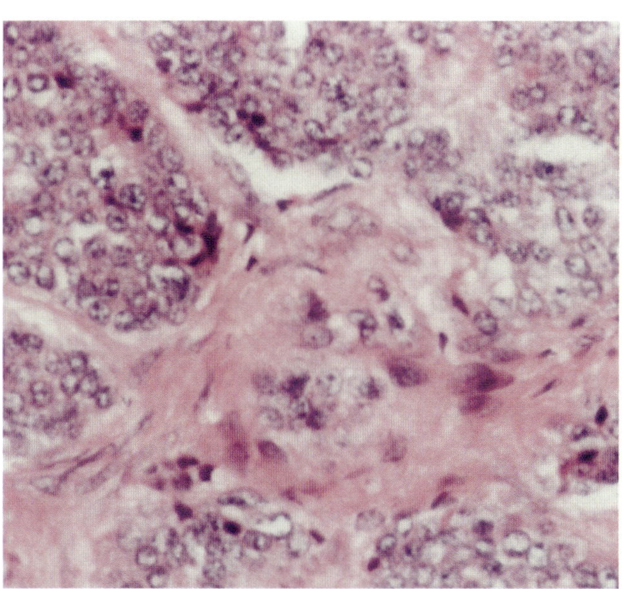

Figure 31.23, H&E x 520

4. Carcinoid Metastatic to Liver (figs. 31.24-31.33)

Figures 31.24-31.33 are taken from
biopsy sections of the anterior edge of the
right lobe of the liver obtained during
exploratory laparotomy from the same
patient as presented in figures 31.8-31.16
and 31.21-31.23.

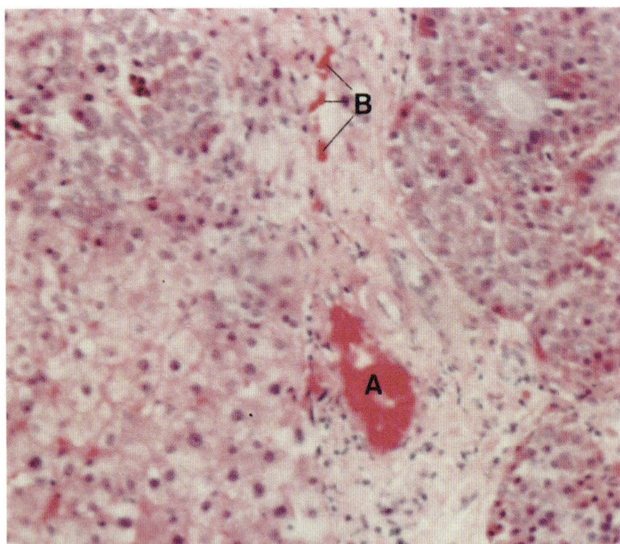

Figure 31.24, H&E x 200

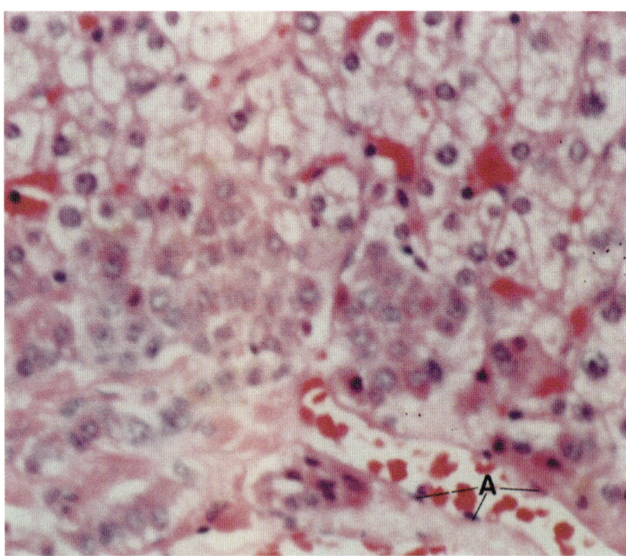

Figure 31.25, H&E x 400

*Blood and blood vessel development
from metastatic carcinoid in the liver
(figs. 31.24-31.28)*

Figure 31.24. This figure is a low-magnification of carcinoid of the ileum metastatic to the liver. In the lower left quadrant, there is normal liver tissue. In the upper left corner, carcinoid tissue has replaced the normal liver tissue. On the right one-third of this figure, there is metastatic carcinoid tissue. Passing through the middle of this figure, there is a thick tract of fibrous connective tissue with a large irregular area of crimson-red hemoglobinization (A). The red cells in this area are not yet well-formed or well-separated. There are also three small linear areas of irregular hemoglobinization (B). H&E x 200

Figure 31.25. The lower half of this figure is occupied by nests of invading carcinoid tissue. In the lower right, there is a blood vessel developing from carcinoid tissue. In the vascular lumen, there are several fragments of crimson-red hemoglobinized particles in which the red cells are not fully formed or well separated. In the vascular wall, three endothelial

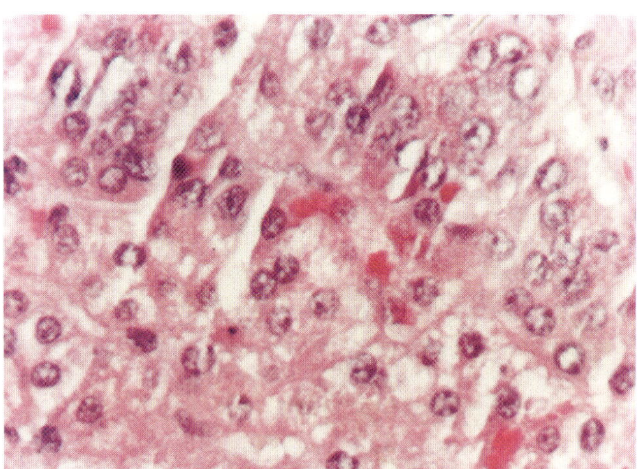

Figure 31.26, H&E x 520

nuclei can be seen (A). The upper portion of this figure is occupied by normal liver tissue in which there are several irregular areas of crimson-red hemoglobinization. The line of separation between the normal liver tissue and the carcinoid cells is irregular but distinct. H&E x 400

Figure 31.26. In this area of the liver biopsy, only carcinoid tissue can be identified. There is crimson-red hemoglobinization of several individual carcinoid cells. H&E x 520

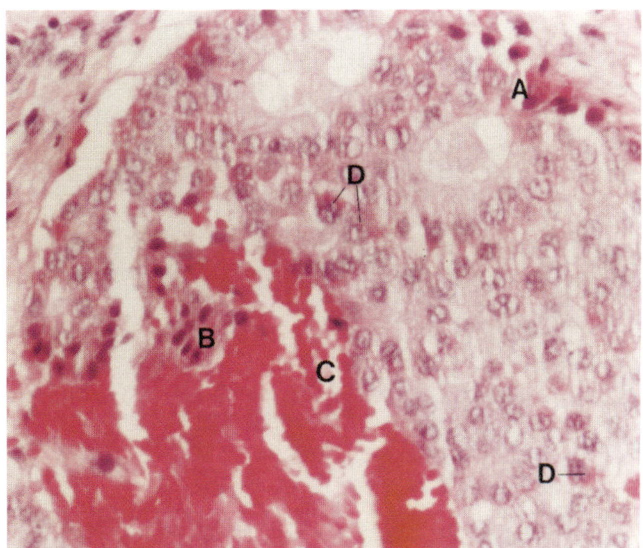

Figure 31.27, H&E x 400

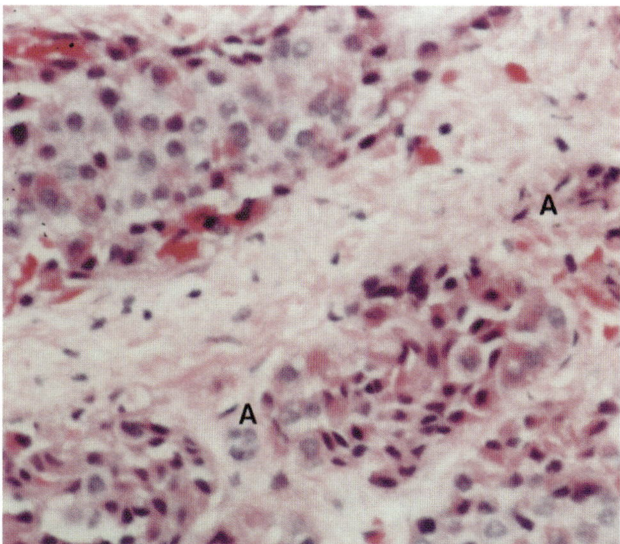

Figure 31.28, H&E x 400

Figure 31.27. The major portion of this figure shows carcinoid tumor tissue. In the lower left corner, there is a large area of carcinoid tissue transforming into a mass of bright red hemoglobinized substance. A few carcinoid cells (A) with pyknotic nuclei show early hemoglobinization. Similar cells can be seen in area (B). In area (C), red cells are starting to form from the hemoglobinized substance. There is partial hemoglobinization of several individual carcinoid cells (D). H&E x 400

Stroma development from metastatic carcinoid in the liver (fig. 31.28)

Figure 31.28. Crossing diagonally across this figure, from lower left to upper right, there is a tract of developing stromal tissue. At the edge of this tract, in areas (A), there is hyalinization of carcinoid tissue with development of stromal tissue with spindle cell nuclei. In this figure there are several areas of hemoglobinization of carcinoid tissue in preparation for red cell and blood vessel formation. The red cells in these areas are not fully formed or well separated. H&E x 400

Extension of metastatic carcinoid into normal liver tissue (figs. 31.29 and 31.30)

Figure 31.29. This is a low-magnification view of this tumor showing extension

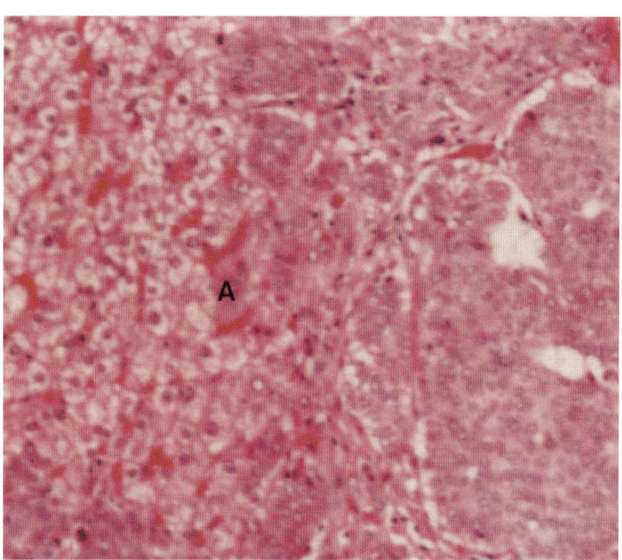

Figure 31.29, H&E x 200

of the carcinoid into the normal liver tissue. Typical carcinoid pattern is seen in the right two-thirds of this figure. In the middle left, extension of tumor tissue into the normal liver tissue can be seen. As usual, there are many developing blood capillaries seen in the liver tissue. On careful examination, many barely recognizable developing nuclei can be seen at the border between the carcinoid and liver tissue. Area (A) is presented in higher magnification in the next figure. H&E x 200

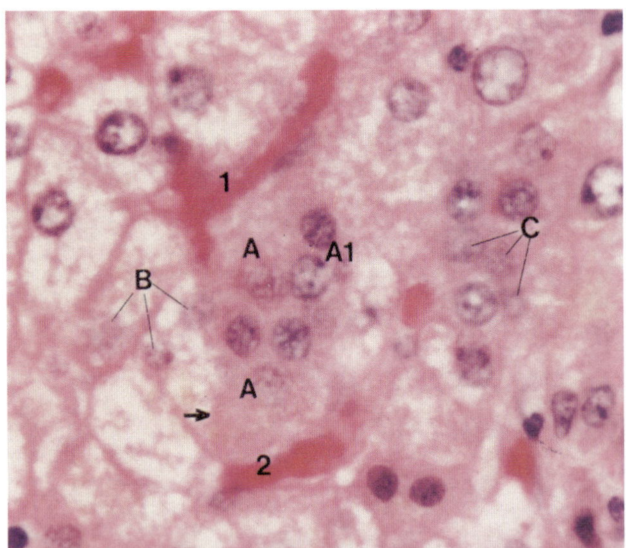

Figure 31.30, H&E x 800

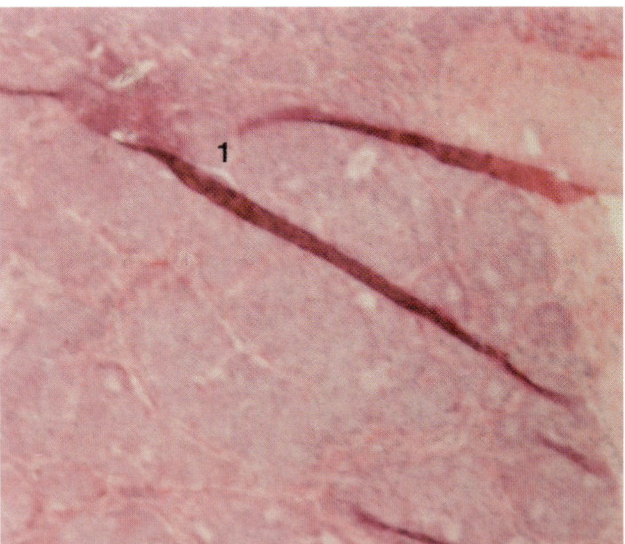

Figure 31.31, H&E x 80

Figure 31.30. In this figure, the appearance of tumor nuclei in various stages, from barely recognizable nuclear structures developing from carcinoid protoplasm (A) to recognizable mature carcinoid nuclei (A1), can be seen. To the left of this group of carcinoid nuclei, the liver cell nuclei are no longer present and there is the appearance of tiny barely recognizable developing carcinoid nuclear structures (B) in the liver cell cytoplasm. As usual for many tumors there is no mitosis seen in the multiplication of these nuclei. A distinct line can be identified where the clear liver cell cytoplasm is transforming into carcinoid cytoplasm (arrow). Partially bordering this area are two developing blood capillaries (1 and 2). Barely recognizable developing carcinoid nuclei are pointed out by (C). H&E x 800

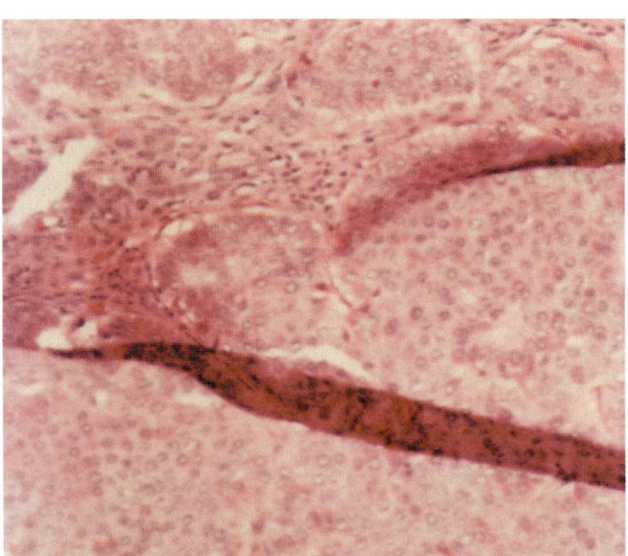

Figure 31.32, H&E x 200

*Development of linear calcifications
by carcinoid metastatic to liver
(figs. 31.31-31.33)*

Figures 31.31-31.33. In these sections of carcinoid of the ileum metastatic to the liver, long, narrow, linear, bright red, thread-like structures filled with pyknotic nuclei are seen crossing the tumor tissue. These structures are the beginning of calcifica-tion of the tumor tissue, which when seen by x-ray, are suppose to be a diagnostic feature of carcinoid. Area (1) is seen in higher magnification in figure 31.32. Similar structures developing in the smooth muscle coat of the ileum of this patient are presented in figures 31.14-31.16; and similar structures developing in uninvolved muscle coat and mucosal tissue are presented in figures 31.40-31.42. Figure 31.31 H&E x 80; figures 31.32 and 31.33 H&E x 200

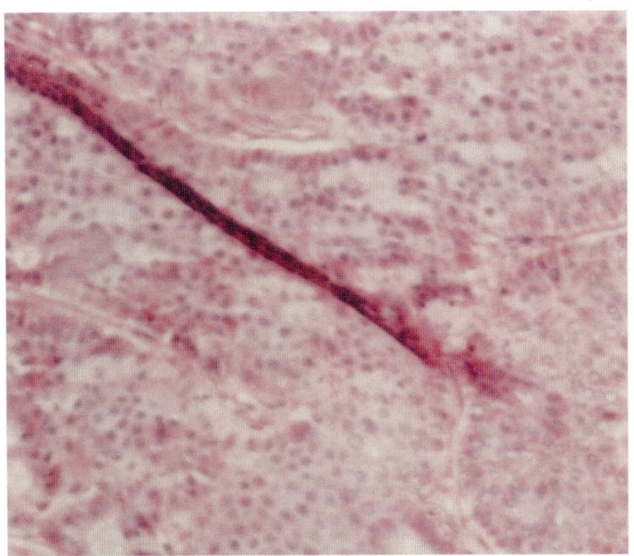

Figure 31.33, H&E x 200

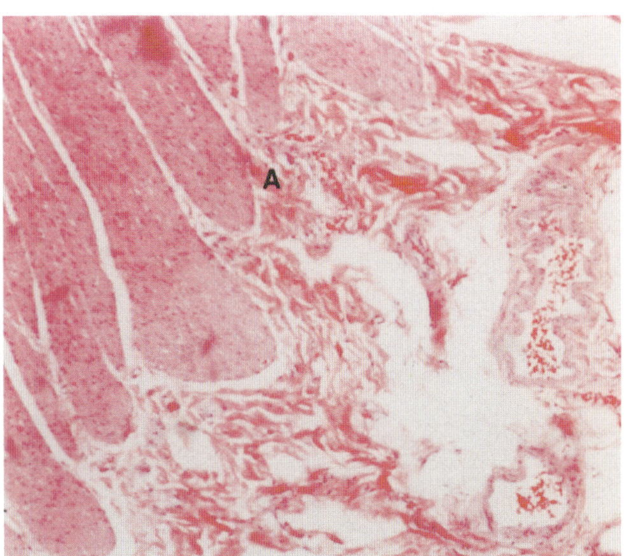

Figure 31.34, H&E x 80

5. Reactions in Intestinal Wall of Ileum to Nearby Carcinoid (figs. 31.34-31.42)

Figures 31.34-31.42 are taken from the apparently uninvolved intestinal wall near the carcinoid of the ileum presented in figures 31.8-31.16. No carcinoid tissue is identified in the figures below.

Fibrinoid degeneration of muscle coat and submucosal tissue with development of blood and blood vessels (figs. 31.34-31.39)

Figures 31.34 and 31.35. These figures show prominent fibrinoid degeneration of muscle coat and submucosal tissue as a reaction to nearby carcinoid of the ileum. These two photomicrographs are from adjoining fields of view. In figure 31.34, lining the wall of the muscle coat there is a large continuous area of fibrinoid degeneration of the hypertropic smooth muscle and the submucosal tissue. On the right, there are some prominently thickened blood vessels in the edematous submucosal tissue. Area (A) is seen in higher magnification in figure 31.36. In figure 31.35, the muscle coat is prominently thickened and within

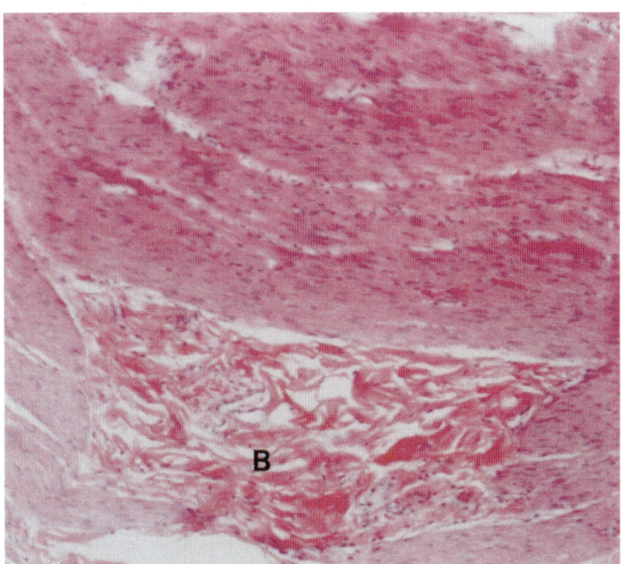

Figure 31.35, H&E x 80

the smooth muscle a considerable portion is undergoing fibrinoid degeneration (B) with replacement of muscle tissue by orange-red fibrinoid material. Area (B) is seen in higher magnification in figure 31.37. H&E x 80

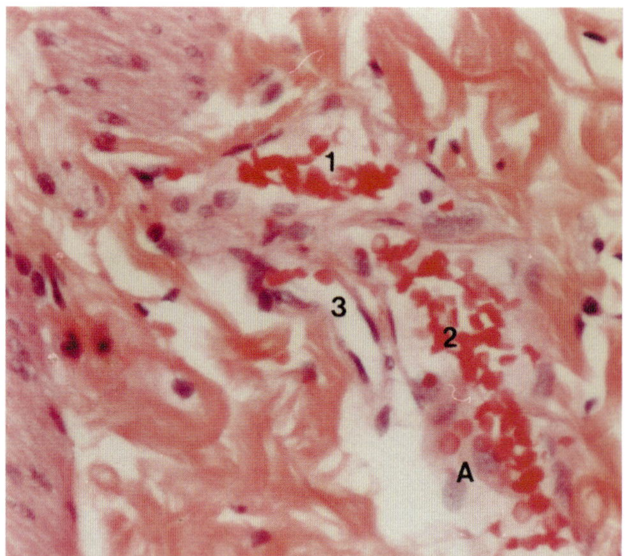

Figure 31.36, H&E x 400

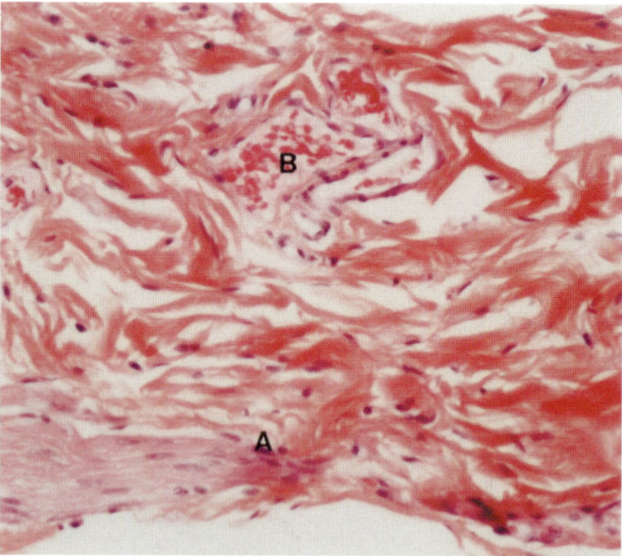

Figure 31.37, H&E x 400

Figure 31.36. This is a magnified view of area (A) in figure 31.34. There are three developing vascular channels (1, 2 and 3) in which the endothelium is barely developed. All three vessels are partially filled with irregular, newly-formed, hyper-chromatic red cells. These red cells are not well separated or well rounded. In vessel (2) remnants of background substance (A) can still be identified. H&E x 400

Figure 31.37. This is a magnified view of area (B) in figure 31.35. In area (A), fibrinoid degeneration is transforming the smooth muscle tissue into fibrinoid material. Arising from this fibrinoid material there is a delicate, newly-formed, vascular channel (B). This blood vessel is loosely filled with locally developed, newly-formed, red cells of different sizes, shapes, and various degrees of hemoglobinization. There are faint remnants of fibrinoid material still present in this developing blood vessel. The endothelial lining is not uniformly developed. H&E x 400

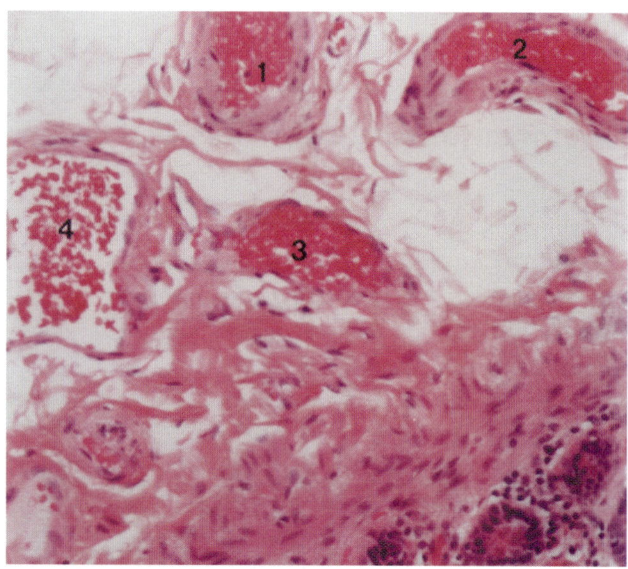

Figure 31.38, H&E x 200

Figure 31.38. There is thickening associated with fibrinoid degeneration and prominent vascularization of the submucosa. Blood vessels (1), (2), and (3) are filled with tightly packed red cells. Blood vessels (1) and (2) have thickened well-formed vascular walls. A small portion of mucosa is included in the lower right corner where two mucosal glands can be identified. Developing vascular channel (4) is shown in higher magnification in figure 31.39. H&E x 200

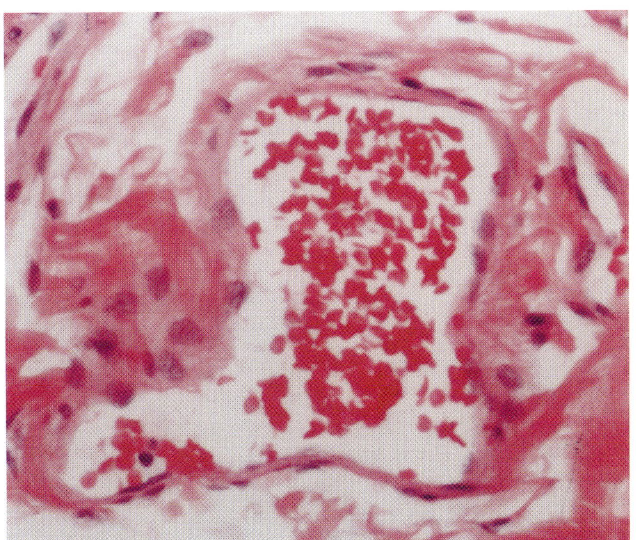

Figure 31.39, H&E x 400

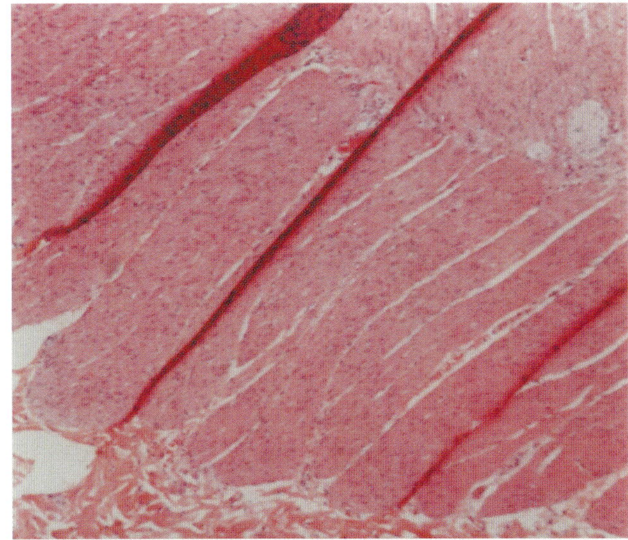

Figure 31.40, H&E x 80

Figure 31.39. This is a cross-section of a dilated developing blood vessel filled with fragments of irregularly-shaped, not well separated, hemoglobin particles. Only a few hemoglobin particles appear as well rounded red cells. This developing blood vessel is only partially lined by endothelial cells. H&E x 400

Development of calcifications in uninvolved intestinal wall near carcinoid (figs. 31.40-31.42)

Figure 31.40-31.42. These are low-magnification views of apparently uninvolved muscle coat and mucosal tissue of the ileum near carcinoid. No carcinoid tissue is identified in these areas. In figure 31.40, there are three long, narrow, linear, slightly-curved, bright red, thread-like structures crossing the muscle coat. In figures 31.41 and 31.42, similar parallel linear structures are seen crossing the lining mucosa. These structures show the beginning of calcifications and when seen by x-ray is considered to be a diagnostic feature of carcinoid. (See also figures 31.14-31.16 and 31.31-31.33.) H&E x 80

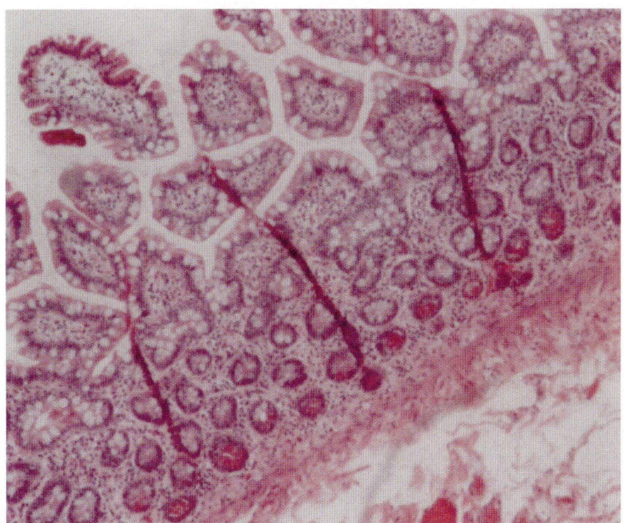

Figure 31.41, H&E x 80

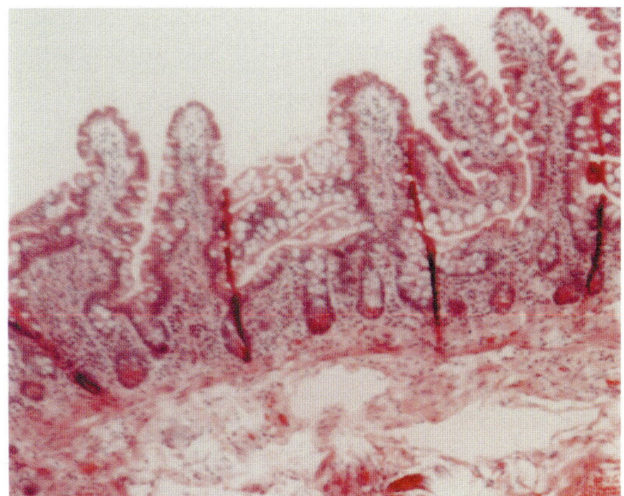

Figure 31.42, H&E x 80

Chapter 32

ADRENAL CORTICAL TUMOR   (figs. 32.1-32.2)

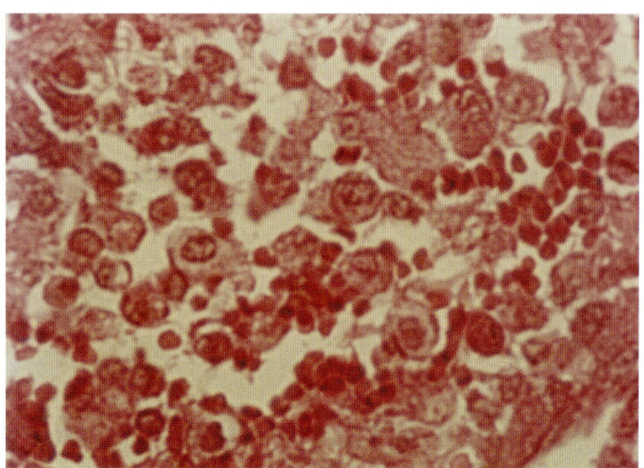

Figure 32.1, H&E x 520

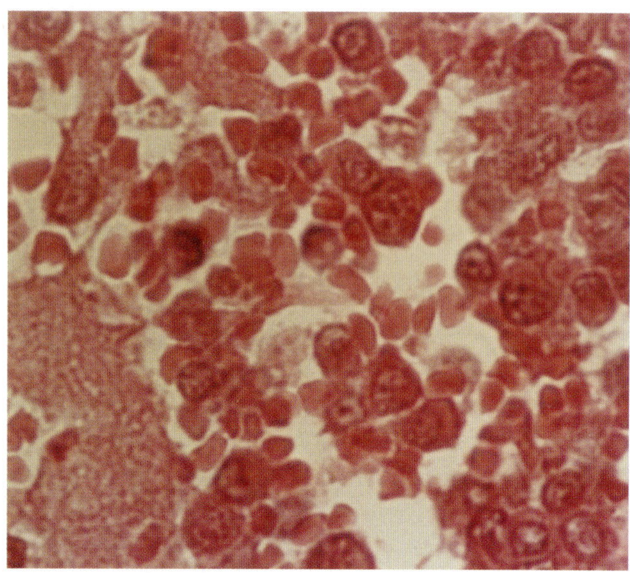

Figure 32.2, H&E x 1040

Figures 32.1 and 32.2 are taken from the left adrenalectomy of a 17-year-old male with adrenal cortical tumor and Cushing's syndrome. For cellular changes in benign adrenal tissue see Chapter 5, subsection 3.

Red cell development directly from cancer cells (figs. 32.1 and 32.2)

Figure 32.1. This is a blood-forming tumor arising from the adrenal cortex. There is prominent presence of red cells develop-ing from and replacing the cortical tumor (seen better in higher magnification in the next figure). H&E x 520

Figure 32.2. This section of adrenal cortical tumor shows extensive replacement of cancer cells by locally developed highly hyperchromatic red cells. The cancer cells and interstitial tissue are directly transforming into the hyperchromatic red cells. Scattered tumor cells can still be identified among the developing red cells. H&E x 1040

Chapter 33

RENAL CANCERS (figs. 33.1-33.12)

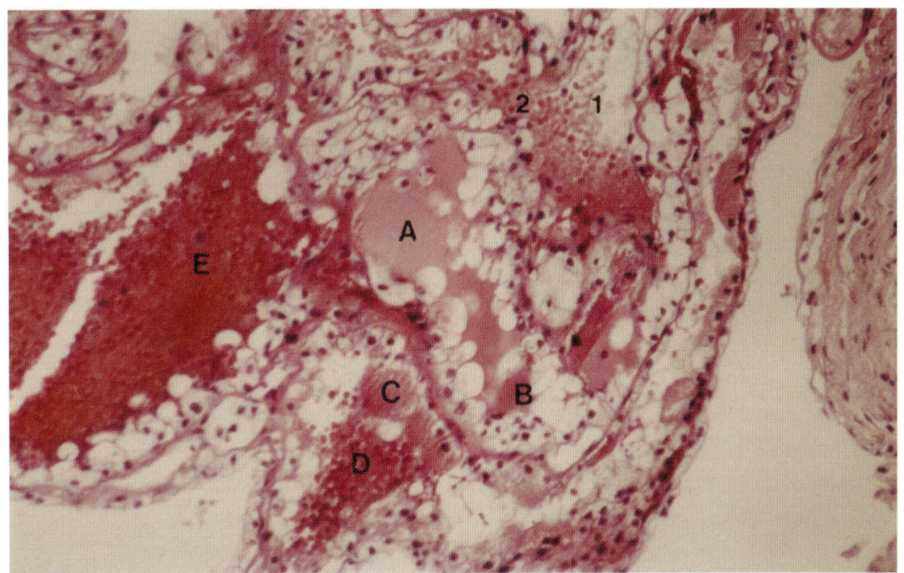

Figure 33.1, H&E x 200

1. Adenocarcinoma (Hypernephroma) (figs. 33.1-33.3)

Figures 33.1-33.3 are taken from a 59-year-old male with a renal carcinoma 1 cm in diameter discovered at autopsy.[1] Cellular changes in benign kidney tissues are presented in Chapter 16.

Development of red cells from liquefied tumor cell product (figs. 33.1 and 33.3)

Figure 33.1. This adenocarcinoma of the kidney (hypernephroma) shows characteristic vacuolated tumor cells with tiny hyperchromatic nuclei and marked "hemorrhagic areas". These so-called "hemorrhagic" areas actually represent the various stages of blood formation from liquefied tumor tissue. Different stages of red cell development from the erythrogenic fluid derived from liquefied tumor tissue can be seen. This erythrogenic fluid shows a progressive degree of hemoglobinization (A, B, C and D). In area (C), there are minute lightly stained hemoglobin particles which show a more mature development into red cells at (D). In area (E), red cells are developing from concentrated hemolysate-like fluid, hemoglobin gel. In area (1), the red cells are developing from clear erythrogenic fluid. Red cells developing from the clear fluid start as fine light pink granules (1) and become slightly darker red as they become more hemoglobinized (2). H&E x 200

---

1. Additional photomicrographs and findings in renal carcinoma are presented in Volume I, figures 96-105.

Figure 33.2. Red cell development (mostly in small groups) from liquefied tumor tissue is seen in different areas in different stages. (A) is an early stage of red cell development from an erythrogenic mesh consisting mostly of small vacuoles with practically no hemoglobin. In the cluster of red cells in the center, developing red cells (B) are barely hemoglobinized and red cells (C) are hyperhemoglobinized. There are also several red cells developing from an irregular area of hyperhemoglobinized tumor tissue (D). The characteristic clear vacuolated tumor cells, each with a tiny single hyperchromatic nucleus, can be clearly seen at this higher magnification. Most of the nuclei are small, round, hyperchromatic, and lymphocyte-like. Some of these tiny tumor cell nuclei show compact lobulated hyperchromatic characteristics (E). There are rare nuclei showing a prominent single nucleolus (arrows). H&E x 800

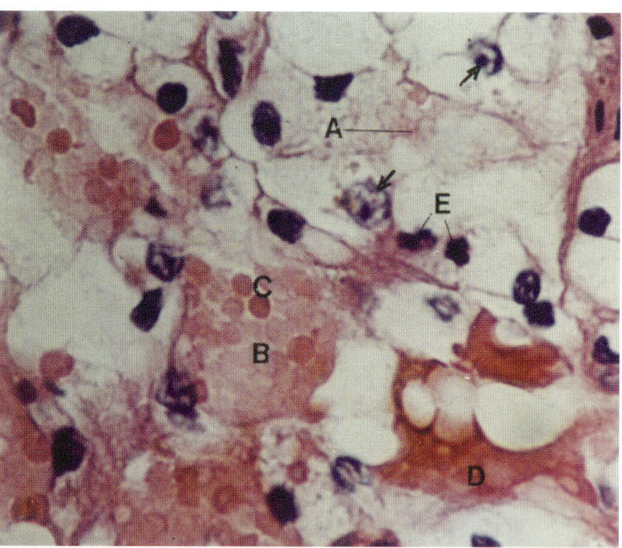

Figure 33.2, H&E x 800

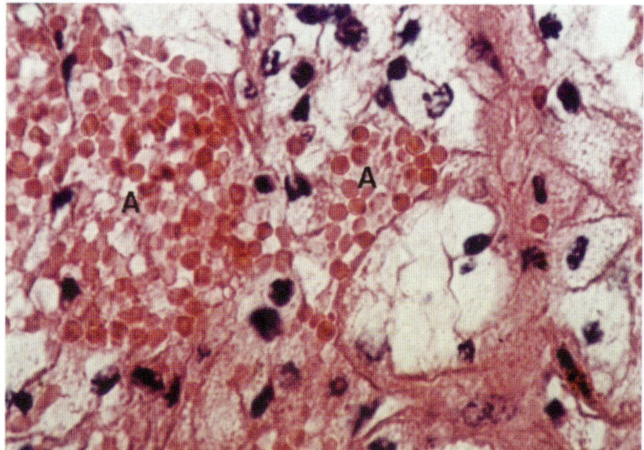

Figure 33.3, H&E x 520

Figure 33.3. Sheets of tumor cells are being replaced by an erythrogenic mesh (A) with developing red cells. The surrounding tumor cells are characteristically composed of single, tiny, hyperchromatic nuclei surrounded by clear cytoplasm, similar to the nuclei described in figure 33.2. H&E x 520

2. Papillary Adenocarcinoma (figs. 33.4-33.6)

Figures 33.4-33.6 are presented here to demonstrate the profuse transformation of tumor tissue into red blood cells.[2] These figures are taken from a 59-year-old male with papillary adenocarcinoma of the right kidney removed by nephrectomy.

Striking example of profuse red blood cell formation by papillary adenocarcinoma of the kidney (figs. 33.4-33.6)

Figure 33.4. A low-magnification view showing the architectural pattern of a papillary adenocarcinoma of the kidney with prominent blood formation is seen in this figure. This is a highly erythrogenic tumor. There are several large areas of red erythrogenic fluid in various stages of hemoglobinization. Development of this red erythrogenic fluid from the tumor tissue and its direct transformation into red blood cells is demonstrated in figures 33.5 and 33.6. H&E x 104

Figure 33.5. This figure shows the characteristic vacuolated tumor cells arranged in papillary pattern and sur-

---

2. A similar process is demonstrated in mammary carcinoma undergoing necrosis in C3H mice in figures 24.1-24.7.

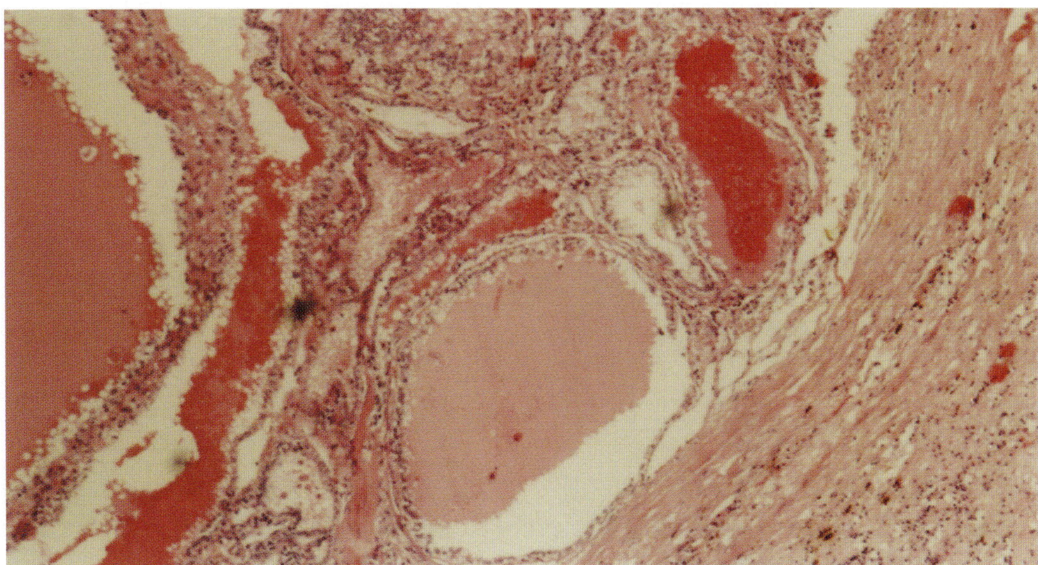

Figure 33.4, H&E x 104

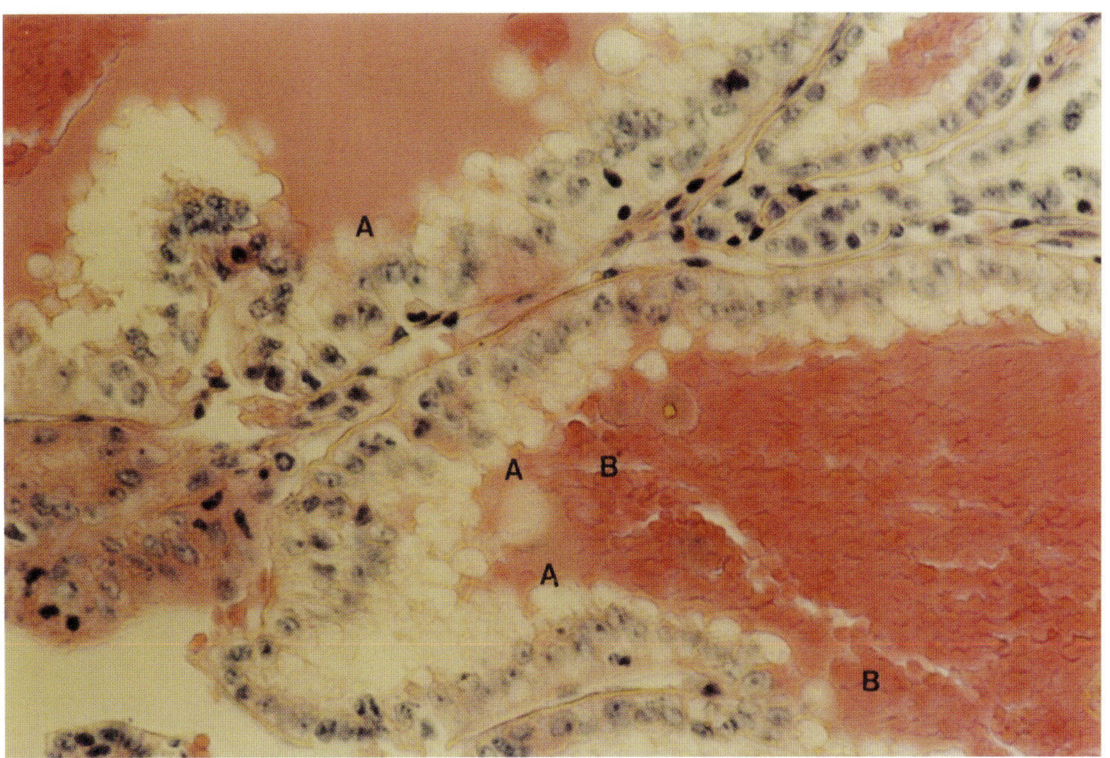

Figure 33.5, Giemsa x 520

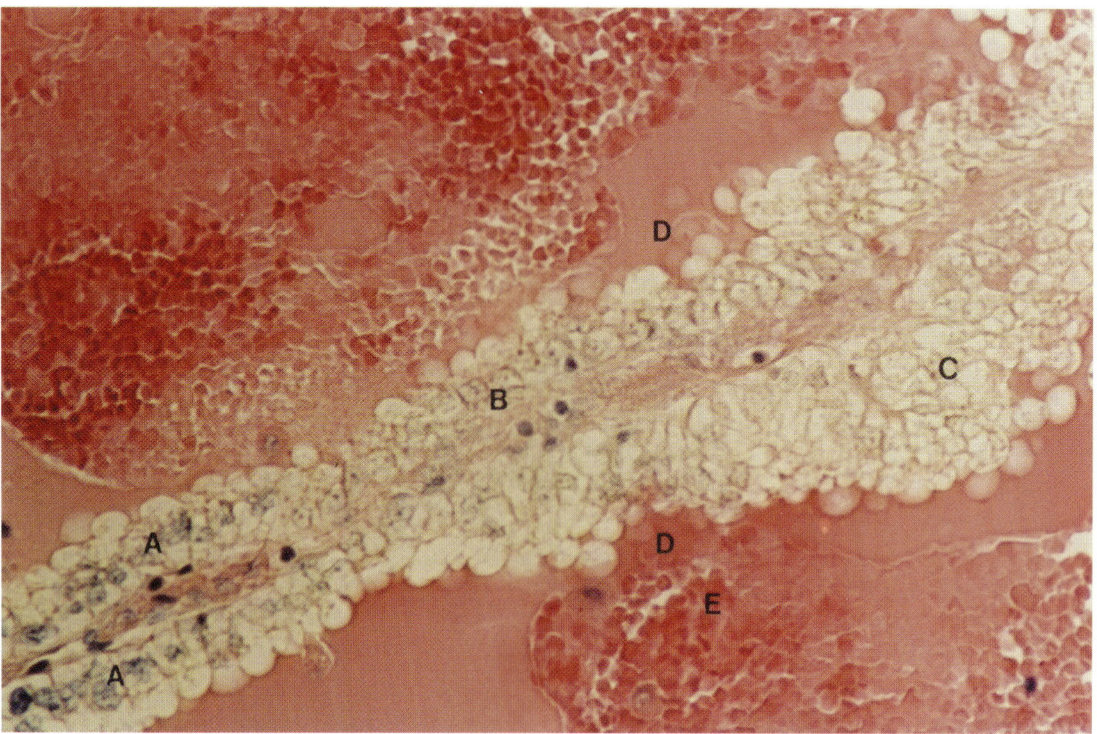

Figure 33.6, Giemsa x 520

rounded by large areas of red erythrogenic fluid. The tumor cells lining the papillary fronds produce vacuoles which transform into the red erythrogenic fluid (A). Direct development of red cells from this erythrogenic fluid is seen here in areas (B) and also in the next figure. Giemsa x 520

Figure 33.6. Prominent vacuolation of the tumor cells of this papillary frond is seen diagonally across this figure. Characteristic tumor cells with nuclei can be seen in areas (A). In area (B), the tumor cell nuclei are vanishing and only fine nuclear dots remain. In area (C), the tumor cells have transformed into multi-layered vacuolated cells with no nuclei. These vacuoles become hemoglobinized (D) and liquefy to form the red erythrogenic fluid. Direct development of red cells from this erythrogenic fluid is seen in area (E). A larger area of red cell development from the erythrogenic fluid is seen in the upper left segment of this figure. Giemsa x 520

3. Wilms' Tumor (figs. 33.7-33.12)

Figures 33.7-33.12 are taken from Wilms' tumor of the left kidney removed by nephrectomy from a four-year-old boy.

Erythrogenic changes in normal kidney tissue as a reaction to nearby Wilms' tumor (figs. 33.7 and 33.8)

Figure 33.7. This is a low-magnification view of the cortical region of a kidney with Wilms' tumor. Replacing the normal kidney tissue, there is development of large markedly dilated tubular spaces filled with red hemoglobinized fluid. H&E x 104

Figure 33.8. This is a higher magnification of an area close to figure 33.7. There are large markedly-dilated tubular spaces filled with hyperhemoglobinized fluid developed from kidney tissue. In the lumen

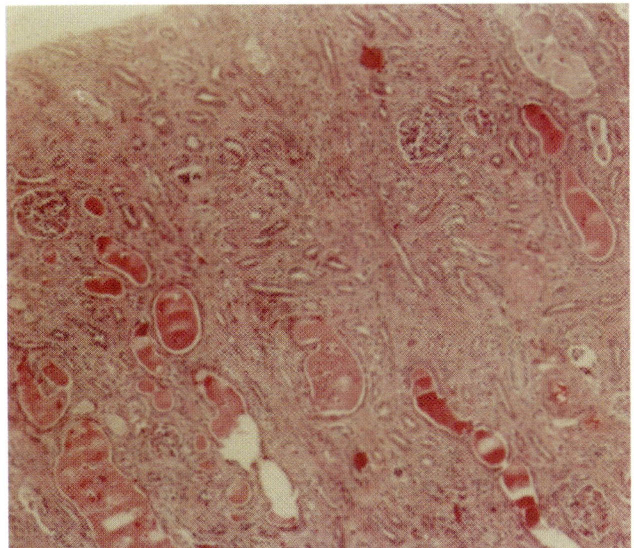

Figure 33.7, H&E x 104

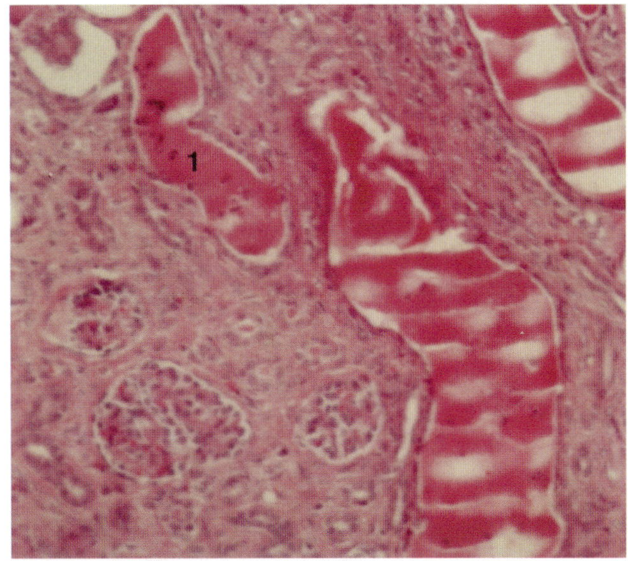

Figure 33.8, H&E x 200

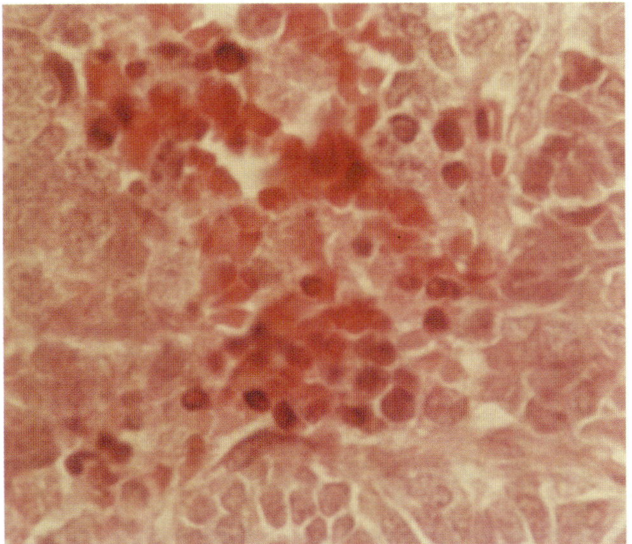

Figure 33.9, H&E x 1040

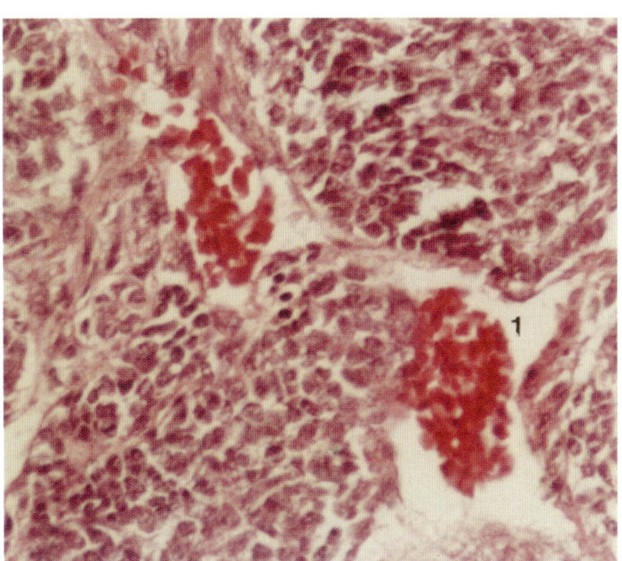

Figure 33.10, H&E x 520

of dilated formative tubular space (1), scattered remnants of fine dark undissolved kidney tissue can still be identified. Hemorrhagic changes can also be seen in the three glomeruli in the left half of this figure. H&E x 200

Blood and blood vessel development directly from tumor tissue (figs. 33.9-33.12)

Figure 33.9. Tumor cells passing through

pyknotic nuclear stages are forming compressed, hyperhemoglobinized red cells. A collection of red cells developed directly from tumor cells is seen in the center of this figure. H&E x 1040

Figure 33.10. In this section of Wilms' tumor of the kidney there are two oval areas where compact, compressed red cells are developing directly from the tumor cells. Area (1) is seen in higher magnification in the next figure. H&E x 520

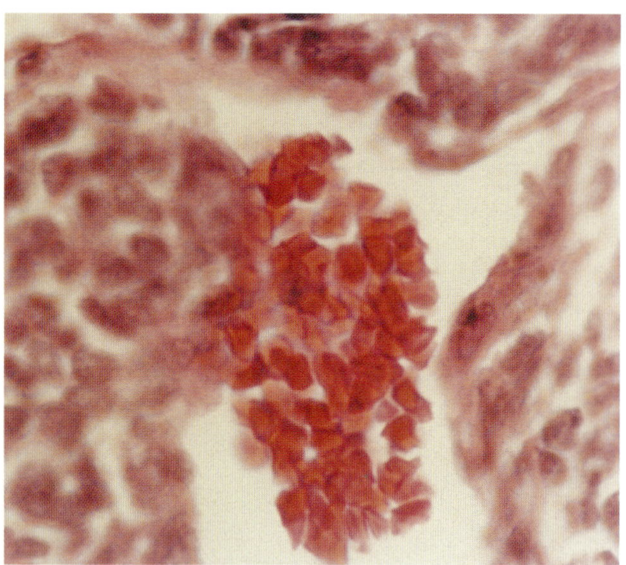

Figure 33.11, H&E x 1040

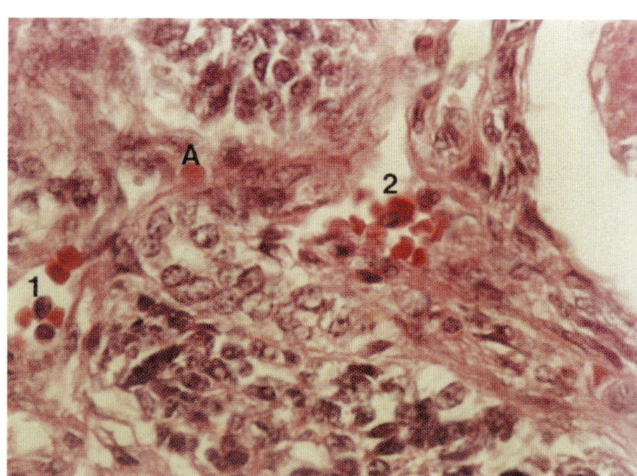

Figure 33.12, H&E x 520

Figure 33.11. Seen along the line of attachment of this cluster of red blood cells with the tumor tissue, is transformation of tumor cells directly into compressed, hyperhemoglobinized red cells. Remnants of tumor tissue can be seen attached to some of the irregularly shaped red cells throughout this cluster. H&E x 1040

Figure 33.12. Red cells developed from disintegrating tumor cells lie within the lumens of these two developing blood vessels (1 and 2). In area (A), one well-hemoglobinized red cell is developing from tumor tissue. H&E x 520

Chapter 34

ADENOCARCINOMA OF THE PROSTATE (figs. 34.1-34.8)

1. Prostatic Carcinoma (figs. 34.1-34.5)

Figures 34.1 and 34.2 are taken from a surgically excised specimen of adenocarcinoma of the prostate. Figure 34.3 is taken from a surgical specimen of adenocarcinoma of the prostate with invasion of seminal vesicles from a second patient. Figures 34.4 and 34.5 are taken from a surgically excised prostatic carcinoma from a third patient. Figure 34.4 is reprinted from Volume I.[1] For cellular changes in benign tissues of the prostate see Chapter 17.

Typical adenocarcinoma of prostate
(figs. 34.1 and 34.2)

Figure 34.1. This is a low-magnification view of a typical picture of adenocarcinoma of the prostate in the center surrounded by normal prostatic glandular tissue. The cancer tissue consists of disorganized glandular structures haphazardly distributed in the stromal tissue. Area (1) is seen in higher magnification in the next figure. H&E x 80

Figure 34.2. In this adenocarcinoma there are several irregular, disorganized, glandular structures separated by hyalinized and collagenous stromal tissue. The atypical glandular structures are made up of small, pyknotic epithelial cells with small irregular pyknotic nuclei. Unlike many other malignant tumors, this cancer does not show large atypical cells with abnormal nuclear structures. A few remaining benign cells are seen at (A). H&E x 400

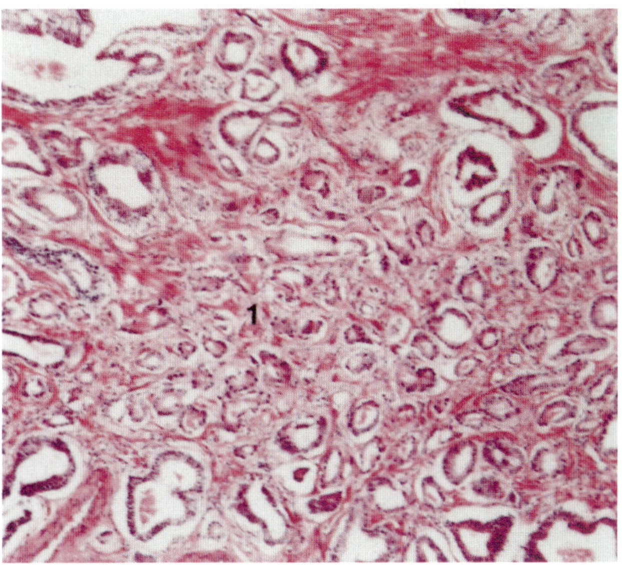

Figure 34.1, H&E x 80

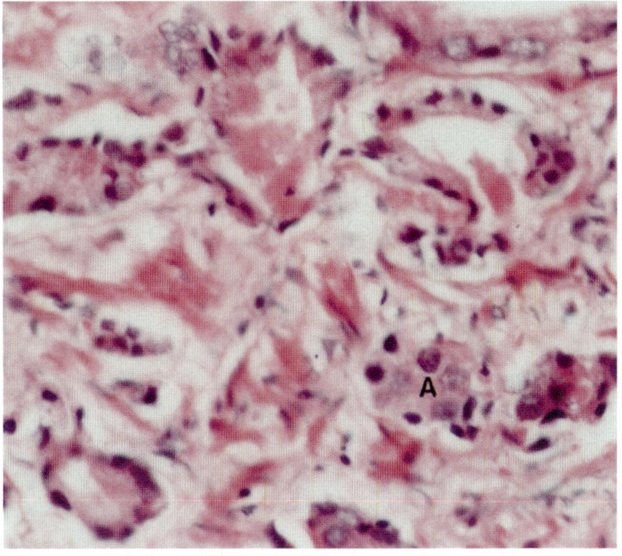

Figure 34.2, H&E x 400

---

1.   Additional photomicrographs and findings in prostatic carcinoma are presented in Volume I, figures 106-112.

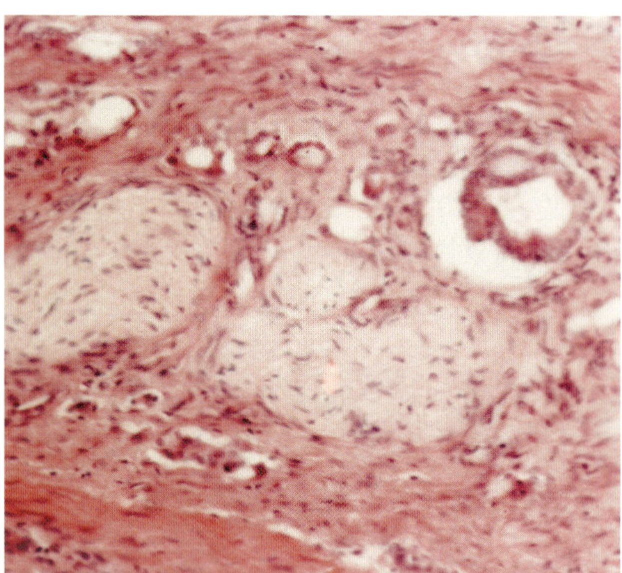

Figure 34.3, H&E x 200

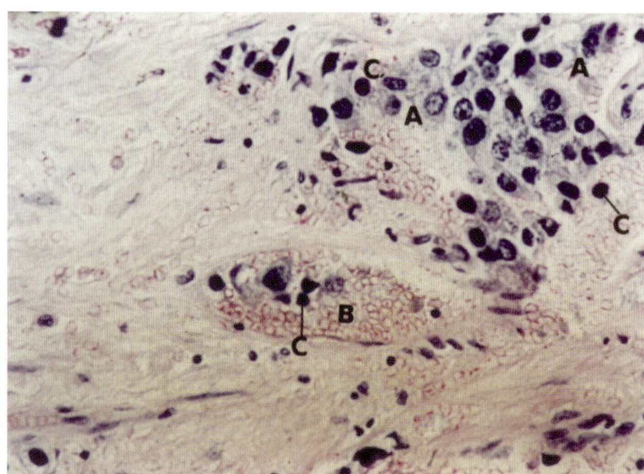

Figure 34.4, Giemsa x 260

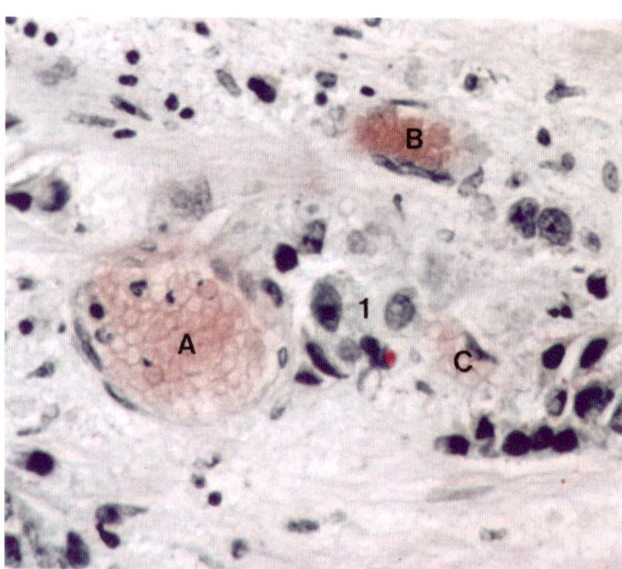

Figure 34.5, Giemsa x 400

*Perineural invasion by adenocarcinoma
of prostate (fig. 34.3)*

Figure 34.3. This figure shows nerve sheath invasion by prostatic cancer. The cancer tissue is formed of atypical cells surrounding and penetrating the perineural tissue. H&E x 200

*Origin of vacuolar red cells (ghost red cells)
and blood vessel formation from malignant
epithelium (figs. 34.4 and 34.5)*

Figure 34.4. Hyalinization of tumor tissue and development of sheets of minute vacuolar red cells (ghost red cells) are seen here over a large area of the figure. Note many ghost red cells contain tiny hemoglobin dots in the center. The tumor cells are also being replaced by sheets of vacuolar red cells without going through hyalinization (A). In the lumen area of the formative blood capillary (B), a few tumor cells have not yet been transformed into ghost red cells. The lower wall of this formative vascular channel is lined by a partially developed protoplasmic membrane derived from peripheral hyalinized cancer tissue with a suggestion of an early stage of endothelial nucleus formation. Arising from vanishing cancer cells, a few lymphocytes (C) are seen in the lumen of formative vascular channel (B) and several in the upper right section, where cancer tissue is transforming into ghost red cells and lymphocytes. Giemsa x 260

Figure 34.5. In this area of the same tissue section as seen in figure 34.4, there are three developing blood capillaries (A, B and C) filled with not yet fully hemoglobinized red cells (ghost red cells) developed from cancer cells. The endothelial linings have not yet completely surrounded the vascular lumens. A few nucleated cells suggesting SN cells and large lymphocytes as well as some unidentified cellular remnants are present in capillary (A). In an early stage of blood vessel formation (C), a column of developing lightly hemoglobinized red cells are present. Only a single endothelial nucleus has developed. In area (1), there is a group of cancer cells, some of which are almost vanished. Giemsa x 400

2. Prostatic Carcinoma Metastatic to Bone Marrow (figs. 34.6-34.8)

Figures 34.6-34.8 are taken from bone marrow clot sections showing a metastatic carcinoma of the prostate. This specimen was obtained from the sternum of a 70-year-old male during a routine bone marrow aspiration procedure for undiagnosed anemia. Figures 34.7 and 34.8 are reprinted from Volume I.[2]

Development of red cells from adenocarcinoma of prostate metastatic to bone marrow (figs. 34.6-34.8)

Figure 34.6. This is a low-magnification view of an aspirated sternal bone marrow clot section with metastatic prostatic carcinoma tissue in the center. H&E x 130

Figure 34.7. Liquefaction of cancer cells and development of red blood cells from the light red liquefied cell product (erythrogenic gel) is shown in the middle and lower right of this figure. The development of lymphocytes (A) and segmented nuclear cells (B) from cancer cells and their subsequent position in the erythrogenic gel, because of the lysis of surrounding cells in formation of such fluid, is also seen in this figure. (Compare with the next figure). H&E x 520

Figure 34.8. Red cells of various sizes and with various shades of red are arising from dissolving cancer cells. These developing red cells lie in the pathway of a formative vascular channel with the gradual dissolution of remaining cancer cells in the formation of plasma. H&E x 1300

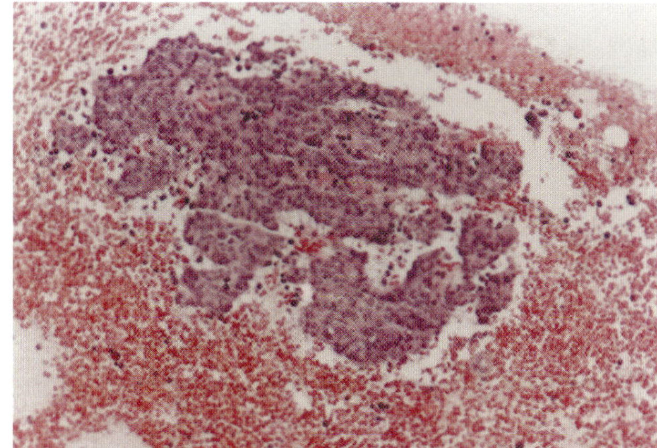

Figure 34.6, H&E x 130

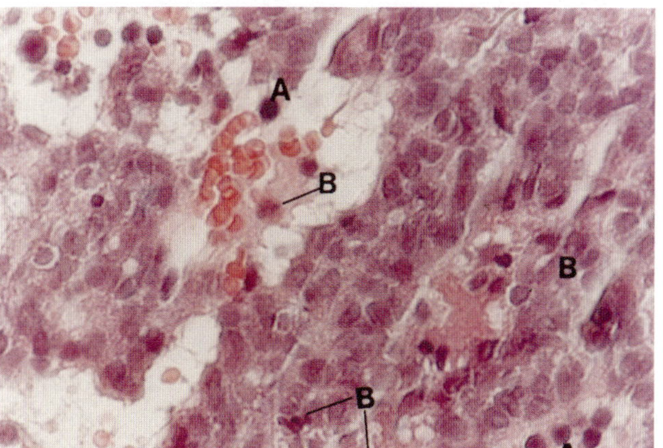

Figure 34.7, H&E x 520

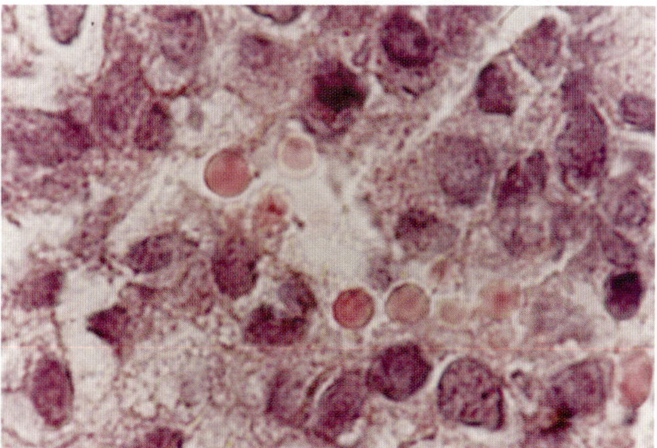

Figure 34.8, H&E x 1300

---

2. Additional photomicrographs of this prostatic carcinoma metastatic to bone marrow are presented in Volume I, figures 113 and 114.

Chapter 35

BONE CANCERS (figs. 35.1-35.6)

1. Giant Cell Tumor (figs. 35.1-35.3)

Figures 35.1-35.3 are from a specimen of giant cell tumor of the distal right femur surgically excised from a 36-year-old female. Cellular changes in cartilage and bone are presented in Chapter 6.

Blood and blood vessel formation from giant cell tumor of the bone (figs. 35.1-35.3)

Figure 35.1. In this section of giant cell tumor of the bone, there is development of red cells and an early stage of blood vessel formation (A) from the tumor tissue. The red cells are developing directly from the tumor tissue with the liquefaction of the latter. The endothelium of this blood vessel has not yet developed. (B) shows a developing irregular tubular space formed by liquefying tumor cells. Only a few red cells have developed from liquefied tumor tissue. H&E x 520

Figure 35.2. The large, multinucleated giant cells and spindle-shaped tumor cells are better seen at this magnification. Three multinucleated giant cells can be seen in this figure. H&E x 1040

Figure 35.3. In this higher magnification the development of red cells directly from the tumor cells can be seen. Early stages of partial hemoglobinization of a cluster of tumor cells can be seen in area (A). Smaller groups of tumor cells (B and C) are also transforming into red cells. H&E x 1040

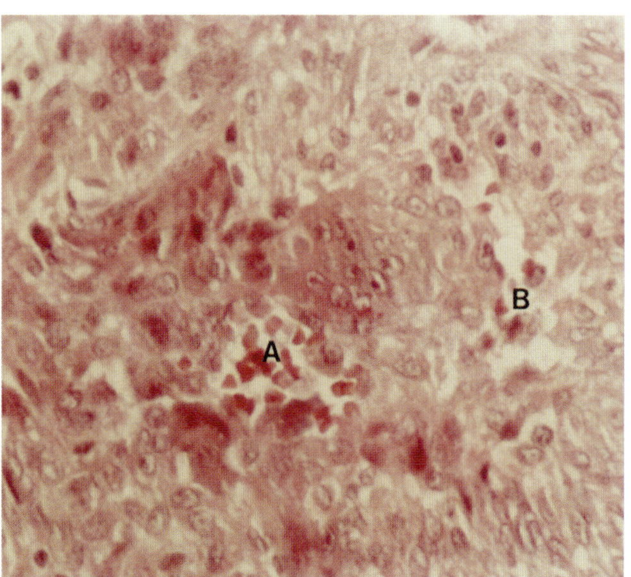

Figure 35.1, H&E x 520

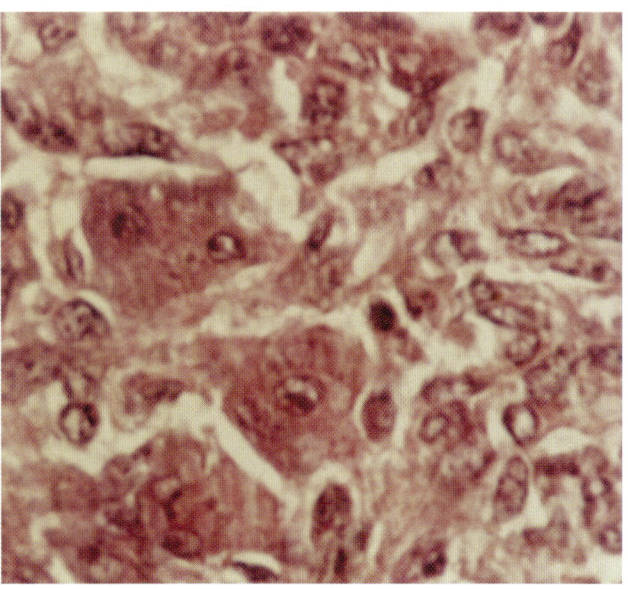

Figure 35.2, H&E x 1040

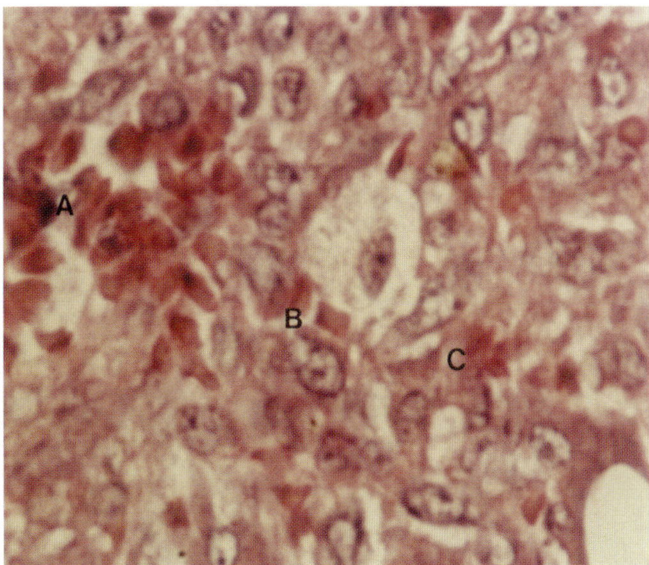

Figure 35.3, H&E x 1040

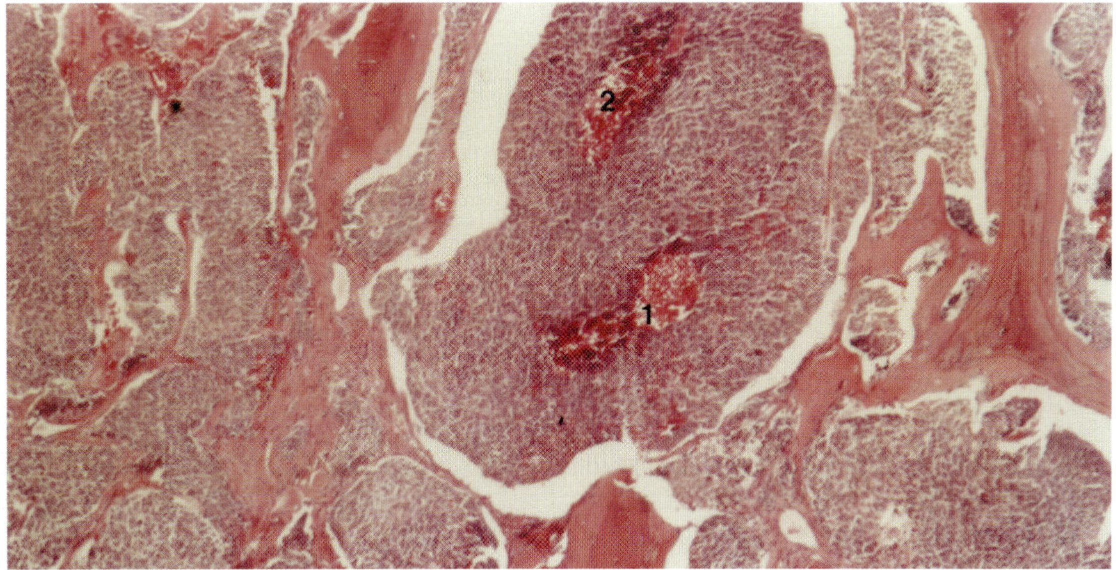

Figure 35.4, H&E x 104

2. **Bone with Metastatic Bronchogenic Carcinoma** (figs. 35.4-35.6)

Figures 35.4-35.6 are taken from a surgical resection specimen from a 50-year-old male with pathological fractures of the seventh and eighth ribs due to metastatic bronchogenic carcinoma.

Red cells developing directly from metastatic cancer cells (figs. 35.4-35.6)

Figure 35.4. This is a low-magnification view of a metastatic bronchogenic carcinoma involving the ribs. The fractured remains of bone tissue can be seen among the bronchogenic small cell carcinoma tumor tissue. There are two large areas (1 and 2) in this tumor from which red cells are arising and replacing the tumor tissue. Area (1) is seen in higher magnification in the next figure. H&E x 104

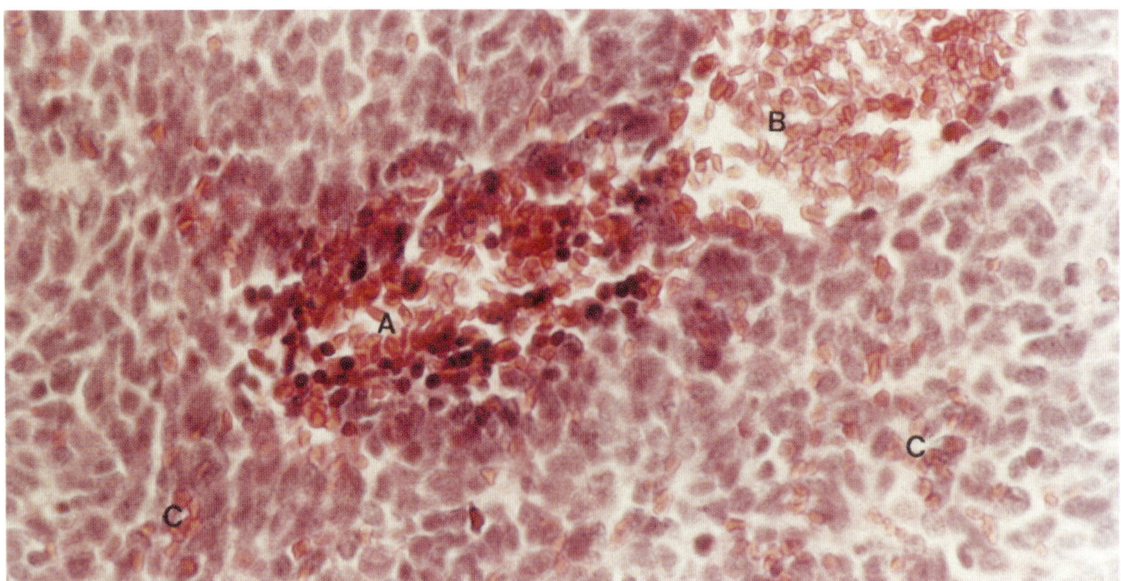

Figure 35.5, H&E x 520

Figure 35.5. In area (A), the tumor cells are transforming into small cells with hyperchromatic pyknotic nuclei and hemoglobinized cytoplasm. These cells transform directly into multiple hyperhemoglobinized red cells which are not yet fully rounded. Remnants of nuclear material can be seen on many of the developing red cells. In area (B), the red cells are well-formed but not well-rounded and no nuclear material remains. The endothelial lining has not yet formed. Scattered throughout this figure, tumor cells are transforming directly into red cells, most prominently seen in areas (C). This process is demonstrated in higher magnification in the next figure. H&E x 520

Figure 35.6. In several areas the tumor cells are directly developing into multiple, hyperhemoglobinized red cells which are not yet fully rounded. H&E x 1040

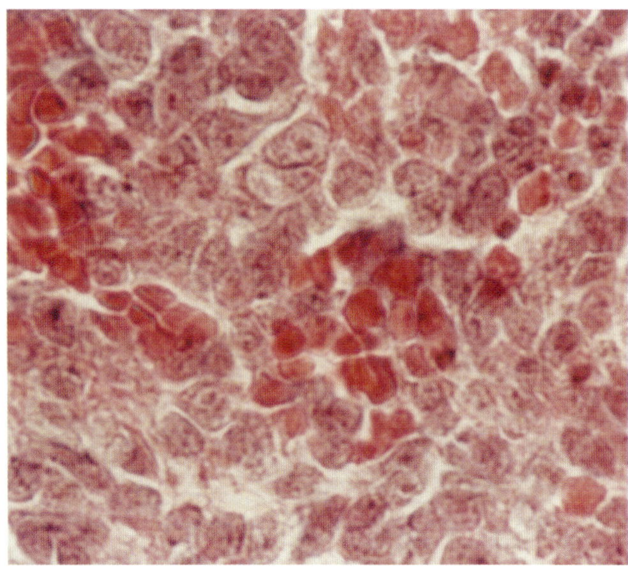

Figure 35.6, H&E x 1040

Chapter 36

BRAIN CANCERS (figs. 36.1-36.6)

1. Astrocytoma (figs. 36.1-36.2)

Figures 36.1 and 36.2 are taken from brain tissue surgically excised from a patient with astrocytoma (Grade III). Cellular changes in brain tissue are demonstrated in Chapter 19.

Blood and blood vessel formation from astrocytoma (fig. 36.1)

Figure 36.1. Seen here is astrocytoma (Grade III) with increased abnormal cellularity compared to normal brain tissue. There is prominent cellular and nuclear pleomorphism of the tumor astrocytes. A large blood capillary filled with hemoglobin particles, in which the red cells are not well-formed, is seen in the center. A single layer of endothelial lining cells can be seen at the periphery of this column of locally developed blood from brain tissue. Except for one area in the center of the vascular lumen, no nuclear structures can be seen. Background reddish color occupying some areas of the brain tissue (A) probably point to an early stage of blood formation. Area (B) shows red cells developing from an area similar to (A). H&E x 200

Red cell development from tumor tissue (fig. 36.2)

Figure 36.2. In this section of astrocytoma scattered hemorrhage can be seen throughout the figure. This so-called hemorrhage is actually developing red cells replacing the tumor tissue. No attempt at vessel formation is noted in this figure. H&E x 400

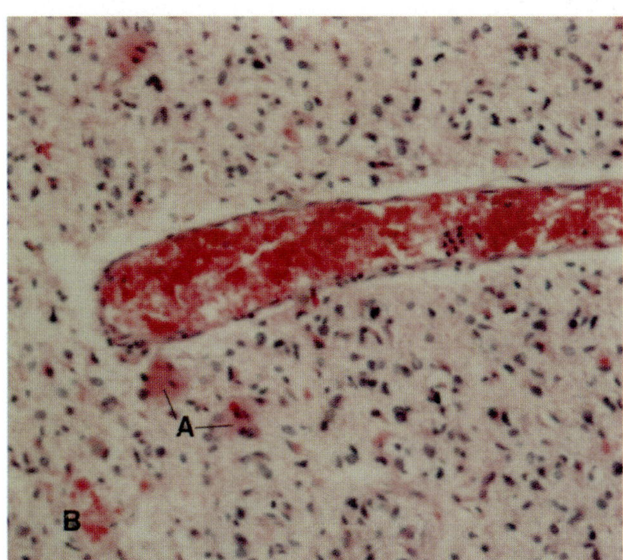

Figure 36.1, H&E x 200

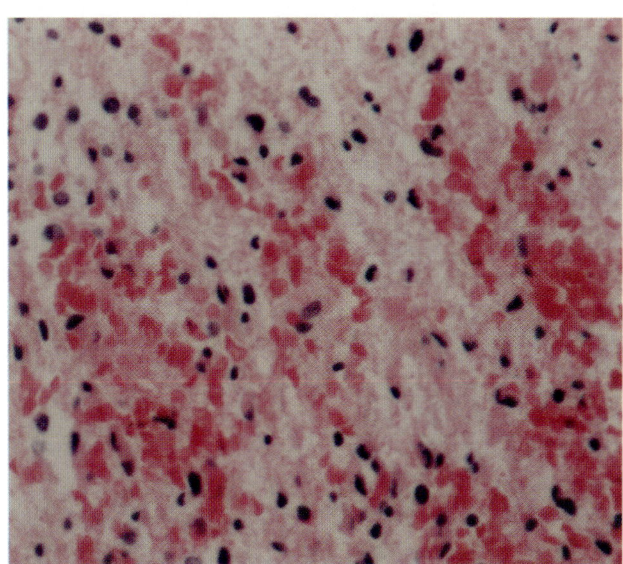

Figure 36.2, H&E x 400

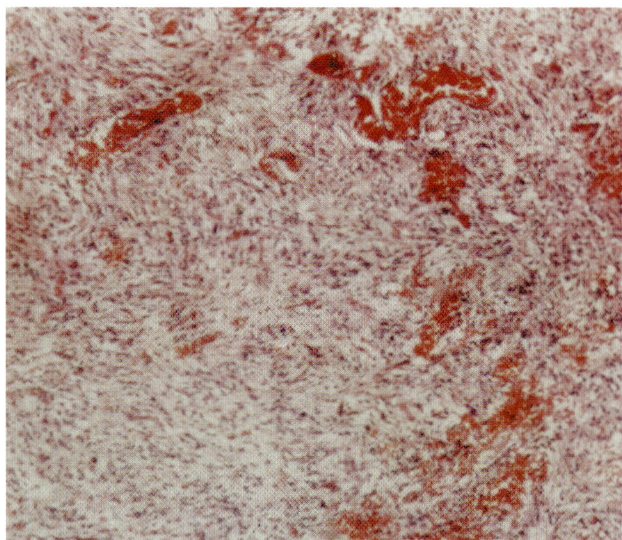

Figure 36.3, H&E x 104

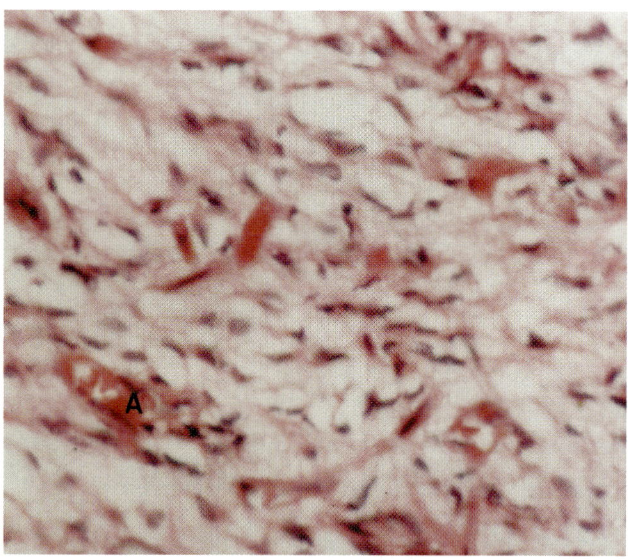

Figure 36.4, H&E x 520

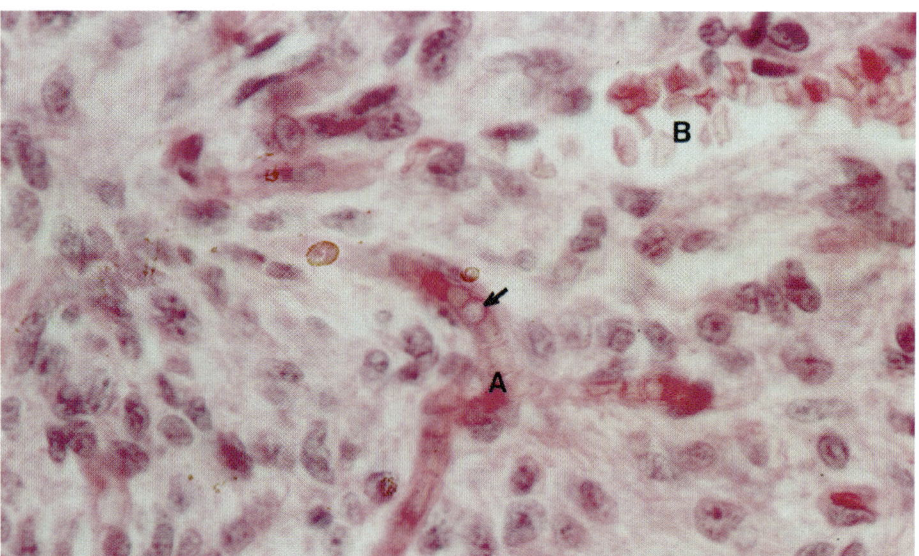

Figure 36.5, H&E x 800

2. Glioblastoma Multiforme (figs. 36.3-36.6)

Figures 36.3 and 36.4 are taken from surgically excised brain tissue from a patient with glioblastoma multiforme. Figures 36.5 and 36.6 are from surgically excised brain tissue from a second patient with glioblastoma multiforme.

Blood and blood vessel development from glioblastoma multiforme (figs. 36.3-36.6)

Figure 36.3. This is a low-magnification view of glioblastoma multiforme of the brain with scattered areas of so-called hemorrhage. H&E x 104

Figure 36.4. The red cells within developing blood capillary (A) are developing directly from the tumor tissue. Other areas of this tumor show streaks of hemoglobinization. The tumor cell nuclei seen here are very narrow, wavy, thread-like and widely separated from each other. This is an atypical variety of glioblastoma multiforme. H&E x 520

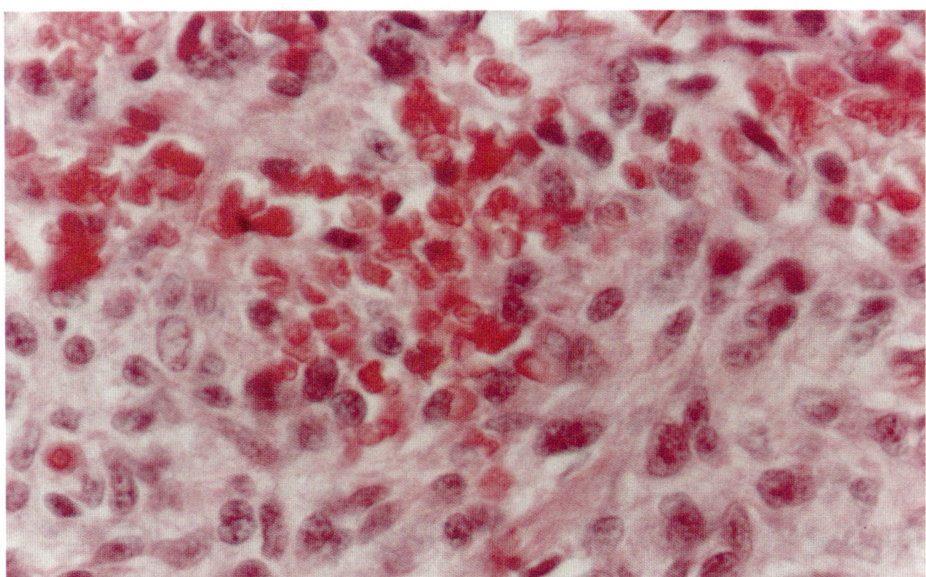

Figure 36.6, H&E x 800

Figure 36.5. In this section of glioblastoma multiforme, single columns of developing red cells are replacing the tumor tissue. In most areas the endothelium has not yet formed. Three narrow columns of developing red cells are joining at the center (A) to form a Y-shaped capillary. The red cells are developing in a single column and some of the red cells are not fully hemoglobinized (arrow). In the irregular, developing blood vessel (B), there are widely-spaced endothelial cells developing from tumor cells. Many of the highly irregularly shaped red cells in the developing lumen are only partially hemoglobinized and no nucleated red cells are seen. H&E x 800

Figure 36.6. In this figure many deformed, not yet rounded, red cells are developing from vanishing tumor cells. No particular attempt at blood vessel formation enclosing these red cells is noted. H&E x 800

Chapter 37

ACTIVE CELLULAR LYSIS,
A PHENOMENON OF GROWTH PROCESSES

Cellular lysis, a phenomenon of growth processes, and its role in the formation of different epithelial patterns was demonstrated earlier[1] in breast carcinomas in mice, and briefly reproduced in Volume I (McDonald, 1989, figs. 3-6). Cellular lysis provides space for the origin and growth of new cells; it also provides the new cells with nutrients (as the lytic products are absorbed by the new cells), vital substance and self-reproducing enzymes systems, as suggested by Shaver (1953). Shaver believed that the substances released by injured cells and tissue extracts contain a whole spectrum of factors, substrates, coenzymes, building blocks and possibly self-reproducing enzyme systems.

Cell lysis is seen in every tissue, benign or malignant. In benign tissues this lytic process proceeds less visibly, unless one examines the tissue sections critically, and may be preceded by diverse tissue changes which may be classified as cell digestion (as if digested by proteolytic enzymes), necrosis and less frequently other types of changes which fall under the various pathological terms: hyaline, hydropic (watery vacuolar), fibrinoid, granular, mucoid, myxoid, amyloid and keratoid degeneration. In normal tissues cellular lysis is usually preceded by cell digestion and/or hydropic degeneration.

In malignant tissues the lytic process is usually obvious, particularly in some tumors. The main cellular changes leading to cell lysis may be categorized as cell digestion (such as digestion by proteolytic enzymes), hydropic degeneration (watery vacuolar changes) and tissue necrosis. The last process (necrosis), usually not seen in benign tissues unless diseased, is common in malignant tumors. The faster the rate of tumor growth the more rapid the lytic process, particularly by cell digestion. New growth out of digested or necrosed cancer tissues is shown in Volume I (figs. 3-6, 29-31, and 58). Following cellular lysis, cell generation, and cell differentiation, a variety of epithelial patterns may be evolved. In this way a large number of fast growing tumor cells are accommodated by the same amount of space within the tumors. In the area of highly rapid cell generation and dissolution (McDonald, see Ghosh, 1959b) there may not be time for cytoplasmic or even nuclear differentiation, as can be seen in many figures in Volume I.

Another role of cellular lysis in morphogenesis is found in the formation of new vascular channels (Volume I, figs. 7 and 8). The lytic product provides the original plasma within the independent primordia of vascular channels. The lining endothelium arises by the transformation of remaining peripheral tumor cells or tumor cell substance. The endothelium may also arise from the acellular membrane derived from lytic tumor cells or from clear fluid of cell lysis (tissue plasma gel). In the latter case, endothelial cells usually develop by the differentiation of tiny dedifferentiated nuclei originating in this membrane (Volume I, figs. 7 and 8). A similar mechanism in the formation of vascular channels following cell liquefaction is demonstrated in livers of chick embryos and adult chickens (McDonald, 1968). Volume I demonstrates a similar way of vessel

---

1. *British Journal of Cancer* 13:200-207, 1959, and abstracted in the *Year Book of Pathology*, 1960, pp. 53-55 (McDonald, see Ghosh, 1959b). These findings were also presented at the Fifty-fifth Annual Meeting of the American Association of Pathologists and Bacteriologists (McDonald, 1958) and at the Tenth International Cancer Congress (McDonald, 1970b).

formation, together with locally developed red cells in the vascular lumen, in various malignant and benign tissues. The textbook theory implies that in early embryonic life independent primordia of vascular channels arise from certain mesodermal tissue, and after the closed vascular system has developed and the circulation begun, new blood vessels always arise by "budding" from pre-existing blood vessels (Maximow and Bloom, 1957, p. 249). However, I agree with Clark and Clark (1932) who believed that under normal conditions new capillaries are continually being formed while others may be retracted or absorbed and that the capillary pattern in any part of the body is plastic and capable of alteration in response to changes in the immediate environment.

In Volume II (McDonald, 1995) the role of cellular lysis of cardiac muscle in the development of new vascular channels in myocardium is explained. Lysis of myofibers creates vascular spaces and original plasma. The lytic product may also produce red cells directly (Volume II, figs. 1-5, 11-13, 83 and 87). Extensive lysis of cardiac muscle resulting in the development of large clefts or sinuses containing clear fluid, which is commonly called edema fluid, is shown in acute rheumatic fever in the third chapter. If these clefts or sinuses are lined by endothelium the fluid is known as plasma (lymph plasma or blood plasma) and the channels are called lymphatic vessels or blood vessels. It is conceivable that fragments of tissue destined for vessel formation may resist complete lysis and be carried away as emboli. Lie (1987) reported myocardial tissue fragments as emboli in the systemic and pulmonary circulation. The possibility of tumor tissue also resisting complete lysis and acting as a source of distant metastasis is demonstrated in breast carcinomas (McDonald, 1962, and Volume I, fig. 8). Human embryonic cardiac muscle undergoing cellular lysis in the formation of the atrio-ventricular passage and atrio-

ventricular valve cusps is demonstrated in figures 22-1 and 22-2 of Volume II.

In this volume, in benign tissues, the role of cellular lysis in direct development of red cells from liquefying tissue is easily seen in embryonic mesenchymal tissue (fig. 1.5), cardiac muscle (fig. 1.6), and liver cells (chick, fig. 1.4, and human, fig 1.7), and adult mouse liver (figs. 12.1-12.3). Cellular lysis of thyroid tissue producing hemoglobinized colloid and red cells is presented in figures 5.9-5.11 and 25.3. The role of cellular lysis in development of blood vessels (including red cells and original blood plasma) is demonstrated in squamous epithelium of the skin (figs. 2.6, 2.8 and 2.9), cardiac muscle (fig. 4.2), skeletal muscle (fig. 4.24), smooth muscle (fig. 13.6), bone (fig. 6.5), and in adult liver of mouse (figs. 12.5 and 12.28), chicken (fig. 12.10) and human (figs. 12.26, 12.27 and 29.5). As the blood vessels develop the endothelium forms from surrounding, remaining local tissues. Similarly, the development of lymphocytes and lymphatic channels from the lysis of dermal tissue is seen in figures 8.7-8.9. Remnants of local tissue resisting lysis and lying within the lumens of the developing vessels are demonstrated in figures 4.24, 12.10, 12.25, 13.6 and 29.5.

In the development of the secretory system, the lytic product (secretion) of specific cells producing various enzymes enters directly or is transported via a ductal system into its site of action. That the lumens of acinar and ductal structures are produced by cellular lysis is shown in well-differentiated mammary carcinomas (McDonald, 1959b). The transformation of glandular tissue into blood and blood vessels with the secretory products (hormones, enzymes, etc.) automatically released into the blood stream is discussed in Chapter 5, Endocrine Glands (figs. 5.2, 5.7, 5.19, 5.26, 5.33, 5.35, 5.41, 5.46, 5.47).

In malignant tissue, the role of cellular lysis in the development of red cells, original plasma, and blood vessels directly from liquefying tumor tissue is easily seen in melanoma metastatic to lung (fig. 22.18), mammary carcinoma in mice (figs. 24.8-24.10), adenocarcinoma of liver (fig. 29.1), and renal adenocarcinoma and papillary adenocarcinoma (33.1 and 33.6). Cellular lysis of tumor tissue leading to the development of inflammatory cells is demonstrated in human mammary carcinoma (fig. 24.15) and carcinoid of colon (fig. 31.19). Liquefaction of thyroid cancer cells with development of hemoglobinized colloid and red cells is described in figures 25.1, 25.2 and 25.4. The dissolution of tumor tissue forming a lacy pattern is seen in carcinoid of the colon (fig. 31.17).

Necrosis leading to cellular lysis is a common finding in malignant tissues. New growth of a variety of cells and cell structures out of digested or necrosed tumor tissue is demonstrated in this volume. Development of blood and blood cysts from necrotic mammary carcinoma in mice is demonstrated in figures 24.1-24.7. New growth of inflammatory cells out of digested or necrosed cancer tissues is shown in bronchogenic squamous cell carcinoma of lung (fig. 27.1) and adenocarcinoma of colon (figs. 30.13-30.15). Necrosis in development of red cells, original plasma, and/or new blood vessels from dissolving tumor tissue is demonstrated bronchogenic squamous cell carcinoma of lung (figs. 27.1 and 27.2), adenocarcinoma of the lung (figs. 27.3 and 27.4), islet cell tumor of pancreas (figs. 28.5 and 28.6), reticulum cell sarcoma of stomach (fig. 30.7), and carcinoid of the ileum (figs. 31.2 and 31.3). Remnants of tumor tissue resisting lysis and lying within the lumens of developing vessels are demonstrated in figures 22.16, 24.8, 27.2, 30.21 and 34.4.

MY OBSERVATIONS AND COMMENTS ON VOLUME I

Mammary Carcinoma in mice and humans, Basal and Squamous Cell Carcinoma, Malignant Melanoma, Carcinomas of Colon, Kidney and Prostate; Brief Examples in Gastric Wall, Spleen, Adrenal Cortex, Renal Medulla, Liver, Epidermis, Dermis, Adipose Tissue, Bone Marrow, and Cardiac, Skeletal and Smooth Muscles

I. Cellular lysis, a phenomenon of growth processes, provides space, nutrients and self-reproducing substance for the origin and growth of new cells. It also plays an important role in the formation of different epithelial patterns. The main cellular changes that lead to cell lysis are cell digestion, hydropic degeneration and tissue necrosis, the last process being particularly notable in malignant tissues. The process of cellular lysis in the development of vascular channels has the ability to provide space for the vessel as well as constitute its initial plasma. The products of cell lysis, as well as cells in the process of liquefaction, may produce red cells, usually within formative vascular channels (Vol. I, figs. 3 to 6).

II. I have observed that changes in malignant epithelium can lead to connective tissue stroma formation with collagen fibers. This view has been supported by several researchers, including Leob (1901), Foulds (1937), Al-Adnani, et al. (1975), Berman and Foidart (1980), and Sakakibara, et al. (1982). Hyalinization of cancer cells has been found to be an important phenomenon in the transformation of malignant epithelium into connective tissue stroma. Hyalinization may be preceded by pyknotic spindling of cancer cells. There is evidence of direct transformation of cancer cells into narrow and broad bands of collagenous fibers. Similar to malignant epithelium, connective tissue stroma, either in stages of transformation or transformed from cancer cells, may give rise to red cells, vascular channels and various reactive cells (Vol. I, figs. 34, 35, 38, 39, 64 to 67, and 70 to 72).

III. Blood and blood vessels may form from various benign and malignant tissues. The endothelium of the developing vessels arises from:

1. Direct transformation of peripheral cells into endothelial cells

2. Development of endothelial nuclei from acellular protoplasmic membrane derived from lysing peripheral cell substance (which may even be liquid)

3. The protoplasmic membrane surrounded by blood or lymph column

4. Structure-less remains of peripheral cells

5. Single-layer columns of vasoformative cells, each cell producing red cells and surrounding endothelial cells (Example: unicellular origin of blood capillaries, common in chorionic villi)

6. The smooth muscle of stomach walls and from cardiac muscle fibers

7. All solid organs

8. Plasma cells

I agree with Clark and Clark (1932), who believe that under normal conditions new capillaries are continually being formed while others may be retracted or absorbed. In addition, the capillary pattern in any part of the body is plastic and capable of alteration in response to changes in the immediate environment (Vol. I, figs. 7, 17, 61, 89, 90, 92, 106, 107, 110 and 120-1).

IV. The various ways of blood vessel formation in the body are as follows:

1. Formation of narrow capillaries,

often with a single column of red cells, as seen in cardiac muscle, liver, etc. (Vol. I, figs. 127-1 and 134 to 137)

2.   Formation of vascular channels by coalescence of small, unicellular vascular units in which the endothelium is formed by the outer cell substance enclosing one or more intracellularly developing red cells (Vol. I, figs. 126 and 127-1)

3.   Transformation of groups of local cells into red cells and plasma surrounded by endothelium developing from peripheral cell substance (Vol. I, fig. 22)

4.   Transformation of local cells *en masse* into red cells (Vol. I, figs. 14, 90, 91, 127-1, 136 and 137)

5.   From sheets or columns of vacuolar red cells arising from local tissues (Vol. I, figs. 106, 107 and 125)

6.   From unilocular or multilocular blood cysts and blood sinuses (Vol. I, figs. 23, 24, 26, 27, 29, 30, 32 and 33)

7.   From erythrogenic reactive cells of local origin (Vol. I, figs. 45 to 47, 63, 65, 69, 83 to 85, 117 to 121 and 123)

8.   Vascular channels containing plasma with or without lymphocytes (categorized as lymphatic channels) may arise from local tissues, such as adipose tissue, dermal tissue, or malignant tissue (Vol. I, figs. 53, 59 and 92)

V.   Red cells develop in various ways without passing through the nucleated stages of erythropoiesis.   They may originate in the following ways:

1.   Intracellular or intranuclear origin as hemoglobin globules (Vol. I, figs. 13 to 17, 22, 61, 83, 89, 100, 106, 110 and 121)

2.   From liquefied cell product, such as tissue plasma gel, erythrogenic gel, or hemoglobin gel; usually, this process is preceded by cell digestion or, less frequently, by hydropic degeneration, necrosis, or other changes (Vol. I, figs. 19, 24, 29, 30, 32, 33 and 33-1)

3.   From erythrogenic mesh (Vol. I, figs. 130 to 132)

4.   From tissue necrosis without going through liquefaction (Vol. I, figs. 58 and 59)

5.   Development of sheets or columns of vacuolar red cells (ghost red cells) upon disappearance of cellular structures, usually by hyalinization (Vol. I, figs. 106, 107 and 134 to 137)

6.   From inflammatory or reactive cells, including endothelial cells, which arise locally in benign and malignant tissues (Vol. I, figs. 45 to 47, 62, 65, 112, 117, 119, 120-1, 122 and 123)

7.   Directly from collagen fibers transformed from malignant epithelium in the development of stromal tissue (Vol. I, figs. 16, 42, 43, 46, 47 and 65)

VI.   Red cells may appear in different sizes, shapes and hemoglobinizations at their origin from bone marrow plasma, most prominently seen in cases of pernicious anemia.   A marked variation of red cell size and hemoglobinization is seen in malignant and benign tissues as shown in prostatic carcinoma, malignant melanoma, basal cell carcinoma, epidermis, renal medulla, etc. Red cells may sometimes arise as "ghost cells," as in renal tubules and cardiac muscle fibers (Vol. I, figs. 43, 88, 106, 107, 116, 125 and 130 to 137).

VII.   Colorless hemolysis is possibly the primary mechanism by which aged and defunct red cells are removed from the circulation.   This reaction involving red cells changing into colorless fluid may also be the main way for the disposal of red cells in the tissues due to an actual hemorrhage caused by the rupture of blood vessels, as occurs in surgical interventions (Vol. I, figs. 138 and 139).   Colorless hemolysis may also be involved in the disposal of additional locally-produced red cells which failed to join the circulation due to lack of vascularization enclosing them.

In addition, hemosiderin pigment development is intimately connected with abortive attempts at red cell generation, as seen in renal cell carcinoma (Vol. I, figs. 101 to 105).   There is also evidence to show that hemosiderin pigment is present in certain benign and malignant tissues.

VIII. Like red cells, inflammatory or reactive cells may arise locally from benign and malignant tissues and may acquire an intravascular position by developing in the lumen area of formative vascular channels. They may also acquire an intravascular position by extension of a vascular lumen by dissolution of the peripheral tissue and realigning of the new endothelium. The development of inflammatory or reactive cells is described in detail in this book.

IX. Benign tissues adjoining malignant lesions often show significant cellular changes. Examples include:

1. Dermis and epidermis adjoining basal cell carcinoma (Vol. I, fig. 69)

2. Dermis and skeletal muscle adjoining squamous cell carcinoma (Vol. I, figs. 74 to 77)

3. Adipose tissue surrounding mammary carcinoma (Vol. I, figs. 52 to 55)

4. Skeletal muscle adjacent to colon carcinoma (Vol. I, figs. 93 to 95)

In skeletal muscles, numerous lymphocytes or dedifferentiated tiny fusiform nuclei may completely fill the muscle fibers, thus possibly stopping further extension of the tumor in some cases. Similar reactive cells that arise from adipose tissue adjacent to mammary carcinoma may also suppress further tumor growth. The enhancement of these natural defense processes for the elimination of cancer cells and creation of an environment which will oppose benign cells from becoming malignant should be considered for an experimental therapeutic study of cancer.

X. The local spread of a malignant lesion need not be contiguous. Benign tissue around a malignant lesion may become cancerous due to some unknown initiating factor coming from the tumor tissue. For example, adipose tissue around breast cancer may develop malignant characteristics similar to the original carcinoma.

Because lymphatic channels and blood vessels can be formed from local tissues including cancer tissues by liquefaction, unliquefied tumor cells or fragments lying loose in the lumen may become a source of metastasis (Vol. I, figs. 8, 92 and 106). For tumor tissue fragments to become intravascular it does not need to invade the vessel wall. This is because the vessel is formed from the tumor tissue, and the latter would already be intravascular, or within the vessel wall, from the beginning of vessel formation.

XI. My observations will be helpful in the solution of some biological problems by providing new knowledge concerning growth processes of blood and blood vessels in benign and malignant tissues. For example:

1. I have shown various instances where blood vessel formation encloses red cells, thus explaining how red cells enter blood vessels. The popularly accepted theory states that after the closed vascular system has developed and the circulation begun in the embryo stages, new blood vessels always arise from preexisting blood vessels. The possibility of blood and blood vessel formation from local tissues (Vol. I, figs. 13 to 22, 42 and 45 to 47) has been erroneously ignored.

2. The formation of single or multiloculated epithelium-lined blood cysts along with the local development of red cells within them (Vol. I, figs. 23, 26, 30 and 32), which are often seen in the mammary carcinoma of C3H mice, explains how blood can exist inside these cysts without any source of entrance.

3. I have demonstrated that red cells can arise locally from various benign and malignant tissues as well as from inflammatory or reactive cells. Thus, I do not agree with the commonly accepted theory that red cells can only develop through the nucleated stages of erythropoiesis. I have also shown that the so-called process of

"erythrophagocytosis" is actually "erythrogenesis," and these red cells do not show any deformity or degeneration, which is generally expected in a phagocytic process.

4. The presence of red cells outside formative vascular channels is due to a lag in the formation of the vessels enclosing these red cells and not due to "hemorrhage" as is commonly believed (Vol. I, figs. 25, 57, 96, 105, 108 and 116).

5. In cases of gastritis red cells may originate locally from gastric mucosa, mainly through reactive cells associated with superficial rupture of mucosal lining cells.

6. In cases of vascular congestion, with engorged blood capillaries, there is usually no evidence of pressure on tissues adjacent to the distended vascular channel. This is because the distended capillaries enclosing many red cells are formed from the local tissues with replacement of the latter. This phenomenon has been demonstrated in the liver (Vol. I, fig. 127-1) and other tissues.

7. Vessel formation from lysing local tissue is found to be an effective way of carrying the tissue product into the circulation. This is true for all types of tissues. I agree with Clark and Clark (1932), who believe that new capillaries are continually forming while others are retracting.

8. The liquefaction of local tissues is involved in the development of the vascular content within formative vascular channels. This explains the occasional presence of intravascular benign tissue fragments; they are present as a result of incomplete tissue lysis during the production of the fluid content of these vessels.

9. My findings point to a new concept concerning metastasis. In malignant tissues, cell lysis and cell growth occurring in rapid succession along with some undissolved tumor tissue fragments can be a source of metastasis (Vol. I, figs. 8, 92 and 106).

10. Like red cells, various inflammatory or reactive cells arise locally from different benign and malignant tissues. Many of these reactive cells give rise to red cells in the form of hemoglobin globules (Vol. I, figs. 45 to 47, 62, 83, 112 and 120 to 123). For example, a large number of lymphocytes are shown transforming into red cells as hemoglobin globules (Vol. I, figs. 65 and 121). This explains why the lymphocyte percentage of blood remains fairly constant despite the fact that a large number of them are poured into the circulation daily through the thoracic duct.

11. The neoplastic origin of stroma explains how it can grow inside malignant tumors and how it shows the same characteristics despite generation after generation of transplantation (when it should originate from different animals in each generation). This also explains how it is possible for the stroma of a metastatic tumor to develop and grow within the cardiac lumen without being connected to the cardiac walls of the host animal (Vol. I, figs. 1 and 2).

12. Benign tissues at the periphery of a growing malignant lesion can themselves become malignant by the influence of the tumor and show characteristics of the primary tumor (Vol. I, figs. 48 to 51). This explains the mechanism of local spread of cancer to adjoining tissues.

13. I believe that hemosiderin pigment is not necessarily associated with hemorrhage because this pigment may not be found in post-surgical incision areas where profuse bleeding has usually occurred. In addition, a large amount of hemosiderin may be found in many tissues and organs of patients suffering from hemochromatosis without any evidence of hemorrhage.

14. Different forms of hemoglobin as well as heterogeneity of serum ferritin is expected as red cells are constantly developing from different cells of different tissues of the body in various ways.

15. The development of blood and blood capillaries from cardiac muscle (Vol. I, figs. 134 to 137) and the possibility of vessel and red cell formation from necrotic tissue may explain the mechanism of establishing collateral circulation passing through infarcted myocardium.

XII. Profuse cellular proliferation may take place with or without mitosis (which is rare). I agree with the existence of mitosis but find that it plays a small role in cellular proliferation. I believe that mitosis is an infrequent way of cell multiplication in benign and malignant tissues. This is evidenced in the following cases, all of which do not involve any mitosis:

1. The development of various inflammatory or reactive cells from benign and malignant tissues

2. The origin of endothelial cells

3. The transformation of cancer tissue into collagenous fibrous stroma

4. The development of collagen directly from cardiac muscle fibers in chronic anoxic conditions

5. The production of a huge number of undifferentiated cells in dissolving muscle fibers (skeletal muscle) near a malignant lesion

6. The development of many tiny dedifferentiated cells from dissolving cardiac muscle fibers in acute rheumatic fever

7. The development of cancer cells from normal adipose tissue close to a malignant lesion

8. The origin of well-differentiated cancer cells from necrosed tumor tissue as well as from the products of cell lysis

XIII. My findings may have a great implication on many branches of biological science. For example, the production of inflammatory cells within benign and malignant tissues as well as the vast degree of diversity in the transformation of malignant epithelium into benign tissues (including fibrous connective tissue, blood and blood vessels) both appear to have a profound effect on tumor growth, and may be considered in experimental and clinical studies on malignancies. In addition, the enormous capacities endowed in living tissues for cell generation, differentiation and transformation can be of great significance in the study of cell life cycles.

XIV. I wish to emphasize that we should not over-rely on technology and curtail our powers of observation and analysis using our greatest tool, the human mind. Preconceived ideas apparently limit one's accurate view of the phenomenon at hand. I hope that based on my findings, further studies by many inquisitive minds will unveil additional pathways of cellular growth. Through my microscopic findings, I wish to depict a "gentle transition" in the development of different cellular and tissue structures guided by nature.

(Volume I, Copyright 1989; ISBN Number 0-9627824-0-8; 129 color and 13 black and white photomicrographs; 111 pages)

I. Red Cell Formation Directly as Hemoglobin Globules from Malignant and Benign Tissues, Including Reactive (Inflammatory) Cells

As demonstrated in this book from various organs and tissues and from reactive cells, the developmental processes of red cells without passing through the nucleated phases of erythropoiesis, as generally believed to be the only way of red cell formation

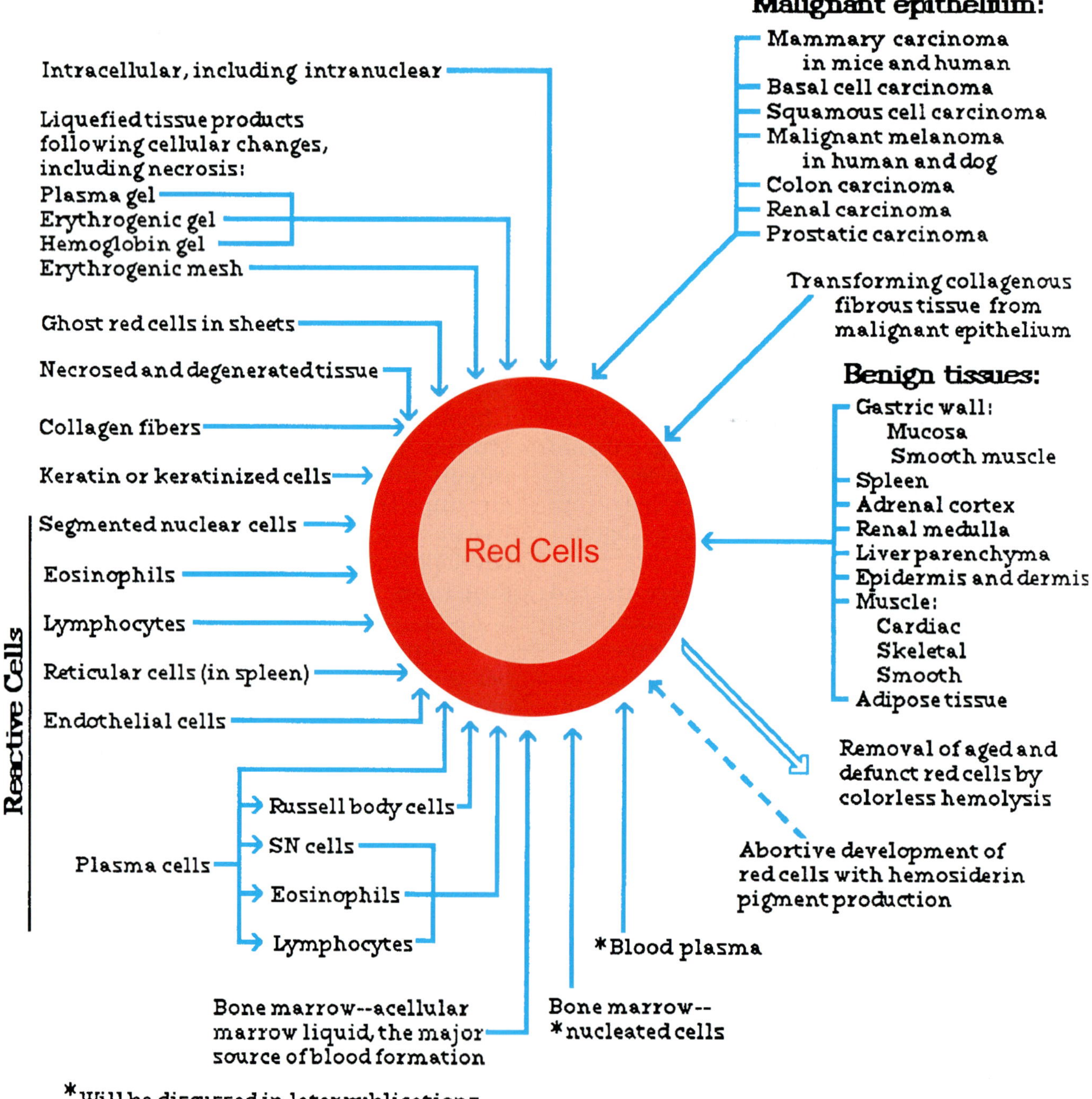

Author's note: This diagram is reproduced from Volume I (McDonald 1989) to demonstrate the variety of ways and different sources involved in the development of red blood cells.

MY OBSERVATIONS AND COMMENTS ON VOLUME II

Cardiac Muscle

Adult and embryonic tissues from humans and animals; and in chronic ischemic conditions, acute rheumatic fever, and coronary occlusion with myocardial infarction.

The vast amount of data that I have accumulated on cardiac muscles through decades of research is presented in Volume II of this series. To demonstrate the broad scope of this study, I present detailed illustrations from the myocardium of humans and a few animals (rabbits, mice and chicks). In addition, Volume II demonstrates the embryonic development of human and chick cardiac muscles, as well as, the development of cardiac muscle of chick embryos *in vitro*.

In the first chapter (figs. 1 to 16-3), the direct development of red cells and blood capillaries from cardiac muscle is shown in humans, rabbits, rats and mice, as well as, in embryonic cardiac muscles of human and chicken. The development of red cells of cardiac muscle origin is also shown in tissue cultures. The muscular coat of blood vessels is shown to arise from cardiac muscle or myogenic fibrous tissue. In addition, the direct development of red cells as hemoglobin globules and the development of endothelium and smooth muscles of larger vessels from cardiac muscles are demonstrated in this chapter.

The second chapter (figs. 17 to 22-2) is devoted to the mechanism of transformation of cardiac muscle into collagenous fibrous tissue with the development of spindle shaped nuclei. The cardiac valves are shown to be of cardiac muscle origin, as demonstrated in the human embryonic stage.

Chapter three (figs. 23 to 55) notes that acute rheumatic fever was an important heart ailment prevalent in children before the antibiotic era. However, even as late as 1980, 40% of cardiac admissions to a major Indian teaching hospital were for rheumatic fever (Agarwal, 1981). At Children's Medical Center in Akron, Ohio in 1986, twenty-three patients from 3 to 16 years of age were treated, as compared to twenty-five cases in the whole preceding decade (Congeni, B., Rizzo, C., et al, 1987). Most shocking is the result of a study done by Primary Children's Medical Center in Salt Lake City, Utah, from January 1985 to June 1986, seventy-six cases of rheumatic fever in children ranging from three to seventeen years of age were reported (Veasy, L.G., Weidmeier, S.E., et al, 1987).

Since the specimens of heart for rheumatic fever are no longer readily available, I studied beautifully preserved autopsy slides and tissues from Barnes Hospital in Washington University's School of Medicine in St. Louis, Missouri. The slides, dating as far back as 1917, were from children who died of acute rheumatic fever. There were hundreds of such cases for which I was given permission to study. I am greatly indebted to Dr. Wilbur A. Thomas, then a senior faculty member in the General Pathology Department of Washington University School of Medicine, for allowing me access to this autopsy material.

This chapter illustrates damaged cardiac tissue and its subsequent reaction in acute rheumatic fever. Through detailed study I have found that in rheumatic fever, cardiac muscle damage is the main culprit and not the collagen (as is commonly believed). The unique findings of my studies are as follows:

1. Acute rheumatic fever causes severe cellular lysis of cardiac muscle along with a lesser extent of fibrinoid and hyaline degeneration.

2. Cellular proliferation and the

development of fibrinoid material from damaged muscle, as well as from regenerating myoplasm, are involved in the production of Aschoff bodies, the diagnostic histological feature of rheumatic fever.

3. Even in fatal cases there is ample evidence of cardiac muscle regeneration without mitosis. This is accomplished by redifferentiation of dedifferentiated cells arising in lytic myofibers.

4. Muscle damage and inadequate muscle regeneration are the causes for the production of collagen fibers in the development of myocardial scar tissue.

5. The acute inflammatory infiltrates that are often in the cardiac valves are shown to be of local origin, either from cardiac muscle or from myogenic fibrous tissue of cardiac valves.

Finally, the fourth chapter (figs. 56 to 90) deals with the damage and transformation of cardiac muscle in coronary occlusion. In acute myocardial infarction I have described the variety of ways in which cardiac muscles, in various stages of coagulation necrosis, attempt recovery by developing fibrous tissue and vascularization. The affected muscle fibers can transform into granulation tissue with the production of blood capillaries and different types of inflammatory or reactive cells. In addition, infarcted muscle may transform into collagenous fibrous tissue, either directly or through chronic reactive cells of muscle origin. The possibility of profuse blood formation (usually classified as hemorrhage) by freshly infarcted muscle, and acute suppurative inflammatory cell generation from noninfarcted myofibers are demonstrated. Both of these phenomena have been attributed to occasional cardiac rupture.

(Volume II, Copyright 1995; ISBN Number 0-9627824-1-6; 105 color and 5 black and white photomicrographs, 117 pages)

MY OBSERVATIONS AND COMMENTS ON VOLUME III

New Discoveries in Hematology

I have been studying the development of red cells, white blood cells, reticulocytes, platelets and megakaryocytes in blood and bone marrow for many years, mainly utilizing the material from patients whom I have seen in hematology consultations, as well as collected material from other sources. As a result, I have discovered many important histocytogenic processes in the development of red cells, white blood cells, reticulocytes, platelets and megakaryocytes. My findings in blood and bone marrow are demonstrated in *Volume III, New Discoveries in Hematology* (McDonald, 2001).

Unlike solid tissues, where cellular structures are not particularly movable, the structures of blood and bone marrow are more difficult to formulate. However, I feel that I have circumvented this difficulty with my various methods of approaching cellular study, including the use of numerous photomicrographs, and taking into consideration each cell's past, present and future possible structural changes.

My findings are in sharp discordance with popularly accepted theories on the development of red and white blood cells, platelets, reticulocytes, megakaryocytes, as well as other cellular and acellular structures which are not mentioned in literature. However, in describing these various cellular and acellular structures, I have used the accepted cellular terminology of erythropoiesis and granulopoiesis in most cases to make it easier for the readers to comprehend the essence of the discussion without getting entangled in a web of revised terms and nomenclature. Such usage in no way diminishes my firm conviction regarding the way various cellular and acellular structures behave in the blood and bone marrow.

Presented below is a brief list of my discoveries in bone marrow and circulating blood that are discussed in Volume III based on demonstrations of many color photomicrographs.

1. Red cells, nucleated cells (segmented nuclear cells, lymphocytes, plasma cells, eosinophils, monocytes, etc.), and megakaryocytes originate directly from bone marrow and blood plasma, as well as through various coagulation products of marrow and blood plasma (fibrinoid bodies, fibrin membrane, fibrin plaques, fibrin particles and loosely woven fibrin particles named "basket cells") (Vol. III, figs. 3.1-3.22, 4.1-4.15, 11.10-11.12, 11.25, 11.26, 13.2, 13.4, 13.10, 13.15, 13.16, 13.19-13.21, 13.28-13.32, 13.37-13.42, 13.54-13.56, 13.58 and 13.59).

2. A type of reticulum cell (nurse cell) and the surrounding bone marrow plasma are involved in the development of a large number of erythropoietic cells, particularly normoblasts, which then further develop into red cells (Vol. III, figs. 5.1-5.6, 13.43 and 13.44).

3. The direct transformation of various adult nucleated cells into red cells is prominently seen in bone marrow clot and biopsy sections. However, bone marrow smears, which dry quickly, give less time for nucleated cells to transform into red cells. Thus, fewer nucleated cells are seen in marrow clot sections than in marrow smears (see Vol. III, Chapter 3). Occasionally, we also see SN cells turning into red cells in peripheral blood smears (Vol. III, figs. 6.9-6.11).

4. Blue fibers, possibly arising from the bluish necrosis of marrow tissue, are intimately involved in the direct development of red cells as hemoglobin globules and various white blood cells, and to a small extent platelets (Vol. III, figs. 7.1-7.22, 13.4, 13.22-13.24, 13.33 and 13.34).

5. The blackish reaction associated with

the disappearance of red cells is directly involved in the production of white blood cells, platelets and megakaryocytes (nucleus and cytoplasm of the latter arising simultaneously), and also various types of colony formation by layers of red cells surrounding nucleated cells in fluid medium (Vol. III, figs. 8.1-8.26, 13.8 and 13.9).

6. New findings on the origin of reticulocytes are presented in Volume III. Clumps of microscopic, finely-beaded, thin fibrillary material (which stain blue with methylene blue or crystal violet) appear in blood or marrow plasma. Tiny fragments of this blue material together with surrounding blood or marrow plasma are directly involved in the development of individual reticulocytes. No nucleated cells are seen taking part in the formation of reticulocytes (Vol. III, figs. 9.1-9.12).

7. I do not agree with the common belief that aged and defunct red cells are removed by phagocytosis in the spleen and lymph nodes. As a pathologist I have examined hundreds of surgical and autopsy specimens of spleen and lymph nodes, but I have hardly found any cells in these tissues which could be definitely called erythrophagocytes. Instead, I found that red cells are developing within the cells, and colorless hemolysis, a process in which red cells hemolyze without turning the plasma red, is the primary way by which red cells make their way out of the circulation. This can be easily observed in cases of hemolytic anemia and thrombocytopenic purpura where massive hemolysis occurs without the plasma turning reddish, as opposed to hemolysis due to mismatched blood transfusion (where the plasma does turn red) (Vol. III, figs. 10.1-10.8, 13.17 and 13.18).

8. In bone marrow smears, necrosis (which is commonly seen and ignored) is followed by the disappearance or fading away of bone marrow cells. This creates space for the regeneration of new marrow tissue, usually consisting of the previous type of cells (Vol. III, figs. 10.9-10.28).

9. Colonies of layers of red cells surrounding nucleated cell(s) are found floating in fluid medium of the bone marrow and appear to be one of the various sources of origin in development of both red and white blood cells. These colonies usually consist of centrally developing nucleated cells (which may be SN cells or other nucleated cells, and rarely megakaryocytes, or even a group of platelets) which are surrounded by an outer layer or layers of developing red cells. The innermost section of these red cell layers usually show blackish reaction (Vol. III, figs. 10.29-10.44).

10. Megakaryocytes originate from marrow plasma in two different ways. In one way, the nucleus is formed by the clumping of fibrinoid or nuclear bodies of marrow plasma origin, and the cytoplasm arises from the surrounding marrow plasma. The other way involves the simultaneous development of the nucleus and cytoplasm through the blackish reaction of red cells (Vol. III, figs. 11.10-11.12, 11.25 and 11.26). I find that megakaryocytes have nothing to do with platelet production (Vol. III, figs. 11.10, 11.13 and 11.14).

11. Platelets develop directly from red cells and from blood and marrow plasma. Also, platelets occasionally develop from fibrin tags and from the marginal cytoplasm of large lymphocytes (Vol. III, figs. 11.1-11.40 and 13.35).

12. Sickle cells originate directly from blood and marrow plasma and, to a small extent, from the coagulation products of marrow plasma (woolly bodies). In autopsy specimens of a patient who died of sickle cell anemia, a direct replacement of solid tissues by sheets of sickle cells is seen in almost all extramedullary organs, most prominently in the liver and spleen (Vol. III, figs. 12.1-12.35).

Brief Synopsis of the Chapters in Volume III

Chapter 1. Embryological Studies of Blood Formation in Chicks and Humans: The present-day more or less accepted theory of erythropoiesis in adults was originally developed by studying red cell formation in the yolk sac membrane of chick embryos following a few hours of incubation. It is believed that hypothetical stem cells, which are originally developed in the yolk sac membrane, are carried to the liver and then to the bone marrow, where in adults they continue to grow passing through various developmental stages until red cells are formed. Because of the popular belief in this theory, I studied blood formation in chick embryos and discovered that not only does the yolk sac membrane form red cells through various nucleated stages, but at the same time, other tissues also form red cells directly as hemoglobin globules. The capacity of red cell development from embryonic tissues, particularly of the heart and liver, is shown *in vitro.*

Chapter 2. Bone Marrow Development from Cartilage and Bone: Because the bone marrow is popularly believed to be the only site of red cell formation, I decided to study how bone marrow is actually formed. I have studied the development of bone marrow in humans, rabbits and chicks, and the role of human bone marrow in the formation of red cells, white blood cells, reticulocytes, platelets and megakaryocytes. I found that bone marrow originates from dissolved cartilage cells as well as from dissolving bone. The marrow plasma then becomes the original mother substance of red cells, white blood cells, platelets and megakaryocytes.

Chapter 3. Red and White Blood Cell Development from Marrow Plasma: A large number of red cells, white blood cells and platelets arise directly from marrow or blood plasma. The majority of red cells develop from marrow and/or blood plasma, which can be called plasma gel, without going through any nucleated stages. A vast number of red cells and nucleated cells appear from clear marrow plasma or through the different shades of red plasma, which can be designated as plasma gel, erythrogenic or hemoglobin gel, and erythrogenic mesh or net.

Chapter 4. Red and White Blood Cell Development from Plasma Coagulation Products: Red and white blood cells are formed through various coagulation products of marrow or blood plasma. These coagulation products include fibrin membrane/film, fibrin plaques, fibrin particles, loosely woven fibrin particles named "basket cells," and solid round fibrinoid bodies. All of these coagulation products may be transformed into red cells, white blood cells, platelets and megakaryocyte nuclei (with the megakaryocyte cytoplasm arising from surrounding clear plasma).

Chapter 5. Development of Commonly Known Erythropoietic Cells in Association with Special Reticulum Cells (Nurse Cells): Rarely, in bone marrow smears, special reticulum cells (nurse cells) are found in association with erythropoietic cells, mainly normoblast type cells. In rare cases, nurse cells may sometimes produce a large number of red cells through small hyperchromatic normoblasts. An earlier author inadvertently came across these special reticulum cells in a puncture of vertebral bone marrow while drawing spinal fluid and named them "nurse cells" (Craver and Carson, 1991).

Chapter 6. Red Cell Development Directly from Nucleated Cells in Bone Marrow Clot and Peripheral Blood: In bone marrow smears, one observes many more nucleated cells as compared to the number seen in marrow clot and biopsy sections of the same sample. This is because in sections, many nucleated cells have transformed into red cells by the time the marrow clot is fixed in formalin. In these sections, one can see many transitional stages between nucleated cells and red cells. Lymphocytes, plasma cells, SN cells and eosinophils are all seen transforming into red cells. Even metastatic cancer cells drawn from aspirated sternal bone marrow can convert into red cells.

Chapter 7. Blue Fibers Found in Circulating Blood and Bone Marrow and Their Direct Participation in the Development of Red and White Blood Cells, and to a small extent Platelets: Even though blue fibers or fragments are present in almost every blood and marrow smear examined, even those taken decades ago, I started discovering them only four/five years ago. Blue fibers play an important role in the origin of red and white blood cells. Although the nature and source of blue fibers remain uncertain, large numbers of SN cells, a few eosinophils, and a few lymphocytes arise from them. Blue fibers take part in the direct production of mature red cells from adjacent plasma and play a very small part in the production of platelets. The possible source of blue fiber origin has been found to be from blue necrosed marrow tissue (see Chapter 10).

Chapter 8. The Blackish Reaction and Disappearance of Red Cells (This Phenomenon is Associated with the Further Development of Red Cells, White Blood Cells, occasionally Platelets and Megakaryocytes): The blackish reaction associated with the disappearance of red cells is a very important phenomenon in the development of colonies of more red cells, white blood cells and platelets in bone marrow and, to a lesser extent, circulating blood. The blackish reaction of red cells is involved in the formation of SN cells, lymphocytes, monocytes, eosinophils, plasma cells, commonly designated erythropoietic and granulopoietic series cells, as well as rarely platelets and megakaryocytes. Generally, the blackish reaction of red cells is the prominent way by which nucleated cells, particularly in bone marrow, are formed. Blackish reaction of red cells is frequently associated with colony formation of red cells and centrally placed nucleated cells.

Chapter 9. Reticulocyte Development from Reticulum (SGF) and Marrow or Blood Plasma:

According to conventional belief, reticulocytes are young red cells that show remnants of nuclear material carried over from the orthochromic normoblast stage. In my study, I could not see any structure which could be labeled as a transformational stage between orthochromic normoblast and reticulocyte. I found that reticulocytes develop from a material of unknown origin, which consists of clumps of thin, finely-beaded, matted fibrils that stain blue with conventional reticulocyte stain (new methylene blue or crystal violet) as seen in reticulocyte preparations of blood smears. This blue material has been previously observed and given the name substantia granulo-filamentosa (SGF) (Griffith, et. al., 1970). I found that the clear plasma surrounding the SGF particles becomes the hemoglobin in the formation of reticulocytes, and the SGF gradually vanishes inside developing red cells.

Chapter 10. Means of Disappearance and Regeneration of Red Cells: I do not agree with the common belief that aged and defunct red cells are removed by "erythrophagocytes" in the spleen and lymph nodes. Instead, I have found that "erythrophagocytes" represent cells with red cell forming capabilities. When hemosiderin pigment is present in "erythrophagocytes," this represents an unsuccessful attempt at red cell production. Aged and defunct red cells are primarily removed by the colorless hemolysis of red cells, a process in which red cells hemolyze without producing red-colored plasma. Red cells can regenerate in various ways in bone marrow; such as directly from marrow plasma, through various plasma coagulation products, or from blue fibers.

In addition, small and large vacuolar spaces, as well as necrosis and regeneration of bone marrow, caught my attention in almost every bone marrow sample examined. With subsequent study, I found this is the way that space is acquired for the regeneration of sheets of new marrow cells. Marrow tissue undergoes necrosis often followed by the development of round vacuolar spaces. Every marrow smear contains clear-cut small and large vacuoles which are great in number, especially in cases of hypoplastic marrow. Somewhat fine-pink woolly bodies of unknown origin are often found floating inside these vacuoles and can produce red and white blood cells, usually in colony formation.

Chapter 11. Various Ways of Platelet Development: My interest in platelet generation started when I could not find any evidence for the development of platelets from megakaryocytes, as is commonly believed. I found that platelets mainly develop directly from red cells and plasma of circulating blood and bone marrow, as well as through the blackish reaction of red cells. Occasionally, platelets arise from fibrin tags or from the marginal cytoplasm of large mononuclear cells which usually fall under the category of large lymphocytes. Rarely, platelets arise from blue fibers. A simple blood clotting experiment shows that red cells and plasma produce a huge number of platelets very quickly outside the body, where megakaryocytes do not exist.

Chapter 12. Nonmedullary and Medullary Development of Sickle Cells in Sickle Cell Anemia: In my study of peripheral blood, bone marrow, and various organs from patients who died of sickle cell crisis, I found that all different tissues and organs have the capacity for sickle cell development. The liver and spleen are the most active sites of sickle cell formation. In the tissues, sickle cells are formed directly without passing through any nucleated stages of erythropoiesis. Local tissues may directly form sickle cells in large and small columns towards the development of vascular channels full of sickle cells. Bone marrow plasma can also form sickle cells directly without passing through any nucleated

stages. In addition, pink woolly bodies of unknown origin floating in large and small vacuolar spaces in bone marrow may form sickle cells, ring-form red cells, and normal red cells.

Chapter 13. Additional Newly Discovered Phenomena Associated with Blood and Bone Marrow Diseases: Throughout my career I have examined many cases of various blood and bone marrow diseases. In Volume III, I discuss a few of these hematological cases, showing newly discovered phenomena. Included are pernicious anemia, macroglobulinemia, idiopathic thrombocytopenic purpura, chronic myelocytic leukemia, chronic lymphocytic leukemia, acute lymphocytic leukemia, acute monocytic leukemia with leukemoid reaction, multiple myeloma, and two cases which fall under the category of Banti's syndrome (one diagnosed as hairy cell leukemia and the other called "atypical hairy cell leukemia").

Chapter 14. Inflammatory or Reactive Cells of Medullary and Nonmedullary Origin: A variety of inflammatory or reactive cells (SN cells, plasma cells, lymphocytes and eosinophils) can all develop not only in the blood and bone marrow, but also in the various nonmedullary organs and tissues of the body. In addition, these reactive cells have the ability to form red cells. Finally, I discuss another new finding concerning cells which I have named Redefined Vasoformative Cells of Ranvier (RVR cells), or *fish-head* cells (because of their appearance), which also fall under the category of reactive cells. RVR cells are the most productive of all reactive cells in the development of red cells, endothelial cells, and other types of reactive cells, prominently SN cells, plasma cells and fibroblasts.

(Volume III, Copyright 2001; ISBN Number 0-9627824-2-4; 392 color photomicrographs, 222 pages)

GLOSSARY

Anitschkow myocytes - Cells with a peculiar nuclear structure showing a serrated bar of chromatin in the center surrounded by clear nucleoplasm. These cells were first described by Anitschkow (1913), for whom they are named. These cells are usually present in Aschoff nodules in cardiac muscle in acute rheumatic fever and commonly believed to be connective tissue cells. Their origin from cardiac muscle fibers in acute rheumatic fever is demonstrated by the author (McDonald, 1963, 1975, 1978; and in Volume II, 1995).

Aschoff bodies or Aschoff nodules - Nodules which consist of a variety of cells including Anitschkow myocytes, Aschoff cells and fibrinoid material, which make up a formation seen in cardiac muscle in cases of rheumatic fever (Aschoff, 1939).

Aschoff cells - Cells with specific characteristics taking part in the formation of Aschoff nodules which are the diagnostic feature of rheumatic fever and commonly believed to be of connective tissue origin. The origin of Aschoff cells from cardiac muscle fibers through various cytogenic processes is demonstrated by the author (McDonald, 1963, 1975, 1978 and 1995). *Cf. Aschoff nodules*

Blackish reaction of red cells - Blackish stained red cell margins in contact with plasma. Their disappearance gives rise to nucleated cells, platelets and megakaryocytes (both nucleus and cytoplasm arising simultaneously).

Blood cysts - This term is preferred to hemorrhagic cysts frequently seen in mammary carcinoma in mice (particularly in C3H mice), since blood contained therein is not derived from a hemorrhage but is locally formed (Volume I, McDonald, 1989, figs. 23-32). *Cf. Hemorrhagic cysts*

Blue fibers – Frequently seen in human blood and bone marrow smears, stained with Wright stain, there are large or small blue fibers or fragments which give rise directly to red and white blood cells. These newly observed phenomena are described in great detail in Chapter 7 of Volume III (McDonald, 2001).

Coagulation necrosis - A type of necrosis commonly produced by cutting off the blood supply (i.e., infarction) and is characterized by a protoplasmic coagulative process. In such necrosis, general architectural features may be preserved for a considerable period, although cellular detail is lost.

Colorless hemolysis - Destruction of red cells without a color change in the plasma, as opposed to regular hemolysis as in a case of mismatched blood transfusion, in which the plasma becomes red. Colorless hemolysis, which may be the normal method of removal of aged and defunct red cells, occurs excessively in hemolytic diseases, such as auto-immune hemolytic anemia.

Dedifferentiation - The process by which cells return to their primitive or undifferentiated stages, the reverse of differentiation. The dedifferentiated cells may redifferentiate into the same kind of cell or into another type of cell. *Cf. Differentiation*

Degeneration - Cellular alterations with or without erasure of local cells. The specific degeneration is named according to the morphological change or the nature of the abnormally accumulated material (For example: hyaline degeneration, hydropic degeneration, fatty change, amyloid deposition, amyloidosis, etc.).

Differentiation - The process of acquiring new, distinctive, individual cellular characteristics from an undifferentiated or primitive stage, as occurs in cells and tissues of the embryo, or from a transformed primitive stage reverted from once well-developed tissue. *Cf. Dedifferentiation*

Erythrogenesis - The development of red cells. The common belief is that red cells can only develop through nucleated stages of erythropoiesis and only in the bone marrow in adults and only from hypothetical (unrecognizable multipotential) cells termed stem cells (Platt, 1969). Recent research by the author has demonstrated various other ways of red cell formation from practically all tissues and organs, as well as red cell development from reactive or inflammatory cells in benign and malignant tissues (McDonald, 1989, 1995 and 2001).

Erythropoiesis - *see Erythrogenesis*

Erythrogenic gel - A liquefied red-shaded cell product as seen in H & E and Giemsa stains. Red cells usually appear in this fluid as if by crystallization or starting as fine red granules.

Erythrogenic inflammatory cells - Inflammatory cells with the capacity for production of hemoglobin globules (red cells). Erythrogenic capacity of segmented nuclear cells, eosinophils, monocytes/histiocytes, lymphocytes, plasma cells, RVR cells, etc., is demonstrated by the author (McDonald, 1989, 1995 and 2001).

Erythrogenic mesh - A fine mesh of delicate fibrin-like material derived from benign (including acellular bone marrow liquid) and malignant tissues in anticipation of red cell development within sieves, which are often red cell-sized.

Erythrophagocytosis - A term usually given to explain red cells being inside other cells. The present study indicates that red cells are usually forming there instead of being phagocytized (engulfed).

Extramedullary - Pertaining to tissues other than bone marrow. *See also Nonmedullary*

Extramedullary hematopoiesis - The formation of blood or blood cells outside the bone marrow. *See also Nonmedullary*

Fibrin film, membrane, particle, plaque, tag and body - *see Products of the Coagulation Processes of Marrow and Blood Plasma*

Fish-head cells - *see Redefined Vasoformative cells of Ranvier*

Floating erythrogenic cells - Cells with erythrogenic properties either desquamated from peripheral tissue or originating in fluid.

Ghost red cell - Red cell of regular size with a well-defined cell membrane and practically no hemoglobin. These cells are in an early stage of one pathway of normal red cell development (McDonald, 1989, 1995 and 2001).

Granulopoiesis - The development of granulocytes, particularly SN cells, basophils and eosinophils. The common belief is that granulocytes can only develop through nucleated stages of granulopoiesis and only in the bone marrow in adults. Recent research has demonstrated various other ways of granulocyte formation from practically all tissues and organs, from marrow or blood plasma, and from reactive or inflammatory cells (McDonald 1989, 1995 and 2001).

Grape cell - *see Russell Body Cell*

Hemoglobin gel - A thick liquefied cell product with concentrated hemolysate-like staining characteristics. It stains orange-red with H & E. This product produces a compact deformed, angular red cell mass in the beginning.

Hemorrhage - The escape of blood from blood vessels. This term is commonly used whenever red cells are seen outside the vascular lumen. Even when there is no evidence for vascular injury, which is commonly the case, a possible leakage in the vascular wall is an explanation given to this occurrence. Erythrogenesis from local tissues is often mistaken as hemorrhage.

Hemorrhagic cysts - *see Blood cysts*

Hemorrhagic necrosis - A common term for the presence of blood (red cells) in necrosed tissue. The development of red cells through the process of necrosis of local tissues is often mistaken as hemorrhage.

Hemosiderin - Iron-containing pigment granules, identified in tissue sections by Prussian blue reaction. With H & E stain, hemosiderin appears as golden-brown pigment granules. The current study shows its association with the abortive development of red cells.

Histogenesis - The development of tissues from undifferentiated cells of the germ layers of the embryo or from adult tissues.

Hyalinization - Erasure of background cellular structures, often with a ground glass effect.

Inflammatory cell - *see Reactive Cell*

In vitro - Grown outside the living body.

In vivo - Grown within the living body.

Keratosis - A general term used in the description of skin lesions with development of increased keratin or keratin-forming cells.

Mott cell - A special kind of plasma cell filled with clear, transparent globules, usually of red cell size.

Necrosis - The histological criteria of necrosis consist of nuclear pyknosis, fragmentation and liquefaction. The cytoplasm of cells is also involved in this process. Regeneration of necrosed tissue is demonstrated in this study and in Volumes I, II and III (McDonald, 1989, 1995 and 2001).

Nonmedullary - The term preferred by the present author, instead of *extramedullary*, for all organs and tissues other than bone marrow, because these tissues are equally important in the direct local development of red and white blood cells without passing through various nucleated stages as seen in bone marrow.

Nucleosis - Marked focal proliferation of nuclei, intracellular or from background substance without evidence of mitosis.

Nurse cell - A special type of reticulum cell which is found to give rise to many erythrogenic cells, mainly normoblasts, which then give rise to red cells (see also Volume III, figures 5.1-5.6, 13.43 and 13.44).

Parenchyma - The functional element of a tissue or organ, as distinguished from its supporting tissue or stroma.

Phagocytosis - *see Erythrophagocytosis*

Plasma - The fluid portion of the blood, bone marrow, or lymph in which the cells are suspended.

Plasma cell - Morphologically, plasma cells are easily distinguished from other cell types. Plasma cells are spherical or ellipsoidal and range from 5 to 30 microns in size. Their cytoplasm is abundant and always basophilic. Plasma cells have a well-defined peri-nuclear clear zone. The nucleus is round or oval, eccentrically placed, and small in relation to cell size. It also contains dense masses of chromatin, often arranged in a wheel-spoke fashion (Wintrope, 1974). Plasma cells can transform into reactive cells and red cells and endothelial cells in development of capillaries (Volume II, McDonald, 1995, figs. 67 and 117).

Plasma cell body - Tiny irregular (plasma cell-like) cytoplasm and a dense nuclear structure with or without characteristics of plasma cells. This name was given by Patterson (1986). These plasma cell bodies can give rise to mature plasma cells (Volume I, McDonald 1989, figs. 37 and 69).

Plasma gel - The clear, colorless liquid product of cell lysis providing tissue fluid, blood plasma and lymph plasma. The current research indicates that this fluid has the capacity for red cell production as well as for other kinds of cell generation (McDonald, 1989, 1995 and 2001). *Cf. Tissue plasma gel*

Platelets - Small blood cells that participate in the coagulation of blood. My studies (McDonald, 2001) indicate that platelets develop from red cells, from marrow or blood plasma, through the blackish reaction of red cells, and occasionally from the peripheral cytoplasm of large lymphocytes, from fibrin tags, and from blue fibers. Platelets do not develop from megakaryocytes.

Products of the coagulation processes of marrow and blood plasma - Includes fibrin film/membrane, fibrin plaque, fibrin tags, irregular fibrinoid particles (including basket cells), round fibrinoid bodies, and nuclear bodies. These products transform into red cells and nucleated cells. Megakaryocyte nuclei arise from clumping of nuclear bodies with cytoplasm arising from surrounding marrow plasma.

Pyknosis - A loss of nuclear and cytoplasmic detail and an intense basophilia.

Reactive cell or inflammatory cell - Cell that appears in tissues as a result of a variety of stimuli (known or unknown) or as a part of the natural growth process. Reactive cells include: Redefined Vasoformative Cells of Ranvier (RVR cells), plasma cells, segmented nuclear cells (SN cells or neutrophils), lymphocytes, eosinophils, multinucleated giant cells; and cells with less specific features, such as monocytes, histiocytes, etc.

Reactive phenomenon - The appearance of reactive cells, red cells and endothelial cells; the proliferation of blood capillaries and lymphatic channels; and the development of collagen fibers associated with spindle shaped nuclei. This phenomenon may occur in response to known or unknown stimuli or may even be a part of the natural growth process.

Red cell - The red cell is a hemoglobin globule. Synonyms for red cell are erythrocyte or red blood corpuscle. The origin of red cells from every tissue and organ is shown by the author in this volume and previous volumes (McDonald, 1989, 1995 and 2001).

Redefined Vasoformative cells of Ranvier (RVR Cells) - Also named *fish-head* cells by the present author due to its appearance (see figs. 21.21 and 21.22). These cells have the capacity to form red cells, white blood cells, blood capillaries, fibroblasts, etc., in extramedullary tissues.

Reticulocyte - In blood and bone marrow smears, round cells that contain finely-beaded, granular, matted fibrils (SGF) of unknown origin which stains blue with Reticulocyte Stain (New Methylene Blue or Crystal Violet). SGF surrounded by bone marrow or blood plasma gives rise to reticulocytes.

Russell body cells or grape cells - Large cells transformed from plasma cells. These cells contain hyaline globules which initially appear like a bunch of grapes; they then become red in color and are often about the size of red cells.

RVR cell - *see Redefined Vasoformative cells of Ranvier*

Segmented nuclear cell (SN Cell) - The author prefers this name instead of segmented neutrophils in the tissues since the origin and structure of SN cells differs from that of neutrophils present in the blood, granules do not form unless SN cells are within capillaries or blood vessels.

SGF - *see Substantia granulo-filamentosa*

Sickle cell - Sickle cells arise from red cells

in deoxygenated conditions in patients with sickle cell anemia. In addition, sickle cells develop directly from every organ and tissue of the body, particularly the liver and spleen, as demonstrated in Chapter 12, Volume III (McDonald, 2001). Occasionally they may develop from woolly bodies in bone marrow in cases of sickle cell anemia (Volume III, figs. 12.3-12.8).

SN cell - *see Segmented nuclear cell*

Stroma - The supporting tissue or matrix of an organ, as distinguished from its functional element or parenchyma.

Substantia granulo-filamentosa (SGF) - Name given to the clumps of finely-beaded, granular, parallel, thin, fibrillary, blue material associated with the development of reticulocytes (Griffith, 1970).

Tissue plasma gel - Plasma gel in tissues. *Cf. Plasma gel.*

Transdifferentiation - The direct conversion of one differentiated cell type into another distinct cell type.

Woolly body - Non-cellular, self-generated body with the capacity for development of sickle cells, ring-form red cells, and regular red cells in bone marrow.

REFERENCES

Agarwal, B.L. Rheumatic heart disease unabated in developing countries. *Lancet* 2 : 910-911, 1981.

Al-Adnani, M. S., et al. "Inappropriate production of collagen and prolyl hydroxylase by human breast cancer *in vivo.*" *Br J Cancer* 31 : 653-660, 1975.

Altschul, R. "Nucleosis of skeletal muscle: its value as a biological test." *Science* 103 : 566-567, 1948.

Anitschkow, N. Experimentelle Untersuchungen uber die Neubildung des Granulationsgewebes im Herzmuskel. *Bietr Path Anat Allg Pathol* 55 : 373-415, 1913.

Aschoff, L. The rheumatic nodules in the heart. *Ann Rheum Dis* 1 : 161-166, 1939.

Badarau, L. "L'angiogenese et l'hemopoiese dans les villosites terminales libres pendant l'evolution du placenta humain." *Rev franc Gynec* 64 : 667-688, 1969.

Bennet, M., and Cudkowicz, G. "Hemopoietic progenitor cells with limited potential for differentiation, erythropoietic function of mouse marrow lymphocytes." *J Cell Physiol* 72 : 129, 1968. Cited by Harris and Kellermeyer 1970.

Berman, J. J., and Foidart, J. M. "Synthesis of collagen by rat liver epithelial cultures." *Ann NY Academy Sci* 349 : 153-164, 1980.

Bjornson, Christopher R. R., et al. "Turning Brain into Blood: A Hematopoietic Fate Adopted by Adult Neural Stem Cells in Vivo." *Science* 283 : 534-537, 1999.

Bloom, W. and Fawcett, D. *A Textbook of Histology,* 10th Ed. Philadelphia: W. B. Saunders, Co., 1975, pp 487, 492.

Broyles, Robert H., and Deutsch, Michael J. "Differentiation of Red Blood Cells in vitro." *Science* 190 : 471-473, 1975.

Clark, E. R., and Clark, E. L. "Observations on living preformed blood vessels in the rabbit ear." *Am J Anat* 51 : 49, 1932.

Congeni, B., Rizzo, C., et al. Outbreak of acute rheumatic fever in northeast Ohio. *J Pediatr* 111 : 176-179, 1987.

Craver, Randall D., and Carson, Tom. "Hematopoietic Elements in Cerebrospinal Fluid in Children." *Am J Clin Path* 95 : 532-535, 1991.

Denys, J. "La cytodiérèse des cellules géantes et des petites cellules incolores de la moëlle des os." *La Cellule* 2 : 245, 1886. Cited by Michels 1931a.

Drummond, James. "On the development of the blood and blood vessels." *Monthly J Med Sci* 19 : 214 and 385, 1854. Cited by Michels 1931a.

Duran-Jorda, F. "Secretion of red blood corpuscles." *Nature* 159 : 293-294, 1947.

—. "The eosinophil cell: studies in horse and camel." *Lancet* 2 : 451-454, 1948.

Foulds, L. "A transplantable carcinoma of a domestic fowl, with discussion of the histogenesis of mixed tumors." *J Path Bact* 44 : 1-18, 1937.

Ghosh, Hemprova. 1957, 1958, 1959a, 1959b. see McDonald, Hemprova Ghosh.

Griffith, D. A., et. al. "Reticulocyte Generation Time and Rate of Discharge from the Marrow." *Acta Haemat* 43 : 13-20, 1970.

Harris, John and Kellermeyer, Robert. *The Red Cell - Production, Metabolism, Destruction: Normal and Abnormal, Revised Edition.* Cambridge: Harvard University, 1970. p 285.

Heitzmann, C. "Studien am Knorpel und Knochen, über die Rück und Neubildung von Blutgefäßen in Knochen und Knorpel. Wiener Mediz. Jahr Untersuchungen über das Protoplasma." *Wiener Akad Bericht* I-V : 67-68, 1872-73. Cited by Michels 1931a.

Hernandez, Jose A., and Steane, Susan M. "Erythrophagocytosis by Segmented Neutrophils in Paroxysmal Cold Hemoglobinuria." *Am J of Clin Path* 86 : 787-789, 1984.

Horner, W.E. *Special Anatomy and Histology,* 8th ed. Philadelphia: Blanchard and Lea, 1851, pp. 37, 170.

Jones, Wharton T. "The blood corpuscle considered in its different phases of development in the animal series." *Mem I-II Philosophic Transact of Royal Soc of London* II : 63, 89, and 103, 1846. Cited by Michels 1931a.

Jordan, H. E. "The Erythrogenic Capacity of Mammalian Lymph Nodes." *Am J Anat* 38 : 255-279, 1926.

—. "The origin and fate of plasmacytes: a comparative histologic study of plasmacytes of normal lymph nodes and tumors of multiple myeloma." *Anat Rec* 119 : 325-348, 1954.

Keasby, L. "On a new form of leucocyte (Schollenleukozyt, Weill) as found in the gastric mucosa of the sheep." *Folia Haemat* 29 : 155-171, 1923.

Larrimer, John H., Mendelson, David S., and Metz, Earl N. "Howell-Jolly Bodies: A Clue to Splenic Infarction." *Arch Intern Med* 185 : 857-858, 1975.

Latta, J. "The histogenesis of the dense lymphatic tissue of the intestine (Lepus). A contribution to the knowledge of the development of lymphatic tissues and blood cell formation." *Am J Anat* 29 : 159-212, 1921. Cited by Michels 1931a.

Le Gross Clark, W.E. *The Tissues of the Body,* 4th ed. London: Oxford Univ. Press, 1958, pp. 181, 186, 205, 223, 250.

Lever, W.F. *Histopathology of the Skin,* 3rd ed. Philadelphia: J.B. Lippincott Co., 1961, p. 8.

Lie, J.T. Myocardium as emboli in the systemic and pulmonary circulation. *Arch Pathol Lab Med* 111 : 261-264, 1987.

Listinsky, Catherine M. "Common Reactive Erythrophagocytosis in Axillary Lymph Nodes." *Am J Clin Path* : 189-192, 1989.

Loeb, L. "On transplantation of tumors." *J Med Res* 6 : 28-38, 1901.

Lopez, M., Chernavvsky, D., Nomasa, T., Wall, L., Yanagisawa, M., and Gomez, R. "The embryo makes red blood cell progenitors in every tissue simultaneously with blood vessel morphogenesis." *Am J Physiol Regul Integr Comp Physiol* 284 : 1126-1137, 2003.

Loutit, J. F. "Versatile haemopoietic stem cells." *Brit J Haemat* 15 : 333, 1968. Cited by Harris and Kellermeyer 1970.

Maniotis, A., Folberg, R., Hess, A., Seftor, E., Gardner, L., Pe'er, J., Trent, J., Meltzer, P., and Hendrix, M. "Vascular channel formation by human melanoma cells in vivo and in vitro: Vascular mimicry." *Am J Pathol* 155(3) : 739-752, 1999.

Maximow, A. A., and Bloom, W. *A Textbook of Histology,* Seventh Edition. Philadelphia: W.B. Saunders, Co., 1957, pp. 73, 249, 271.

McDonald, Hemprova Ghosh. "Observations on the histogenesis of rheumatic lesions of the heart." (Abstract). *Am J Path* 33 : 598-599, 1957.

—. "The mechanisms of formation of various histologic patterns as observed in mammary tumors in mice." (Abstract). *Am J Path* 34 : 599, 1958.

—. "Demonstration and significance of the independently growing mammary carcinomata within the cardiac lumina in mice, with epithelial and stromal differentiation." *Brit J Cancer* 13 : 115-120, 1959a.

—. "Active cellular lysis, a phenomenon of growth processes and its role in the formation of different epithelial patterns as shown in mammary carcinomas in mice." *Br J Cancer* 13 : 200-207, 1959b. (Abstract - *Year Book of Pathology,* 1960, pp. 53-55)

—. "Formation of new vascular channels from local tissues by cellular lysis as shown in mammary carcinomas in mice." *J Indian Med Assoc* 39 : 115-123, 1962.

—. "Origin of Anitschkow's myocytes from cardiac muscle fibers." *Texas State J Med* 59 : 1062-1067, 1963.

—. "Origin of vascular channels from epithelial tissue as shown in livers." *J Am Med Women's Assoc* 23 : 545-553, 1968.

—. "Structural changes in malignant epithelial cells suggesting stroma formation as shown in mammary carcinomas in mice." *J Am Med Women's Assoc* 25 : 493-501, 1970a.

—. "Mechanism of formation of various epithelial tumor patterns and possible origin of stroma including vascular channels from cancer tissue as shown in mammary carcinoma in mice." (Abstract). *Tenth International Cancer Congress*, p. 299, 1970b.

—. "Myocardial lysis in acute rheumatic fever followed by regeneration of cardiac muscle and origin of Aschoff bodies." *J Clin Path* 28 : 568-75, 1975.

—. "Mechanism of epithelial pattern formation and transformation of cancer tissue into vascular channels and connective tissue stroma." *Indian J Med Res* 78 (Suppl.) : 29-38, 1983.

—. *New Concepts in Blood Formation and Cell Generation in Malignant and Benign Tissues: Volume I.* Waco, Texas: Diagnostic and Cell Research Institute, 1989, pp. 7-111.

—. *New Concepts in Blood Formation and Cell Generation in Malignant and Benign Tissues: Volume II, Cardiac Muscle.* Waco, Texas: Diagnostic and Cell Research Institute, 1995, pp. 8-117.

—. *New Concepts in Blood Formation and Cell Generation in Malignant and Benign Tissues: Volume III, New Discoveries in Hematology.* Waco, Texas: Diagnostic and Cell Research Institute, 2001, pp. 15-213.

McDonald, Hemprova Ghosh and Moore, Melba M. "Primary gastric amebiasis superimposed on reticulum-cell sarcoma." *JAMA* 193 : 971 and 972, 1965.

McDonald, Hemprova Ghosh and Calkins, H.E. "Degeneration of cardiac muscle followed by cell transformation, regeneration and fibrogenesis in rheumatic fever." *Experimental and Toxicological Pathology* (previously *Experimental Pathology*) 15 : 185-195, 1978.

Michels, Nicholas A. "Erythropoiesis. A critical review of the literature." *Folia Haemat International Magazin Fuer Klinische und Morphologische Blutforschung* 45 : 75-128, 1931a.

Moffat, D. J., Rosse, C., and Yoffey, J. M. "Identity of the hemopoietic stem cell." *Lancet* 2 : 547, 1967. Cited by Harris and Kellermeyer 1970.

Murphy, M.J., Jr., Gordon, A. S., and Bertles, J. F.   "The stem cell: A functional and morphological identification by electron microscopy." *J Clin Invest* 48 : 59a, 1969. Cited by Harris and Kellermeyer 1970.

Neumann, Earnest. "Über die Bedeutung des Knochenmarks für die Blutbildung." *Zentralbl f d med Wiss* Nr. 44 : 1868. Cited by Michels 1931a.

Niimi, G., Usuda, N., Shinzato, M., and Nagamura, Y. "Erythrocyte-like globules observed in mouse and hamster visceral yolk sac endodermal cells." *J Analytical Bio-Science* 24(4) : 324-326, 2001.

—.   "Visceral yolk sac and placenta of hamster: Light and transmission electron microscopic study." *J Creative Approach for Health* 1(1) : 46-49, 2002a.

—.   "A light and electron microscopic study of the mouse visceral yolk sac endodermal cells in the middle and late embryonic periods, showing the possibility of definitive erythropoiesis." *Ann Anat* 184 : 425-429, 2002b.

—.   "Appearance of erythrocyte-like globules in the mouse visceral yolk sac endodermal cells on embryonic day 12, with special reference to blood islands." *Ann Anat* 185 : 201-205, 2003.

Patterson, J. W. "An extracellular body of plasma cell origin in inflammatory infiltrates within the dermis." *Am J Dermatopath* 8 : 117-123, 1986.

Platt, William R. *Color Atlas and Textbook of Hematology.* Philadelphia: J.B. Lippincott Co., 1969, plate 1.

Ranvier L.   "De développement et de l'accroissement de vaisseau sanguins." *Arch de Physiol* 6 : 429-445, 1874.

Reynolds, R. C., and Montgomery, P. O'B. "Nucleolar Pathology Produced by Acridine Orange and Proflavine." *The American Journal of Pathology* 51 : 323-339, 1967.

Rindfleisch, E. *Experimentalstudien zur Histologie des Blutes.* Leipzig, 1863. Cited by Michels 1931a.

Robin, (first initial not available). "Note sur les éléments anatomiques appelés myeloplaxes." *J de l'Anat et de Physiol.* 1 : 88, 1874. Cited by Michels 1931a.

Rollet, L. "Entwicklung und Neubildung der Blutkörperchen." In *Hermanns Handbuch der Physiologie, Bd.14,* 1862. Cited by Michels 1931a.

Sabin, F. R. "Studies on the origin of blood vessels and of red blood corpuscles as seen in the living blastoderm of chick during the second day of incubation." *Contrib Embryol* 9 : 213-262, 1920.

Sakakibara, K., et al. "Biosynthesis of an interstitial type of collagen by cloned human gastric carcinoma cells." *Cancer Res* 42 : 2019-2027, 1982.

Schaefer, E. "The intracellular development of blood corpuscles in mammals." *Mon Micr J* 11 : 261, 1874.

Schridde, H. 1905. Cited by Michels 1931b.

Schwann, T. *Microscopical Researches into the Accordance in the Structures and Growth of Animals and Plants.* Translated by Henry Smith and published by Sydenham Soc., London: 1847, pp. 39, 66, 158.

Shaver, J.R. "Studies on the initiation of cleavage in the frog egg." *J Exp Zool* 122 : 169-192, 1953.

Springer, M., Brazelton, T., and Blau, H. "Not the usual suspects: the unexpected sources of tissue regeneration." *J Clin Invest* 107(11) : 1355-1365, 2001.

Veasy, L.G., Weidmeier, S.E., et al. Resurgence of acute rheumatic fever in the intermountain area of the United States. *N Engl J Med* 316 : 421-427, 1987.

Weber, E. H., and Kölliker, A. "Über die Bedeutung der Leber für die Bildung der Blutkörperchen der Embryonen." *Zeitschr f rat Med* 4 : 160-167, 1845. Cited by Michels 1931a.

Wedl, C. *Rudiments of Pathological Histology.* Translated by George Busk and published by Sydenham Soc., London: 1853, pp. 59-61, 82, 83, 530, 610.

Weiss, Paul. "Perspectives in the field of morphogenesis." *Quart Rev Biol* 25 : 177-98, 1950.

Wintrobe, Maxwell H. *Clinical Hematology, Seventh Edition.* Philadelphia: Lea and Febiger, 1974, p. 289.

Yoffey, J. M., Smith, N. C. W., and Wilson, R. S. E. "Studies on hypoxia. V. Changes in the bone marrow during hypoxia at 10,000 and 20,000 feet." *Scan J Haemat* 4 : 145, 1967. Cited by Harris and Kellermeyer 1970.

List of Original Scientific Papers Contributed by

Hemprova Ghosh McDonald, MD, FCAP

(See also Hemprova Ghosh)

1.* A preliminary note on the complement-fixation reaction in Kala-azar with a specific antigen as an aid to diagnosis. Hemprova Ghosh, N.N. Ghosh, and J.C. Ray. *Ann Biochem & Exp Med* 5:153-158, 1945.

2.* Complement-fixation reaction in sera of rabbits actively immunized with living culture of Leshmania donovania. Hemprova Ghosh and N. N. Ghosh. *Ann Biochem & Exp Med* 7:1-2, 1947.

3.* Agglutination reaction in sera of rabbits immunized with different strains of Leishmania donovani. Hemprova Ghosh and N. N. Ghosh. *Ann Biochem & Exp Med* 7:3-6, 1947.

4.* A preliminary note on complement-fixation reaction in amoebiasis. N. N. Ghosh and Hemprova Ghosh. *Ann Biochem & Exp Med* 8:3-10, 1948.

5.* Further studies on complement-fixation reaction in Kala-azar with a specific antigen as an aid to diagnosis. Hemprova Ghosh, N. N. Ghosh, and J. C. Ray. *Ann Biochem & Exp Med* 9:173-178, 1949.

6. Tuberculous lymphadenitis. Hemprova Ghosh. *Am J Clin Path* 24:1044-1049, 1954.

7. Chronic pyelonephritis with xanthogranulomatous change. Hemprova Ghosh. *Am J Clin Path* 25:1043-1049, 1955.

8. Nuclear abnormalities in epidermal carcinogenesis in mice. Hemprova Ghosh. *J Am Med Women's Assoc* 11:417-425, 1956.

9. Observations on the histogenesis of rheumatic lesions of the heart. Hemprova Ghosh. (Presented at the Fifty-fourth Annual Meeting of the American Association of Pathologists and Bacteriologists held in Washington, D.C., April 1957.) Abstract — *Am J Path* 33:598-599, 1957.

10. The mechanisms of formation of various histologic patterns as observed in mammary tumors of mice. Hemprova Ghosh. (Presented at the Fifty-fifth Annual Meeting of the American Association of Pathologists and Bacteriologists held in Cleveland, Ohio, April 1958.) Abstract — *Am J Path* 34:599, 1958.

11. Demonstration and significance of the independently growing mammary carcinomas within the cardiac lumen in mice, with epithelial and stromal differentiation. Hemprova Ghosh. *Brit J Cancer* 13:115-120, 1959.

12. Active cellular lysis, a phenomenon of growth processes and its role in the formation of different epithelial patterns as shown in mammary carcinomas in mice. Hemprova Ghosh. *Brit J Cancer* 13:200-207, 1959. (It also appeared in the Year Book of Pathology, 1960, pp. 53-55.)

---

*Research work carried out at the Indian Institute for Medical Research, Calcutta.

13. Active cellular lysis of local tissues and transformation of peripheral remaining tissues into endothelium leading to vessel formation as shown in mammary carcinomas in mice. (Presented at the Annual Session of the Indian Science Congress Association held in Roorkee, India, 1961.) Hemprova Ghosh. Abstract — *Proc 48th Indian Sc Cong: Part IV*, pp. 16-17, 1961.

14. Myocardial lysis and regeneration of cardiac muscle fibers through the stages of redifferentiation of dedifferentiated cells arising from cardiac muscle fibers as shown in acute rheumatic heart disease. (Presented at the Seventy-fifth Annual Session of American Association of Anatomists held in Minneapolis, Minnesota, March 1962.) Hemprova Ghosh McDonald. Abstract — *Anat Record* 142:257, 1962.

15. Evidences in favor of fibrous metaplasia of cardiac muscle fibers in cases of gradual diminution of vascular supply. (Read by title at the Seventy-fifth Annual Session of American Association of Anatomists held in Minneapolis, Minnesota, March 1962.) Hemprova Ghosh McDonald. Abstract — *Anat Record* 142:318, 1962.

16. Cellular lysis of local tissue in the formation of vascular lumens and the peripheral remaining tissue transforming into endothelium shown in chick embryos and in humans. Hemprova Ghosh McDonald. (Presented at the 13th Annual Veterans Administration Medical Research Conference, Cincinnati, Ohio, Nov 1962.) Abstract — *13th Annual Veterans Administration Medical Research Conference* Nov. 1962, p.111.

17. Formation of new vascular channels from local tissues by cellular lysis as shown in mammary carcinomas in mice. Hemprova Ghosh McDonald. *J Indian Med Assoc* 39:115-123, 1962.

18. Origin of Aschoff nodules from damaged cardiac muscle fibers. Hemprova Ghosh McDonald. (Presented at the Texas Medical Association Annual Session, Dallas, April 1963.) Abstract — *Texas State J of Med* 59:273, 1963.

19. Origin of Anitschkow's Myocytes from cardiac muscle fibers. Hemprova Ghosh McDonald. *Texas State J of Med* 59:1062-1067, 1963.

20. Primary gastric amebiasis superimposed on reticulum-cell sarcoma. Hemprova Ghosh McDonald and Melba M. Moore. *J Am Med Assoc* 193:971-972, Sept. 13, 1965.

21. Origin of Aschoff nodules from damaged cardiac muscle fibers as shown in acute rheumatic fever. Hemprova Ghosh McDonald. (Presented at the Veterans Administration Regional Medical Conference in Houston, April 1965.) Abstract — *Program of Veterans Administration Regional Medical Conference*, April 1965.

22. Cellular lysis and differentiation in cancer. Hemprova G. McDonald. (Presented at the Veterans Administration Regional Medical Conference in Houston, April 1965.) Abstract — *Program of Veterans Administration Regional Medical Conference*, April 1965.

23. Origin of vascular channels from epithelial tissue as shown in livers. Hemprova G. McDonald. *J Am Med Women's Assoc* 23:545-553, 1968.

24. The mechanism of formation of various epithelial tumor patterns and possible origin of stroma including vascular channels from cancer tissue as shown in mammary carcinomas in mice. Hemprova Ghosh McDonald. (Presented at the Tenth International Cancer Congress held in Houston, April 1970.) Abstract — *Tenth International Cancer Congress, Abstracts*, p.299.

25. Structural changes in malignant epithelial cells suggesting stroma formation as shown in mammary carcinomas in mice. Hemprova G. McDonald. *Woman Physician* 25(4):493-501, 1970.

26. Myocardial lysis followed by regeneration of cardiac muscle and origin of Aschoff bodies as observed in acute rheumatic fever. Hemprova G. McDonald. (Presented at the Texas Medical Association Annual Session held in San Antonio, Texas, May 1975.) Abstract — *Program of Texas Medical Association Annual Session,* p. 128.

27. Myocardial lysis in acute rheumatic fever followed by regeneration of cardiac muscle and origin of Aschoff bodies. Hemprova G. McDonald. *J Clin Path* 28:568-575, 1975.

28. Tissue lysis and its significance in composition of plasma and formation of new vascular channels. Hemprova G. McDonald. (Presented at the American Society of Hematology Eighteenth Annual Meeting, Dallas, Dec 1975.) Abstract — *American Society of Hematology, Eighteenth Annual Meeting,* p.127.

29. Degeneration of cardiac muscle followed by cell transformation, regeneration, and fibrogenesis in rheumatic fever. H. G. McDonald and H. E. Calkins. *Exp Path* 15:185-195, 1978.

30. The mechanism of formation of various epithelial patterns and evidence in favor of transformation of malignant tissue into benign histological structures as shown in mammary carcinomas in mice. Hemprova G. McDonald. (Presented at the Fifth Asia Pacific Cancer Conference held in Colombo, Sri Lanka, Sept 1981.) Abstract — *Fifth Asia Pacific Cancer Conference Abstracts and Proferred Papers,* p 188..

31. Mechanism of epithelial pattern formation & transformation of cancer tissue into vascular channels & connective tissue stroma. Hemprova Ghosh McDonald. *Indian J Med Res (Suppl)* 78:29-38, July 1983.

32. The possible transformation of malignant tissue into benign forms as shown in mammary carcinomas in mice. Hemprova Ghosh McDonald. (Presented at the 7th Annual Breast Cancer Symposium held in San Antonio, Texas, Dec 1984.) Abstract — *Breast Cancer Research and Treatment* 4(4), p. 342, 1984.

33. New concepts in origin of blood and blood vessels from local tissues. Hemprova Ghosh McDonald. (Presented at the Tenth Annual Meeting of the University of Calcutta Medical Association of America held in Boston, August 1986.) Abstract. — *Commemorative Volume Tenth Annual Meeting of the University of Calcutta Medical Association of America* .

34. **New Concepts in Blood Formation and Cell Generation in Malignant and Benign Tissues, Volume I.** Hemprova Ghosh McDonald, Diagnostic and Cell Research Institute, Waco, Texas, 1989, 111 pages, 129 color and 13 black and white photomicrographs.

35. **New Concepts in Blood Formation and Cell Generation in Malignant and Benign Tissues, Volume II, Cardiac Muscle.** Hemprova Ghosh McDonald, Diagnostic and Cell Research Institute, Waco, Texas, 1995, 117 pages, 105 color and 5 black and white photomicrographs.

36. **New Concepts in Blood Formation and Cell Generation in Malignant and Benign Tissues, Volume III, New Discoveries in Hematology.** Hemprova Ghosh McDonald, Diagnostic and Cell Research Institute, Waco, Texas, 2001, 222 pages, 392 color photomicrographs.

37. **New Concepts in Blood Formation and Cell Generation in Malignant and Benign Tissues, Volume IV, Blood and Blood Vessels Developing Locally from Every Tissue.** Hemprova Ghosh McDonald, Diagnostic and Cell Research Institute, Waco, Texas, 2004, 268 pages, 476 color and 11 black and white photomicrographs.

INDEX

A

Abscess, vascularization of pericolic abscess **21.33, 21.34;** 155-156

Adenocarcinoma, *see Carcinomas*

Adipose tissue **3.1-3.16, 24.16, 24.17, 30.12;** 43-47, 174, 194, 232, 233, 235; *see also Liposarcoma*
Breast **24.16, 24.17**
Dissolution of skeletal muscle giving false appearance of adipose tissue **4.33**
Fat vacuoles developing from cardiac muscle *in vitro* **4.4**
Hemoglobinization of fat cells **3.5, 3.6, 3.12-3.14**
Hyalinization **3.4**
Lipoma **3.3, 3.4**
Origin of
    Blood and blood capillaries **3.1-3.6, 3.10, 3.11, 30.12**
    Endothelium **3.3, 3.5-3.7**
    Erythrogenic SN cells and blood capillaries **3.7-3.9**
    Lymphocytes **3.11**
    Plasma cells **3.11, 3.14-3.16**
    RVR cells **3.10, 3.11**
    Vascular channels 232
Pericolic **3.1, 3.2, 3.7-3.16, 30.12**
Progressive malignant changes **24.16, 24.17**
Throat **3.5, 3.6**

Adipose tissue cancer, *see Liposarcoma*

Adrenal cortical tumor, developing red cells **32.1-32.2;** 212

Adrenal gland tissue **5.30-5.37;** 64, 73-76, 91
Developing
    Erythrogenic SN cells **5.32**
    Lymphocytes **5.31, 5.33, 5.35;** 91
    Red cells and blood vessels **5.30-5.37**
    Sickle cells **5.37**
Transforming into plasma cells **5.36**

Anitschkow myocytes developing in acute rheumatic fever from cardiac muscle **4.5, 4.8, 4.9;** 50-52, 244

Aschoff cells developing in acute rheumatic fever from damaged cardiac muscle **4.5-4.9;** 17, 48, 50-52, 237-238, 244
Type A **4.5, 4.8**
Type B **4.6, 4.9**
Type C **4.7**

Astrocytoma, developing blood and blood vessels **36.1-36.2;** 225

B

Basal cell carcinoma **22.1-22.6;** 91, 158-159, 232, 233
Developing red cells **22.3, 22.5, 22.6**
Plasmacytoid cells transforming into erythrogenic SN cells and red cells **22.4**
Transformation into plasmacytoid cells **22.2-22.4**

Benign tissues, *see Adipose tissue, Adrenal gland tissue, Blood and bone marrow, Bone, Brain tissue, Cardiac muscle, Cartilage, Colon, Embryonic tissue, Eye, Gall bladder wall, Inflammatory tissue, Intraductal papilloma of breast, Kidney tissue, Liver tissue, Lung, Lymphatic channels, Lymph node, Mammary gland, Pancreas, Parathyroid gland tissue, Pituitary, Prostate, Rectal tissue, Salivary gland, Skeletal muscle, Skin, Smooth muscle, Spleen, Stomach, Thymus, and Thyroid*

Blackish reaction of red cells **7.12-7.14;** 84, 88, 239-240, 242, 243, 244
Developing
    Granulocytic nucleated cells **7.12-7.14**
    Megakaryocytes **7.14**
In combination with blue fibers **7.12**

Blood and bone marrow **7.1-7.20;** 84-90, 239-243; *see also Blue fibers, Blackish reaction, Bone, Bone marrow cells, Bone marrow plasma, Peripheral blood, Plasma coagulation products, and Reticulocytes*
Bone marrow clot sections **7.1, 7.2**
Bone marrow smears **7.4-7.8, 7.11, 7.13, 7.14, 7.19, 7.20**
Peripheral blood, plasma clot section **7.3**
Peripheral blood smears **7.9, 7.10, 7.12, 7.15-7.18**

Blue fibers **7.9-7.12;** 84, 87-88, 239, 242, 243, 244
Developing
    Red cells **7.9-7.11**
    White blood cells in association with blackish reaction **7.11-7.12**

Bone **6.3, 6.5;** 81-83, 241
Developing from cartilage **6.3**
Developing red cells and capillaries **6.5**
Origin of bone marrow 241

Bone cancers **35.1-35.6;** 222-224
Bone with metastatic bronchogenic carcinoma, developing red cells **35.4-35.6**
Giant cell tumor, developing red cells **35.1-35.3**

Bone marrow cells developing from
Blackish reaction of red cells **7.12-7.14;** 239-240, 242-244

Figure numbers are denoted by **bold-faced** type.

Cartilage and bone **6.1-6.4;** 81, 241
Coagulated marrow plasma **7.4-7.6;** 239, 241

Bone marrow plasma
    Developing from cartilage and bone **6.1-6.3;** 241
    Developing platelets **7.19;** 243
    Developing red cells and nucleated cells in
        association with
        Blackish reaction of red cells **7.13, 7.14;** 239, 240,
          242
        Blue fibers **7.11;** 84, 239, 242-244
        Coagulation products **7.4-7.6;** 84, 239, 241
    Direct origin of red cells through
        Crystallization of erythrogenic plasma gel **7.2**
        Erythrogenic mesh **7.1;** 84

Brain cancer, *see Astrocytoma and Glioblastoma
multiforme*

Brain tissue, cerebrum **19.1-19.8;** 139-141
    Developing red cells through erythrogenic SN cells
      19.3-19.5
    Directly developing red cells **19.1, 19.2, 19.6-19.8**

Breast cancer, *see Mammary carcinoma*

Breast tissue, *see Adipose tissue, Intraductal papilloma
of breast, Mammary carcinoma, and Mammary gland*

Bronchogenic carcinoma metastatic to bone, *see Bone
cancers*

Bronchogenic squamous cell carcinoma, *see Lung
cancers*

C

Capillary hemangioma, developing blood and blood
    vessels **2.16, 2.17;** 39

Carcinoid **31.1-31.42;** 198-211, 230
    Colon **31.17-31.20;** 203-204
        Lacy pattern **31.17**
        Plasmacytoid cells developing stroma **31.19**
        Tumor tissue developing
            Erythrogenic plasma cells **31.19**
            Plasma **31.18, 31.20**
            Plasmacytoid cells **31.19**
            Red cells **31.18**
            Stromal tissue **31.20**
    Ileum **31.1-31.16;** 198-203
        Extension of tumor into smooth muscle coat
          31.12, 31.13
        Hemorrhagic necrosis of tumor tissue producing
          red cells and plasma **31.2, 31.3**
        Tumor tissue developing
            Blood vessels **31.4, 31.5**
            Calcifying streaks **31.14-31.16**
            Connective tissue stroma **31.6, 31.7, 31.9,
              31.10**

Crimson red hemoglobin-like substance and red
    cells **31.2-31.5, 31.11**
Endothelium **31.4, 31.5**
Typical carcinoid tumor **31.1, 31.8, 31.9**
    Metastatic to liver **31.24-31.33;** 205-209
        Extension of tumor into normal liver **31.29, 31.30**
        Tumor tissue developing
            Blood vessels **31.24-31.28**
            Calcifying streaks **31.31-31.33**
            Crimson red hemoglobin-like substance and red
              cells **31.24-31.28**
            Endothelium **31.25**
            Stromal tissue **31.28**
    Metastatic to lymph nodes **31.21-31.23;** 204-205
        Tumor tissue developing
            Red cells **31.22**
            Stromal tissue **31.21, 31.23**
    Reactions in intestinal wall to nearby carcinoid
      31.34-31.42; 209-211
        Blood and blood vessel development **31.34-31.39**
        Development of calcifying streaks **31.40-31.42**

Carcinomas, *see Adrenal cortical tumor, Astrocytoma,
Basal cell carcinoma, Bone cancers, Carcinoid, Colon
carcinoma, Glioblastoma multiforme, Liposarcoma, Liver
cancer, Lung cancers, Malignant melanoma, Mammary
carcinoma, Pancreatic cancers, Prostate cancer, Renal
cancers, Salivary gland, Squamous cell carcinoma,
Stomach cancers, and Thyroid cancers*

Cardiac muscle **1.6, 1.9, 4.1-4.21;** 17, 30, 32, 33, 48-
    56, 229, 237-238
    Developing red cells and blood capillaries **4.1-4.3,
      4.11-4.15;** 30, 229, 231, 232, 234, 237-238
    Embryonic, chick, *in vitro* developing
        Fat **4.4;** 48
        Red cells **1.9, 4.3;** 29-30, 33, 48
    Embryonic, human, developing red cells and blood
      capillaries **1.6;** 30, 31-32
    In acute myocardial infarction
        Coagulation necrosis **4.10-412;** 238, 244
        Developing
            Blood and blood capillaries **4.11, 4.13**
            Lymphocyte-like cells **4.11**
            Plasma cells **4.13**
            SN cells **4.13**
         Infarcted myofibers changing into bluish-gray,
          ground glass-like substance **4.11**
        Massive red cell formation **4.12**
    In acute rheumatic fever, giving rise to Anitschkow
      myocytes and Type A, B & C Aschoff cells **4.5-
      4.9;** 17, 48, 237-238, 244
    In chronic ischemia, developing blood and blood
      vessels **4.14, 4.15;** 237-238
    Mammary carcinoma emboli within cardiac cavities
      and coronary vessels **4.16-4.21**

Cartilage **6.1-6.4;** 81-83, 241
    Developing
        Bone **6.3**

Figure numbers are denoted by **bold-faced** type.

Marrow cells **6.1-6.4**
Original marrow fluid **6.1**
Red cells **6.4**

Cellular lysis 18, 228-230, 231-235
In development of
Hemoglobinized colloid and red cells from
Thyroid tissue **5.9-5.11, 25.3**
Thyroid tumor tissue **25.1, 25.2, 25.4**
Inflammatory cells **24.15, 31.19**
Lacy pattern of carcinoid of colon **31.17**
Lymphocytes and lymphatic channels **8.7-8.9**
Red cells, original plasma, and blood vessels **1.4-1.7, 2.6, 2.8, 2.9, 4.2, 4.24, 6.5, 12.5, 12.10, 12.26-12.28, 13.6, 22.18, 24.8-24.10, 29.1, 29.5, 33.1, 33.6**
Of glandular epithelium releasing secretory products into the blood stream **5.2, 5.7, 5.19, 5.26, 5.33, 5.35, 5.41, 5.46, 5.47**
Of necrotic tumor tissue in development of
Blood and blood cysts **24.1-24.7**
Inflammatory cells **27.1, 30.13-30.15**
Red cells, original plasma, and blood vessels **27.1-27.4, 28.5, 28.6, 30.7, 31.2, 31.3**
Remnants of local tissue resisting lysis and lying within the lumens of developing vessels **4.24, 12.10, 12.25, 13.6, 22.16, 24.8, 27.2, 29.5, 30.21, 34.4**

Cerebellum, *see Brain tissue*

Cerebrum, *see Brain tissue*

Chick embryonic tissue, *see Embryonic tissue and Tissue culture*

Choroid, *see Eye*

Coagulation necrosis, *see Necrosis*

Coagulation products, *see Plasma coagulation products*

Collagen (collagenous fibers), *see also Connective tissue and Stromal tissue*
Developing blood and blood vessels **2.16, 2.17, 2.24, 2.25, 3.1, 3.2, 5.17-519, 21.11**
Developing from
Cardiac muscle **4.6, 4.9**
Colloid of thyroid **5.17, 5.18, 5.21**
Dermal tissue **8.9**
Epidermal cells **2.12**
Glandular epithelium **5.28, 5.29**

Colon
Hemoglobinization of pericolic fat **3.12-3.14**
Pericolic adipose tissue developing
Erythrogenic SN cells **3.7-3.9**
Plasma cells **3.14-3.16**
Red cells and blood capillaries **3.1, 3.2, 3.7-3.11**
RVR cells **3.10, 3.11**

Pericolic abscess, developing
Blood and blood vessels **21.33, 21.34**
Plasma cells **21.34**
Smooth muscle developing red cells and capillaries through locally formed plasma cells **4.37**

Colon carcinoma, **30.13-30.21**; 194-197, 230; *see also Carcinoid*
Developing
Connective tissue stroma **30.19, 30.20**
Lymphocytes **30.20**
Mucin **30.18**
Red cells **30.21**
Necrotic tumor tissue developing
Blood vessels **30.13-30.17**
Endothelium **30.15**
Erythrogenic SN cells **30.14-30.17**
Hemoglobin particles **30.14, 30.15**
Red cells developing from erythrogenic SN cells **30.14-30.17**

Colorless hemolysis **7.16**; 232, 240, 242, 244

Connective tissue, *see also Collagen and Stromal tissue*
Developing from
Cardiac muscle **4.6, 4.9**
Epidermal cells **2.7, 2.12**
Developing red cells and blood vessels **5.28, 5.29, 13.7, 13.8**
In spleen, developing reactive cells **10.8**

D

Dermis, *see Skin*

Desquamated epithelium 144-147
Developing
Eosinophils **13.9, 21.8, 21.9**
Red cells
Directly **13.9**
From locally formed erythrogenic SN cells **21.3-21.6**
Through granular changes **21.2**
Through vacuolated mesh **21.1, 21.6**
Gall bladder wall **13.9**
Gastric mucosa **21.9**
Nasal epithelium **21.8**
Vaginal **21.1-21.7**

E

Emboli of mammary carcinoma in cardiac cavities and coronary vessels **4.16-4.21**; 48, 54-56, 234

Embryonic tissue **1.1-1.9, 4.3, 4.4, 12.11-12.13, 4.25**; 29-33, 237, 241
Chick
Cardiac muscle *in vitro* **1.9, 4.3-4.4**
Liver *in vitro* **1.8, 12.12**

Figure numbers are denoted by **bold-faced** type.

Liver parenchyma **12.11**
Yolk sac membrane **1.1-1.4**
Chick, *in vitro,* developing
  Red cells **1.8, 1.9, 4.3, 12.12**
  Fat **4.4**
Developing erythrogenic nucleated cells **12.13**
Directly developing red cells **1.1-1.7, 4.25, 12.11, 12.13**
Human
  Cardiac muscle **1.6**
  Liver **1.7, 12.13**
  Mesenchymal tissue **1.5**
  Skeletal muscle **4.25**

Emphysematous lung **11.1**

Endocrine glands **5.1-5.50**; 64-80; *see Adrenal, Pancreas, Parathyroid, Pituitary, and Thyroid*

Endothelium 18, 20, 21, 48, 64, 94, 104, 119, 128, 150-151, 228-229, 231-233, 235, 237
Developing locally from
  Adipose tissue **3.3, 3.5, 3.6**
  Adrenal **5.30, 5.31, 5.33, 5.35, 5.36**
  Bronchogenic squamous cell carcinoma of lung **27.1, 27.2**
  Carcinoid tissue **31.4, 31.5, 31.25, 31.36, 31.37, 31.39**
  Cardiac muscle **4.1, 4.2, 4.11, 4.13, 4.15**
  Chick embryonic liver tissue **1.4**
  Colon carcinoma **30.15**
  Connective tissue **13.7, 13.8, 21.11**
  Dermal collagen **2.16, 2.17**
  Dermal tissue **2.8, 8.7, 8.8**
  Epidermis **2.5, 2.6, 2.8**
  Eye, choroid and retina **20.2-20.4**
  Gall bladder wall **13.1-13.8**
  Glioblastoma multiforme **36.5**
  Human embryonic mesenchymal tissue **1.5**
  Kidney **16.1-16.4, 16.6**
  Liver **12.4-12.6, 12.8, 12.9, 12.26, 12.28**
  Liver carcinoma **29.1, 29.2**
  Liposarcoma **23.2, 23.3**
  Lymphocytes **8.5 13.4**
  Malignant melanoma **22.16**
  Mammary carcinoma in mice **24.7-24.10, 24.15**
  Nevus cells **2.21-2.23**
  Pancreas, Islets of Langerhans **5.46-5.50**
  Parathyroid **5.28, 5.29**
  Pituitary **5.41-5.43**
  Plasma cells **4.36, 14.1, 14.4, 24.15**
  Prostate **4.34, 17.3**
  Prostate carcinoma **34.4, 34.5**
  Rectal tissue **15.1, 15.4**
  RVR cells **3.10, 3.11, 21.22, 21.24**
  Salivary gland tissue **26.4, 26.5**
  Skeletal muscle **4.25**
  Smooth muscle **4.34, 4.36, 4.38, 14.13**
  SN cells **3.7**
  Squamous cell carcinoma **22.10**

  Stomach **4.36, 4.38, 14.1, 14.4, 14.5, 14.13**
  Thyroid **5.4, 5.7-5.11, 5.16, 5.19**

Eosinophils 19, 20, 48, 98, 119, 144, 151, 239, 242
Developing red cells **4.27, 4.39, 4.40, 10.1, 10.2, 10.5, 10.8, 13.2, 13.3, 13.8, 13.9, 20.4**; 19, 20
Origin from
  Blackish reaction of red cells in bone marrow **7.13**
  Desquamated epithelial cells **13.9, 21.8, 21.9**
  Gall bladder wall **13.1, 13.2, 13.3, 13.8, 13.9**
  Gastric mucosa **14.1, 14.14, 21.9**
  Labium majora **21.27, 21.28**
  Nasal epithelium **21.8**
  Plasma cells **4.28, 14.1**
  RVR cells **21.22, 21.29, 21.31, 21.32**
  Retina of eye **20.4**
  Skeletal muscle **4.27, 4.29**
  Smooth muscle **4.39, 4.40**
  Spleen **10.1, 10.5, 10.8**

Epidermis, *see Skin*

Erythrogenic gel 232, 241, 245
Developing red cells in
  Bone marrow **7.2, 34.7**
  Dermis **2.8**
  Mammary gland **18.3**

Erythrogenic mesh 245
Developing red cells **7.1, 12.14, 12.15, 16.2, 21.1, 21.6, 33.2, 33.3**

Erythrogenic net developing red cells **2.18**

Esophagus, *see Squamous cell carcinoma*

Eye, choroid and retina **20.1-20.4**; 142-143
Developing blood and blood vessels **20.1-20.4**
Developing erythrogenic SN cells **20.4**

F

Fat vacuoles, *see Adipose tissue*

Fibrin films/membranes, *see Plasma coagulation products*

Fibrinoid particles, *see Plasma coagulation products*

Fibroblast origin from RVR cells **21.23**

Fibrocyte
Developing red cells **2.18-2.20**
Origin from
  Epidermal cells **2.12**
  Skeletal muscle **4.32**

Fish head cells, *see Redefined Vasoformative Cells of Ranvier*

Figure numbers are denoted by **bold-faced** type.

Follicular adenocarcinoma of thyroid, *see Thyroid cancers*

G

Gall bladder wall **13.1-13.9;** 116-118
  Origin of erythrogenic reactive, inflammatory cells
    Eosinophils **13.1, 13.2, 13.3, 13.9**
    Lymphocytes **13.4**
    SN cells **13.2, 13.3**
  Origin of red cells and blood capillaries from
    Connective tissue **13.7, 13.8**
    Desquamating epithelial cells **13.9**
    Smooth muscle **13.2, 13.5-13.8**

Gastric mucosa, submucosa, serosa, and smooth muscle, *see Stomach and Stomach cancers*

Ghost red cell formation **4.1, 4.2, 12.16, 16.2, 16.4, 16.5, 34.4, 34.5;** 232, 245

Giant cell tumor, *see Bone cancers*

GI tract, *see Colon, Ileum, Rectal tissue, and Stomach*

GI tract tumors, *see Carcinoid, Colon carcinoma, and Stomach cancers*

Glioblastoma multiforme, developing blood and blood vessels **36.3-36.6;** 226-227

Granulation tissue **21.10-21.17;** 148-150
  Development of red cells and blood capillaries from locally formed reactive, inflammatory cells
    Grape cells **21.16, 21.17**
    Lymphocytes **21.17**
    SN cells **21.11-21.15**

Grape cells, *see Russell body cells*

H

Hashimoto's disease, origin of lymphocytes and lymphatic tissue from thyroid tissue **5.22-5.24**

Heart, *see Cardiac muscle*

Hemoglobin gel developing red cells **33.1, 34.7;** 232, 241, 245

Hemoglobin globules developing in nonmedullary tissues in formation of red blood cells **1.4, 1.6-1.8, 2.1, 2.5, 4.22-4.25, 5.1-5.9, 5.35, 6.4, 7.1, 7.10, 10.2, 12.1, 14.7, 14.9, 14.12, 17.1, 18.2, 21.2, 21.5, 21.6, 21.13, 21.15, 21.16, 21.25-21.27, 22.16, 24.2, 30.4;** 18, 19-20, 29-30, 31, 56, 81, 104, 119, 148, 232, 234, 237, 239, 241, 247

Hemorrhagic cysts in mammary carcinoma in mice **24.1-24.7;** 18, 168-170, 230, 232, 233, 246

Hemorrhagic dermatofibroma, developing red cells and hemosiderin pigment from fibrocytes **2.18-2.20;** 39-40

Hemorrhagic necrosis developing blood and blood capillaries **12.18, 24.1-24.7, 27.3, 27.4, 31.2, 31.3;** 230, 246

Hemorrhoids, *see Rectal tissue*

Hemosiderin pigment 39, 232-233, 234, 242, 246
  Development in association with abortive attempts at red cell formation from
    Erythrogenic mononuclear cells of lymph node **8.2**
    Fibrocytes of dermis **2.18-2.20**
    Histiocytes in lung **11.7**
    Histiocytes of thyroid origin **5.14-5.16**

Histiocytes 20, 56, 101
  Developing red cells **11.7, 11.8**
  Origin from
    Colloid of thyroid **5.14-5.16**
    Lung **11.7, 11.8**
    Skeletal muscle **4.30, 4.31**

Human embryonic tissue, *see Embryonic tissue*

Hyalinization 64, 91, 228, 231, 232, 244, 246
  In development of
    Colloid **5.10, 5.11, 5.15**
    Inflammatory cells **2.14, 4.13, 5.15, 5.23**
    Red cells and blood capillaries **2.11, 3.4, 5.11, 5.28, 5.29, 5.42, 16.4, 16.5, 17.1, 20.2, 31.4, 31.5, 34.4**
  Of adipose tissue in the local spread of mammary carcinoma **24.16, 24.17**
  Of tumor tissue in development of stroma **24.11, 24.12, 30.19, 30.20, 31.8-31.10, 31.23, 31.28**

Hypernephroma, *see Renal cancers*

Hypophysis Cerebri, *see Pituitary*

I

Ileum, *see Carcinoid*

Infarcted cardiac muscle, *see Cardiac muscle*

Inflammatory exudates **21.1-21.9;** 144-147
  Origin of erythrogenic SN cells from vaginal epithelium **21.3-21.7**
  Origin of eosinophils from
    Gastric mucosa **21.9**
    Nasal epithelium **21.8**
  Red cells developing from
    Desquamated vaginal epithelium **21.1-21.6**
    Erythrogenic SN cells **21.3-21.7**

Figure numbers are denoted by **bold-faced** type.

Inflammatory tissue **21.1-21.34;** 18, 20, 34, 43, 48, 56, 61, 116, 119, 144-156, 230, 232, 233, 234, 235, 238, 243, 245, 246; *see also Abscess, Eosinophils, Granulation tissue, Inflammatory exudates, Lymphocytes, Plasma cells, RVR cells, and Segmented nuclear cells*
   Erythrogenic **4.36-4.40, 13.2, 13.3-13.5, 13.8, 14.2-14.4, 21.4-21.6, 21.12-21.17, 21.20-21.22, 21.24-21.27, 21.33, 21.34, 30.1-30.4, 30.15**
   Origin from
      Bronchogenic squamous cell carcinoma of lung **27.1, 27.2**
      Cardiac Muscle **4.13**
      Colon carcinoma **30.15**
      Gall bladder wall **13.1-13.9**
      Gastric mucosa **14.1**
      Gastric serosa **21.20-21.22**
      Lung **11.7**
      Malignant Melanoma **22.13**
      Nasal epithelium **21.9**
      Pancreas **28.2**
      Skeletal muscle **4.26-4.32**
      Skin **2.11-2.15, 21.10-21.17**
      Smooth muscle **4.36-4.40**
      Squamous cell carcinoma of skin **22.7, 22.8**
      Stomach adenocarcinoma **30.1-30.4**
      Vaginal exudate **21.3-21.7**

Intercellular bridges **2.6, 2.14, 22.15**

Intraductal papilloma of breast **18.4-18.9;** 136-138
   Typical papillary structures **18.4, 18.5**
   Developing red cells and blood vessels **18.6-18.9**

In vitro, chick embryonic tissue, *see Embryonic tissue*

Ischemic cardiac muscle, *see Cardiac Muscle*

Islets of Langerhans, *see Pancreas*

K

Kidney cancer, *see Renal cancers*

Kidney (renal) tissue **16.1-16.13;** 128-132, 231, 232
   Developing
      Blood and blood vessels **16.1-16.13**
      Lymphocytes **16.1, 16.3-16.5**
      Plasma cells **16.5**
   Directly developing sickle cells **16.12, 16.13**
   Transformation of glomeruli into blood and blood vessels **16.7-16.10**

L

Labia majora granulation tissue **21.23-21.32;** 152-155
   Origin of RVR cells **21.23**

RVR cells developing
   Eosinophils **21.31, 21.32**
   SN cells **21.28-21.30**
   Red cells and blood vessels **21.24-21.27**

Lipoma, *see Adipose tissue*

Liposarcoma, developing blood and blood vessels **23.1-23.3;** 166-167

Liver cancer, adenocarcinoma **29.1-29.5;** 188-189
   Glandular and stromal structures of tumor **29.1, 29.2**
   Cancer cells developing
      Blood vessels **29.2**
      Clear plasma **29.2**
      Red cells **29.1, 29.2**
      SN cells **29.2, 29.4**
      Stroma **29.3**
   Cancer cells invading normal liver tissue **29.4**
   Normal liver tissue near approaching cancer developing blood and a blood vessel **29.5**

Liver tissue **1.4, 1.7, 1.8, 12.1-12.28, 28.8, 29.4, 29.5, 31.24, 31.25, 31.29, 31.30;** 19, 21, 30, 31, 33, 91, 104-115, 228, 229, 230, 231, 232, 234, 240, 241, 243
   Chick embryonic liver developing red cells **1.4, 12.11**
   Chick embryonic liver *in vitro* developing red cells **1.8, 12.12**
   Cords developing red cells and blood vessels **12.15, 12.16, 12.18-12.20**
   Developing
      Endothelial cells **1.4, 12.4-12.9, 12.26-12.28**
      Erythrogenic nucleated cells **1.7, 12.13**
      Large central vein **12.18-12.20**
      Lymphocytes **12.4, 12.8, 12.28;** 91
      Sickle cells **12.21-12.25**
   Directly developing red cells and blood vessels **1.4, 1.7, 1.8, 12.1-12.28**
   Extension of tumor into normal liver **31.29, 31.30**
   Human embryonic liver developing red cells and nucleated cells **1.7, 12.13**
   Necrosed liver tissue, stimulated by nearby cancer, developing red cells **28.8**
   Normal liver tissue
      Invaded by cancer **29.4, 31.24, 31.25, 31.29, 31.30**
      Near approaching cancer developing blood and a blood vessel **29.5**

Lung **11.1-11.8;** 101-103
   Emphysematous lung **11.1**
   Developing hemosiderin pigmented cells as an abortive attempt at red cell development **11.7**
   Origin of erythrogenic SN cells **11.4, 11.5**
   Origin of red cells from
      Alveolar cells **11.6**
      Alveolar fluid **11.2, 11.3**
      Erythrogenic SN cells **11.4, 11.5**

Figure numbers are denoted by **bold-faced** type.

Histiocytes/phagocytes **11.8**

Lung cancers **27.1-27.7;** 181-184
  Adenocarcinoma, necrosed tumor cells developing
      red cells, blood vessels, and plasma **27.3, 27.4;**
      182
  Bronchogenic squamous cell carcinoma **27.1, 27.2;**
      181-182
    Necrosed tumor cells developing
        Red cells, blood vessels, and endothelium **27.1,**
        27.2
      SN cells and other inflammatory cells **27.1**
  Mesothelioma **27.5-27.7;** 183-184
    Developing
        Blood capillaries **27.5, 27.7**
        Plasma **27.5**
        Red cells **27.6, 27.7**
      Undergoing fibrinoid changes **27.5, 27.7**

Lymphatic channels developing from papillary dermis
    2.12, 8.7-8.9; 34, 43, 94-95, 229, 232, 233

Lymph node (lymphatic tissue) **5.22-5.24, 8.1-8.6,**
    22.17, 30.10, 30.11, 30.22; 91-95
  Origin from thyroid tissue in Hashimoto's disease
      5.22-5.24
  Developing red cells and blood vessels from
      Erythrogenic nuclear cells **8.1-8.3**
      Grape cells **8.6**
      Lymphocytes **8.5**
      Lymphoid tissue **8.1-8.6**
      Metastatic tumor tissue **22.17, 30.10, 30.11,**
      31.22
  Lymphocytes transforming into plasma cells **8.6**
  With metastatic
      Carcinoid **31.21**
      Malignant melanoma **22.17**
      Reticulum cell sarcoma **30.10, 30.11**

Lymphocytes 91-95
  Developing platelets 240, 243
  Developing red cells **8.5, 9.1-9.4, 10.1, 10.3, 10.5,**
      10.8, 10.9, 13.4, 22.17; 19, 20, 96, 98, 119, 148,
      234, 241
  Forming stromal tissue **24.11**
  Origin from local tissues 20, 34, 41, 56, 64, 74, 91-
      95, 119, 148, 229, 232, 233, 234, 237, 242, 243
      Adrenal cells **5.31, 5.33, 5.35**
      Blackish reaction of red cells in bone marrow **7.13**
      Cardiac muscle **4.11**
      Colon carcinoma **30.20**
      Dermis, epidermal cells **2.11-2.15, 8.7-8.9**
      Gastric mucosa **14.1**
      Liver parenchyma **12.4, 12.8, 12.28**
      Mammary carcinoma in
          Humans **24.14**
          Mice **24.8, 24.11**
      Nevus cells **2.21-2.23**
      Plasma cells **14.1**
      Prostatic carcinoma **34.4, 34.5, 34.7**

Renal epithelial cells **16.1, 16.3-16.5;** 128
RVR cells **3.10, 3.11;** 151
Salivary gland tissue **26.3, 26.5**
Skeletal muscle **4.27, 4.29**
Skeletal muscle, as a reaction against advancing
    cancer **4.31**
Smooth muscle **4.36**
Spleen **10.8**
Thyroid tissue **5.8**
Thyroid tissue in Hashimoto's disease **5.22-5.24**
Transforming into plasma cells **4.36, 8.6, 26.3;** 98

M

Malignant melanoma **22.13-22.19;** 21, 162-165, 91,
    230, 231, 232
  Of skin
      Blood and nucleated cells developing from keratin
          layer **22.13, 22.14**
      Developing blood and blood vessels **22.15, 22.16**
      Malignant changes in squamous epithelium **22.15**
      Prominent intercellular bridges **22.15**
  Metastatic to lung, developing
      Plasma cells **22.19**
      Red cells **22.18**
  Metastatic to lymph node, developing red cells **22.17**

Mammary (breast) carcinoma **24.1-24.17;** 168-174,
    228, 229, 230, 231, 233
  In humans **24.13-24.17;** 173-174
      Developing
          Collagen fibers, **24.13, 24.14**
          Lymphocytes **24.14**
          Plasma cells **24.15**
          Red cells **24.13**
      Malignant changes in nearby adipose tissue
          24.16, 24.17
      Plasma cells of cancer cell origin developing red
          cells and capillaries **24.15**
      Red cells and capillaries developing from
          transforming collagen fibers **24.14**
  In mice **4.16-4.21, 24.1-24.12;** 17, 18, 168-172, 228
      Necrotic tumor tissue developing
          Blood cysts **24.1, 24.2, 24.5-24.7**
          Endothelium **24.7**
          Hemoglobin globules **24.2**
          Red cells **24.1-24.7**
      Origin from cancer cells of
          Blood vessels **24.8-24.10**
          Connective tissue stroma **24.11, 24.12**
          Endothelium **24.8-24.10**
          Lymphocytes **24.8, 24.11**
          Plasma **24.9, 24.10**
          Red cells **24.8-24.10**
      Emboli within cardiac cavities and coronary vessels
          4.16-421; 54-56

Mammary gland tissue, developing red cells **18.1-**
    18.3; 135

Figure numbers are denoted by **bold-faced** type.

Marrow plasma, *see Bone marrow plasma*

Megakaryocytes 239, 240, 241
   Developing from blackish reaction of red cells in bone marrow **7.14;** 240, 242
   Showing no tendency to form platelets **7.20;** 84, 240, 243

Melanoma, *see Malignant melanoma*

Mesenchymal tissue developing red cells and capillaries **1.5;** 229

Mesothelioma, *see Lung cancers*

Metastatic cancer, *see Bone cancers, Carcinoid, Malignant melanoma, Pancreatic cancers, Prostate cancer, and Stomach cancers*

Multinucleated giant cells, developing from skeletal muscle **4.27**

Muscle tissue **4.1-4.40;** 48-63; *see also Cardiac, Skeletal, and Smooth muscle*

N

Nasal lining, *see Desquamated epithelium*

Necrosis 18, 228, 230, 231, 232, 234, 239, 240, 242, 243, 246
   Coagulation necrosis of myocardium **4.10-4.12;** 48, 238, 244
   Hemorrhagic necrosis developing blood and blood capillaries **12.18, 24.1-24.7, 27.3, 27.4, 31.2, 31.3;** 246
   Of colon carcinoma forming blood and blood vessels **30.13-30.15**

Nevus cells, transformation into red cells, blood capillaries and lymphocytes **2.21-2.23;** 41

P

Pancreas (Islets of Langerhans) **5.44-5.50, 28.2-28.4;** 64, 79-80
   Developing red cells and blood capillaries **5.46-5.50**
   Normal pancreatic tissue near adenocarcinoma
      Becoming hemoglobinized and developing red cells **28.2-28.4**
      Transforming into inflammatory cells, erythrogenic inflammatory cells, and spindle cells towards stroma formation **28.2**
   Transformation of acinar structures into islet cells **5.44, 5.45**

Pancreatic cancers **28.1-28.8;** 185-187
   Adenocarcinoma, with malignant changes **28.1**
   Anaplastic carcinoma head of pancreas

   Metastatic to liver **28.7**
   Nearby necrosed liver tissue developing red cells **28.8**
   Islet cell tumor, necrosed tumor tissue developing red cells **28.5, 28.6**
   Normal pancreatic tissue near adenocarcinoma
      Becoming hemoglobinized and developing red cells **28.2-28.4**
      Transforming into inflammatory cells, erythrogenic inflammatory cells, and spindle cells towards stroma formation **28.2**

Papillary dermis, *see Skin*

Parathyroid gland tissue **5.25-5.29;** 64, 72-73
   Glandular epithelium developing blood and blood capillaries **5.25-5.27**
   Stromal tissue developing red cells and blood capillaries **5.28, 5.29**

Peripheral blood **7.3, 7.9, 7.10, 7.12, 7.15-7.18;** 84, 239-243
   Blood plasma developing
      Platelets in association with red cells **7.17, 7.18**
      Red cells and nucleated cells **7.3**
      Reticulocytes in association with SGF **7.15**
   Blue fibers developing red cells **7.9, 7.10**
   Colorless hemolysis **7.16**
   Combination of blue fibers and blackish reaction **7.12**

Phagocytes 68, 69, 91, 98, 101

Pituitary **5.38-5.43;** 64, 76-78
   Chromophobe cells developing red cells **5.40, 5.42**
   Oxyphil cells developing red cells and blood capillaries **5.40-5.43**
   Three zones of pituitary **5.38, 5.39**

Plasma cells 20, 64, 74, 98, 119, 144, 155, 231, 239, 241, 242, 243, 246; *see also Russell body cells*
   Developing red cells **4.28, 4.36, 4.37, 8.1-8.3, 8.6, 10.7, 10.8, 14.1, 14.3, 14.4, 14.9-14.11, 21.33, 21.34, 24.15, 30.4, 31.19;** 19, 20, 123
   Origin from
      Adipose tissue **3.14-3.16**
      Adrenal epithelium **5.36**
      Basal cell carcinoma **22.2-22.4**
      Carcinoid of colon **31.19**
      Esophageal wall **22.12**
      Gastric mucosa **14.1, 14.3, 14.4**
      Lymphocytes **4.36, 8.6, 26.3;** 91, 92
      Mammary carcinoma in humans **24.15**
      Melanoma metastatic to lung **22.19**
      Renal epithelium **16.5**
      Salivary gland tissue **26.3-26.6**
      Skeletal muscle **4.28**
      Smooth muscle **4.37, 30.4**
      Spleen **10.8**
      Squamous epithelium of skin **2.15**

Figure numbers are denoted by **bold-faced** type.

RVR cells **3.11**
Transformation into
  Endothelial cells **4.36, 14.1, 14.4, 24.15**
  Eosinophils **14.1**
  Lymphocytes **14.1**; 91
  Russell body cells **14.1, 14.9-14.12**; 19, 123
  Segmented nuclear cells **14.1, 22.4**
  Stromal tissue **26.6, 31.19**; 91

Plasma coagulation products of bone marrow (fibrin bodies) developing nucleated cells **7.4-7.6**; 84, 85-86, 239, 240, 241, 247

Platelets, origin from red cells and plasma **7.17-7.19**; 84, 89-90, 239, 240, 241, 242, 243, 247

Prostate, developing red cells and blood capillaries **4.34, 17.1-17.3**; 61, 133-134

Prostate cancer **34.1-34.8**; 91, 219-221, 231
Adenocarcinoma **34.1-34.5**
  Cancer cells developing vacuolar (ghost) red cells, blood vessels, and lymphocytes **34.4, 34.5**
  Perineural invasion **34.3**
  Typical malignant structures **34.1, 34.2**
Metastatic to bone marrow, cancer cells developing red cells, lymphocytes, and SN cells **34.6-34.8**

R

Rectal tissue (hemorrhoids), developing red cells and blood vessels **15.1-15.4**; 126-127

Redefined Vasoformative Cells of Ranvier (RVR cells) **3.10, 3.11, 21.18-21.32**; 46, 144, 150-155, 243, 247
Developing reactive cells **3.10, 3.11, 21.22, 21.28-21.32**
Developing red cells and blood capillaries **3.10, 3.11, 21.20-21.27**
Origin from
  Adipose tissue **3.10, 3.11**
  Edematous subserosa of stomach **21.18-21.22**
  Labium majora **21.23-21.32**

Renal (kidney) cancers **33.1-33.12**; 91, 213-218, 230, 231
Adenocarcinoma, developing red cells from liquefied tumor cell product **33.1-33.3**
Papillary adenocarcinoma, developing red cells from liquefied tumor cell product **33.4-33.6**
Wilms' tumor **33.7-33.12**
  Erythrogenic changes in nearby normal kidney tissue **33.7, 33.8**
  Tumor tissue developing blood and blood vessels **33.9-33.12**

Renal tissue, *see Kidney tissue*

Reticulocytes 20, 84, 239, 240, 242, 247
Colorless hemolysis **7.16**
Origin from SGF and plasma **7.15**

Reticulum cell sarcoma, *see Stomach cancers*

Retina, *see Eye*

Rheumatic fever, *see Cardiac muscle*

Russell body cells (grape cells) 19, 20, 119, 123, 247
Developing red cells **8.6, 14.1, 14.10-14.12, 21.19, 21.20**
Origin from plasma cells **14.1, 14.9-14.12**

RVR cells, *see Redefined Vasoformative cells of Ranvier*

S

Salivary gland **26.1-26.6**; 177-180
Papillary adenocarcinoma **26.1**
Reactions to nearby cancer, development of
  Atypical changes in glandular epithelium **26.4**
  Endothelium **26.4, 26.5**
  Lymphocytes **26.3, 26.5**
  Plasma cells **26.3-26.6**
  Red cells and blood vessels **26.2-26.6**

Segmented nuclear (SN) cells 20, 30, 34, 56, 64, 74, 98, 101, 119, 128, 139, 144, 148, 151, 239, 240, 241-242, 243, 247
Erythrogenic **1.7, 3.7-3.9, 4.26, 4.27, 4.38, 5.32, 10.5, 10.7-10.9, 11.4, 11.5, 13.3, 14.1, 14.2, 16.5, 19.3-19.5, 21.5-21.7, 21.13-21.15, 21.28-21.30, 22.4, 22.16, 30.2-30.4, 30.14, 30.16, 30.17**
Origin from
  Adipose tissue **3.7-3.9**
  Adrenal tissue **5.32**
  Blackish reaction of red cells **7.12-7.14**
  Blue fibers **7.11, 7.12**
  Brain tissue **19.3-19.6**
  Bronchogenic squamous cell carcinoma **27.1**
  Cardiac muscle **4.13**
  Carcinoid tumor **31.19**
  Colon carcinoma **30.14-30.17**
  Embryonic liver **1.7**
  Epidermal cells (skin) **2.13-2.15, 21.10-21.15**
  Fibrinoid particles in bone marrow **7.7, 7.8**
  Gall bladder wall **13.3**
  Gastric wall **14.1, 14.2**
  Lung tissue **11.4, 11.5**
  Mucocutaneous tissue of the lower lip **21.16, 21.17**
  Plasma cells **14.1, 22.4**
  Prostatic carcinoma metastatic to bone marrow **34.7**
  Renal tissue **16.5**

Figure numbers are denoted by **bold-faced** type.

RVR cells **21.28-21.30**
Skeletal muscle **4.26, 4.27**
Smooth muscle **4.38, 13.2, 30.3, 30.4**
Spleen **10.5, 10.7, 10.8**
Stomach adenocarcinoma **30.2**
Vaginal epithelium **21.3-21.7**

Senile ecchymosis, developing red cells from collagen of upper dermis **2.24, 2.25;** 42

SGF, *see Substantia granulo-filamentosa*

Sickle cells 74, 104, 139, 240, 243, 247
Origin from
Adrenal tissue **5.37**
Brain tissue **19.7, 19.8**
Kidney tissue **16.12, 16.13**
Liver tissue **12.21-12.25**
Spleen tissue **10.9**

Skeletal muscle **4.22-4.33;** 21, 48, 56-60, 91, 229, 231, 233, 235
Direct origin of red cells and blood capillaries **4.22-4.25**
Dissolution of muscle fibers giving false appearance of adipose tissue **4.33**
Human embryonic **4.25**
Nucleosis as a reaction against advancing cancer **4.31, 4.32**
Origin of fibrocytes **4.32**
Origin of reactive/inflammatory cells **4.26-4.32**

Skin (dermis, epidermis, and papillary dermis) **2.1-2.25, 8.7-8.9;** 21, 34-42, 148, 229; *see also Capillary hemangioma, Granulation tissue, Hemorrhagic dermatofibroma, Nevus, and Senile ecchymosis*
Developing
Connective tissue **2.7, 2.12**
Endothelial cells **2.5, 2.6, 2.8**
Lymphocytes and lymphatic channels **2.11, 2.12, 2.15, 8.7-8.9**
Reactive/inflammatory cells **2.11-2.15**
Red cells and blood capillaries **2.1-2.11**
Transformation of epidermis into papillary dermis **2.7-2.10**

Skin cancers, *see Basal cell carcinoma, Malignant melanoma, and Squamous cell carcinoma*

Small intestine (ileum), *see Carcinoid*

Smooth muscle **4.34-4.40, 13.1 13.2, 14.13, 14.14, 30.3, 30.4, 31.12-31.14, 31.35, 31.36, 31.40;** 21, 48, 61-63, 91, 116, 119, 126, 229, 231
Developing
Calcifying streaks **31.14, 31.15, 31.40**
Erythrogenic reactive cells **4.36-4.40, 13.1, 13.2, 14.14, 30.3, 30.4**
Direct origin of red cells and blood capillaries **4.34, 4.35, 14.13**

Extension of carcinoid tumor into smooth muscle coat of ileum **31.12, 31.13**
Fibrinoid degeneration developing blood and blood vessels **31.34-31.39**

SN cells, *see Segmented nuclear cells*

Spleen **10.1-10.9;** 91, 98-100, 231, 240, 242, 243
Origin of
Reactive cells from fibrous connective tissue **10.8**
Red cells from reticular cells **10.1-10.3, 10.6**
Sickle cells **10.9;** 240, 243, 247-248
Origin of red cells from locally developed
Eosinophilic granules **10.1, 10.2**
Eosinophils **10.1, 10.5**
Lymphocytes **10.1, 10.3, 10.5, 10.8, 10.9**
Plasma cells **10.7, 10.8**
SN cells **10.5, 10.7, 10.8**

Squamous cell carcinoma **22.7-22.12;** 160-162, 231, 233
Esophagus, developing
Blood and blood vessels **22.10-22.12**
Plasma cells **22.12**
Skin
Chronic inflammatory reaction **22.7, 22.8**
Keratinized tissue developing red cells **22.9**
Transforming into reactive/inflammatory cells **22.7, 22.8**

Stomach **4.35, 436, 4.38, 14.1-14.14, 21.9, 21.18-21.22, 30.3, 30.4, 30.12;** 91, 119-125, 144, 151, 231, 234
Developing red cells and blood capillaries from
Gastric gland epithelial cells **14.6**
Gastric lining epithelium **14.7, 14.8**
Gastric mucosa and submucosa **14.1-14.5, 14.9-14.12, 21.9**
Gastric subserosa **21.18-21.22**
Perigastric adipose tissue **30.12**
Smooth muscle coat **4.35, 4.36, 4.38, 14.13, 14.14, 30.3, 30.4**
Origin of
Eosinophils **14.1, 14.14, 21.9**
Endothelium **4.36, 14.1, 14.4, 14.5, 14.13**
Lymphocytes **4.36, 14.1**
Origin of erythrogenic
Plasma cells **4.36, 14.1, 14.3, 14.4, 14.9-14.11, 30.3, 30.4**
Russell body cells **14.1, 14.10-14.12**
RVR cells **21.18-21.22**
SN cells **4.38, 14.1, 30.3, 30.4**
Red cells and endothelium developing directly from smooth muscle **4.35, 14.13**
Transformation of locally developed plasma cells into
Endothelial cells **4.36, 14.1, 14.4**
Eosinophils **14.1**
Lymphocytes **14.1**
Red cells and blood capillaries **4.36, 14.3, 14.4**

Figure numbers are denoted by **bold-faced** type.

Russell body cells **14.1, 14.9-14.11**
SN cells **14.1**

Stomach cancers **30.1-30.12;** 190-194, 230
   Adenocarcinoma **30.1-30.4;** 190-191
      Blood vessel development **30.1-30.3**
      Developing erythrogenic SN cells **30.2-30.4**
      Erythrogenic SN cells developing red cells **30.2-30.4**
   Reticulum cell sarcoma with necrosis **30.5-30.12;** 191-194
      Developing blood vessels **30.5-30.9**
      Metastatic to lymph nodes, developing red cells **30.10, 30.11**
      Reaction in nearby perigastric adipose tissue, developing red cells and blood capillaries **30.12**
      Superimposed with amebiasis **30.8, 30.9**

Stromal tissue, 18, 48, 64, 72, 92, 133, 231, 232, 234, 235, 248; *see also Collagen and Connective tissue*
   Developing from
      Carcinoid tumor tissue **31.6-31.10, 31.19, 31.20, 31.21, 31.23, 31.28**
      Colon carcinoma **30.19, 30.20**
      Liver carcinoma **29.3**
      Mammary carcinoma **24.11-24.14**
      Pancreatic glandular tissue **28.2**
      Parathyroid gland tissue **5.28, 5.29**
      Plasma cells **26.6**
   Developing red cells and blood vessels **5.28, 5.29, 17.1, 24.14**
   Independently growing mammary carcinoma emboli within the cardiac cavities and coronary vessels questioning the theory that stromal tissue originates only from host tissue **4.16-4.21**

Stupendocyte **14.12**

Substantia granulo-filamentosa (SGF) developing reticulocytes **7.15;** 242, 248

T

Thymus, lymphocytes developing red cells and blood capillaries **9.1-9.4;** 96-97

Thyroid cancers **25.1-25.4;** 175-176, 230
   Adenocarcinoma, developing hemoglobinized colloid and red cells **25.1, 25.2**
   Follicular adenocarcinoma, developing erythrogenic colloid, red cells and blood vessels **25.4**
   Normal thyroid tissue developing hemoglobinized colloid and red cells **25.3**

Thyroid tissue **5.1-5.24, 25.3;** 64-71, 91, 229
   Colloid
      Direct origin of red cells **5.1-5.13, 5.15, 5.19, 25.3**

      Developing erythrogenic nucleated cells **5.11-5.13**
      Developing hemosiderin pigmented cells **5.14-5.16**
   Glandular epithelium of follicles developing
      Erythrogenic colloid **5.10, 25.3**
      Lymphocytes **5.8**
      Red cells and blood vessels **5.4, 5.6-5.11, 5.16**
   Origin of lymphatic tissue in Hashimoto's disease **5.22-5.24**
   Transformation into erythrogenic collagen **5.17-5.21**

Tissue culture, chick embryonic nonmedullary tissue *in vitro* **1.8, 1.9, 4.3, 4.4, 12.12;** 25, 26, 33, 237, 241
   Fat vacuoles developing from cardiac muscle **4.4**
   Red cells developing from
      Cardiac muscle **1.9, 4.3**
      Liver cells **1.8, 12.12**

U

Urinary bladder, smooth muscle developing erythrogenic eosinophils **4.39, 4.40;** 63

V

Vaginal epithelium, *see Desquamated epithelium*

Vacuolar changes 228
   Developing red cells and vascular channels from
      Colloid of thyroid **5.5**
      Desquamated vaginal epithelium **21.1, 21.2**
      Erythrogenic nuclear cells **5.12, 5.13, 5.16**
      Fibrocytes **2.18**
      Kidney tissue **16.1-16.3**
      Liver tissue **12.3, 12.9, 12.15, 12.16**
      Mammary carcinoma **24.14**
      Nevus cells **2.21-2.23**
      Papillary adenocarcinoma of kidney **33.4-33.6**
      Perigastric adipose tissue **30.12**
      Reticulum cell sarcoma **30.10, 30.11**
   In epidermal cells developing SN cells **2.14**

Vacuolar red cells, *see Ghost red cells*

Vacuolated mesh, developing red cells **5.16, 16.2, 21.1, 21.2, 21.6, 33.2;** 232, 241

Y

Yolk sac membrane of chick embryo developing red cells **1.1-1.3;** 19, 29-31

Figure numbers are denoted by **bold-faced** type.